Oxide Spintronics

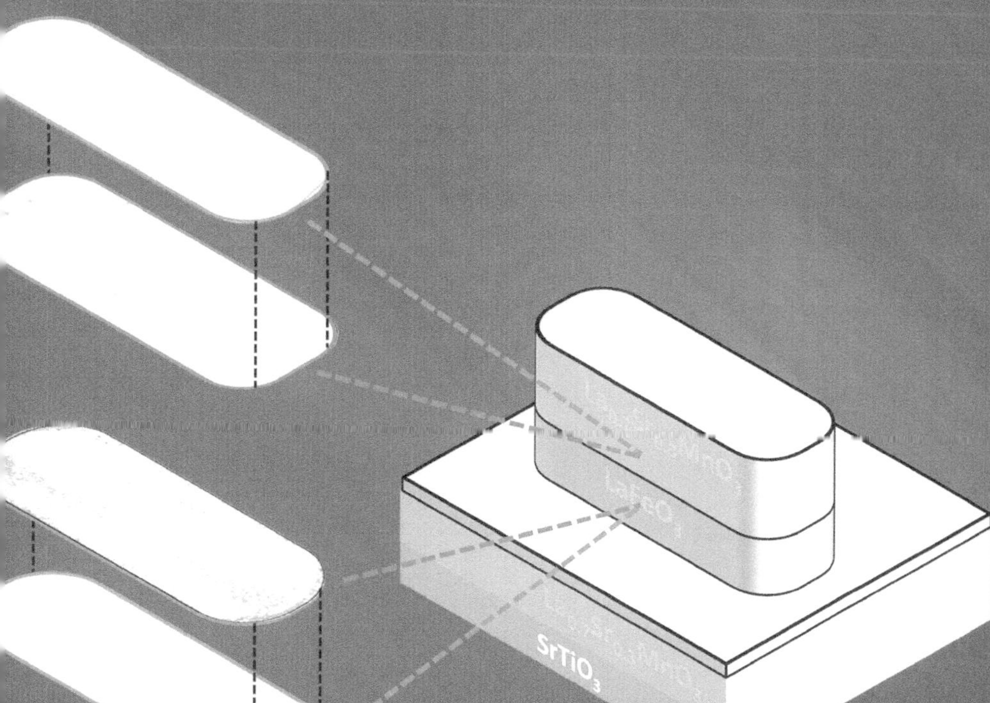

Oxide Spintronics

edited by

Tamalika Banerjee

JENNY STANFORD
PUBLISHING

Published by

Jenny Stanford Publishing Pte. Ltd.
Level 34, Centennial Tower
3 Temasek Avenue
Singapore 039190

Email: editorial@jennystanford.com
Web: www.jennystanford.com

British Library Cataloguing-in-Publication Data
A catalogue record for this book is available from the British Library.

Oxide Spintronics

ISBN 978-981-4774-99-4 (Hardcover)
ISBN 978-0-429-46819-3 (eBook)

Contents

Preface

Oxide materials have been used in mainstream semiconductor technology for several decades, serving as important components such as gate insulators or capacitors in integrated circuits. However, in recent years this material class has emerged in its own right as a potential contender for alternative technologies, generally designated as 'beyond Moore'. The 2004 discovery by Ohtomo and Hwang was a global trendsetter in this context—the observation of a two-dimensional high-mobility electron gas at the heterointerface between two insulating oxides ($LaAlO_3$ and $SrTiO_3$), supported by the rise of nascent deposition and growth monitoring techniques, was an important direction in materials science research. The quest to understand the origin of this unparalleled physical property and for finding other emergent properties has been an active field of research in condensed matter, uniting researchers with diverse expertise from thin film growth, defect control, advanced microscopy, semiconductor technology, computational experts, magnetism and electricity, spintronics, nanoscience, and nanotechnology. This book is an attempt to showcase the important scientific advances that have been made in this direction with new oxide materials interfaces or with new techniques.

Complex oxide materials discussed in this book are commonly perovskites with the general formula ABO_3, where A and B represent the two different cations. Heterointerfaces designed by choosing different cations causes elastic strain at the interface as discussed in Chapter 1 and leads to new structural magnetic or electronic phases. This is triggered not only by the mismatch in the in-plane lattice parameter but also by a mismatch of the oxygen octahedral rotation and of the polar interface charge. This chapter considers coherently strained films of $SrRuO_3$, manganites, and nickelates and shows how elastic strain can be used as a design tool for tuning their electronic and magnetic properties.

The significance of interface engineering at the interface between the half-metal $La_{0.67}Sr_{0.33}MnO_3$ on $SrTiO_3$ is discussed extensively in Chapter 2. The formation of a magnetic dead layer at this interface has been reported earlier in several studies and is deleterious for device applications. The authors have systematically performed compositional engineering at the interface to investigate the different mechanisms that are responsible for its occurrence. They conclude that the tunability of the oxygen octahedral rotation is an important parameter that influences the magnetic dead layer at such interfaces.

Electronic transport (both charge and spin) are strongly influenced by defects, orbital reconfiguration, interdiffusion, and local inhomogeneities at the interface and govern device performance. The field of complex oxide devices gained momentum in recent years by the inclusion of a unique technique that has the ability to probe, at the nanometer scale, and with a high lateral resolution sub-surface features and buried interfaces that are fundamental to the analysis of electronic transport at (non-)engineered interfaces. Chapter 3 discusses one such probe, known as the ballistic electron emission microscope, and demonstrates its first application to the study of complex oxide heterointerfaces on $SrTiO_3$. Using a current-perpendicular-to-plane device geometry, transport parameters such as the mean free path across and close to the magnetic phase transition in ferromagnetic $LaSrMnO_3$ and $SrRuO_3$ are determined.

The advances that have been made in the theoretical understanding of material properties on the atomic scale have been captured in Chapter 4 by means of quantum mechanical calculations within density functional theory (DFT). Several important technological material interfaces are considered such as $SrRuO_3$/$SrTiO_3$, $YMnO_3$/$LaMnO_3$ and Fe/$BaTiO_3$ and experimental findings related to the unusual magnetic and electronic transport are analyzed using DFT-based ab initio electronic structure calculations.

Chapter 5 introduces ferroic and multiferroic materials, in particular, $BiFeO_3$. Displaying magnetism and ferroelectricity in the same material phase, such materials are of tremendous technological importance since they allow control of magnetization (polarization) of devices with an electric (magnetic) field. Demonstrations of domain wall conductivity in thin films of such multiferroics and others have ignited intense research with these materials. The broadly accepted phenomena responsible for the conductivity at the domain walls are the reduction of the band gap at the walls and the electrostatic potential changes due to structural variations. The chapter also discusses the different mechanisms such as octahedral rotations, increased carrier density, and role of oxygen vacancies that are responsible for the observed conductivity at the domain wall in these material systems.

Spintronic functionalities such as magnetoresistance, anisotropic magnetoresistance, and giant or tunnel magnetoresistances are interesting to investigate in heterostructures based on multiferroics and have been discussed in Chapter 6. After describing the most commonly used phenomena in spintronics, the chapter discusses these effects as exhibited in different types of multiferroic tunnel junctions using the external control knobs of magnetic and electric field and suggests research directions for electrical control of spintronic functionalities.

Electronic and orbital reconstructions at oxide heterointerfaces provide an opportunity to trigger magnetic interactions that can be tailored by strain engineering. Chapter 7 reviews such interfacially induced magnetism in different spintronic devices. Spin reconstruction at the interfaces drives them into a novel magnetic state that acts as a spin filter and can lead to an increase in the tunneling magnetoresistance in tunnel junction devices. The chapter discusses different combinations of cuprate/manganite interfaces and observes a magnetoelectric effect that can be used to electrically switch the magnetization of the magnetic layers without an applied magnetic field. This alternative route of electric field control of magnetism is a functionality that is being actively pursued for the development of low-dissipation spintronics.

Finally, Chapter 8 reviews recent advances that have been made in the exploration of the most archetypal two-dimensional electron system (2DES) of $LaAlO_3/SrTiO_3$. This review focuses on the electronic properties of 2DESs such as 2D superconductivity and Rashba spin–orbit coupling at such heterointerfaces. The combined coexistence of 2D-superconductivity with a sizable spin–orbit coupling of the Rashba type, which is tunable by large electric fields, opens up new avenues in oxide electronics and oxide spintronics that utilize such engineered oxide heterointerfaces. The chapter also provides an outlook for future developments in the study of 2D superconductivity and spin–orbit fields using such oxide heterointerfaces.

In summary, the book provides a modest perspective on the vast scientific advances realized using complex oxide heterointerfaces and demonstrates their recognition as a potential material class for defining new technologies. To drive complex oxides as the next frontier materials in electronics, combined efforts in different directions are needed such as (i) the availability of high-quality large-area substrates for integration with the existing semiconductor technology as well as compatible fabrication methods, (ii) tools for quantifying defects and their control, (iii) analytical tools that are non-destructive and nanoprobes that can non-destructively probe the device interface, (iv) predictive modeling and design of heterostructures, thereby establishing a strong link between experiments and theory, and last but not least (v) continuation of our exploration of new emerging properties across such heterointerfaces that can host new topological phases such as skyrmions and design novel methods to probe them.

The authors of this book expect that this fascinating class of materials will soon define future technologies of its own and play a prominent role in current research and technological programs involving quantum and neuromorphic computing. They also believe that the book will be of interest to researchers, engineers, and technologists working in academia or in industries and will propel many scientific explorations utilizing together the complex oxide materials and their devices. I would like to acknowledge the authors for their extensive contributions and patience and the Jenny Stanford Publishing team for their interest and support in this field.

Tamalika Banerjee

Groningen, The Netherlands

Spring 2019

Chapter 1

Elastic Control of Magnetic Order at Oxide Interfaces

Kathrin Dörr
Institute of Physics, Martin Luther University Halle-Wittenberg, 06099 Halle, Germany
kathrin.doerr@physik.uni-halle.de

I wish to devote this chapter to the memory of Michael D. Biegalski who contributed a lot to our present understanding of structural aspects in complex oxide films. He was an inspiring and painstaking colleague, in particular as a staff scientist caring for users and their projects at the Centre for Nanophase Materials Sciences (CNMS) at Oak Ridge National Laboratory. He also was a warm-hearted friend who is missed by many of us.

1.1 Oxide Interfaces

In this chapter, general structural, electronic, and magnetic characteristics of coherent and atomically abrupt (ideal) interfaces between two oxides ABO_3 and $A^*B^*O_3$ are introduced with an emphasis on the role of elastic strain. Real interfaces may form defects

Oxide Spintronics
Edited by Tamalika Banerjee

ISBN 978-981-4774-99-4 (Hardcover), 978-0-429-46819-3 (eBook)
www.jennystanford.com

such as an interdiffused region at the interface, thus broadening the range of observable phenomena. Here, the role of defects is only occasionally mentioned. Further, metal ions on the A sites of the ABO_3 structure (e.g., rare earth ions) may order magnetically at low temperatures. This type of magnetism and coupling to B site magnetic order are also not considered.

1.1.1 Chemical Composition of the Interface

Based on advances of thin film growth techniques, atomically abrupt interfaces of complex oxides have been fabricated in recent years. Scanning transmission electron microscopy (STEM) [1–4] provides tools to inspect elemental concentrations within lateral dimensions of about 1 Å^2 and with a precision down to a few percents. Often, there are several compositional variants of an ideal interface between two complex oxides. For instance, the ABO_3 lattice can be cut along a (001) plane either through the A or the B atoms (Fig. 1.1). Hence, the ABO_3 crystal can end with an AO or a BO_2 face, called an AO or a BO_2 termination. Joining two such crystals (ABO_3 and $A^*B^*O_3$) forms either an AO–B^*O_2 or a BO_2A^*O interface. Interface terminations can profoundly change the electronic and magnetic properties of an oxide interface. A prominent example is the two-dimensional electron system observed at the TiO_2–LaO interface of $LaAlO_3/SrTiO_3$ (001), which is not found for the other termination [5].

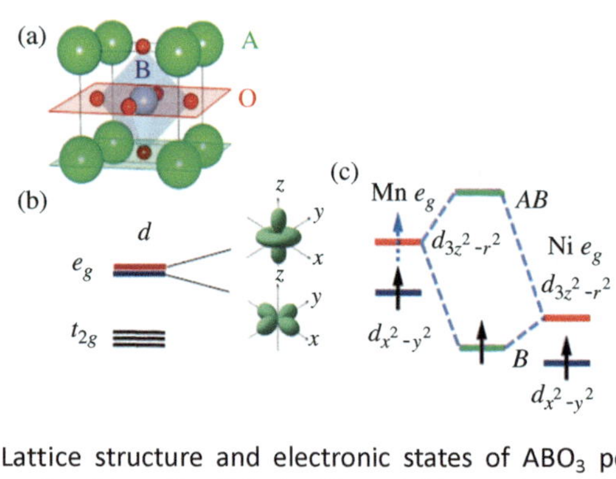

Figure 1.1 Lattice structure and electronic states of ABO_3 perovskite-type oxides: (a) Pseudocubic unit cell (uc) with metal ion A, transition metal ion B and oxygen (O) octahedron. A (001) surface can be an AO plane (green) or a BO_2 plane (red). (b) Electronic d levels of the B ion in cubic coordination and shapes of e_g orbitals. (c) 3 z^2–r^2 hybrid orbital formation at a Mn–O–Ni interface.

Instead of being atomically sharp, an interface may show some intermixing of the components. Intermixing can originate from the growth process if energies of deposited species are large. Otherwise, there may be a chemical driving force to form a solid solution of another chemical phase from the two interface partners. Further, vacancies may find lower energy positions near an interface where lattice sites lack neighboring atoms or where regions with larger elastic strain lower the chemical potential for vacancies. For instance, an enrichment of oxygen vacancies at the $LaAlO_3/SrTiO_3$ (001) interface [6] and the $La_{0.7}Sr_{0.3}MnO_3/SrRuO_3$ interface [7] has been reported. As one more origin of interdiffusion, electrically polar interfaces between insulating oxides may compensate the interface charge by shifting ions instead of electrons (Section 1.1.3) [8].

1.1.2 Interfacial Lattice Structure

The perovskite-type lattice is a three-dimensional network of corner-sharing oxygen octahedrons. Inside each octahedron, a metal B cation is located and forms six B–O–B bonds with neighboring B sites governing the electronic structure (e.g., the electronic bandwidth) and the magnetic interactions. Elastic constraints to this network arise from the ionic radius of the A site ion (chemical pressure) or from hydrostatic pressure or elastic strain of films. The network reacts by collective octahedral rotations around the three coordinate axes and typically less easily, by altering the B–O bond lengths. Hence, every perovskite-type oxide has a specific rotation pattern, which can be classified using the *Glazer* notation [9]. For example, $a^0b^+c^-$ denotes the absence of rotations about the crystallographic *a*-axis, in-phase rotations of octahedrons about the *b*-axis and alternating (out-of-phase) rotations about the *c*-axis. Rotation angles of several degrees are common. The elastic strain of a thin film modifies the angles or even the type of the rotation pattern in a non-trivial way [10–14], since charge, spin, orbital, and lattice degrees of freedom are interlinked (see Section 1.2).

Connecting two such oxides at a coherent interface raises the question of how the rotation pattern at the interface accommodates the elastic stress. Specific interfacial rotation patterns have been discovered (Fig. 1.2), which can deviate from the (strained) bulk patterns of both connected oxides [1, 15, 16]. If the stiffness of rotations in an oxide exceeds strongly that of the interface partner, rotations can be transferred from one oxide into the other. In general, the interfacial rotation pattern depends on both coupled oxides and returns to the (strained) bulk pattern within a range of <10 uc away from the interface [1, 15–17].

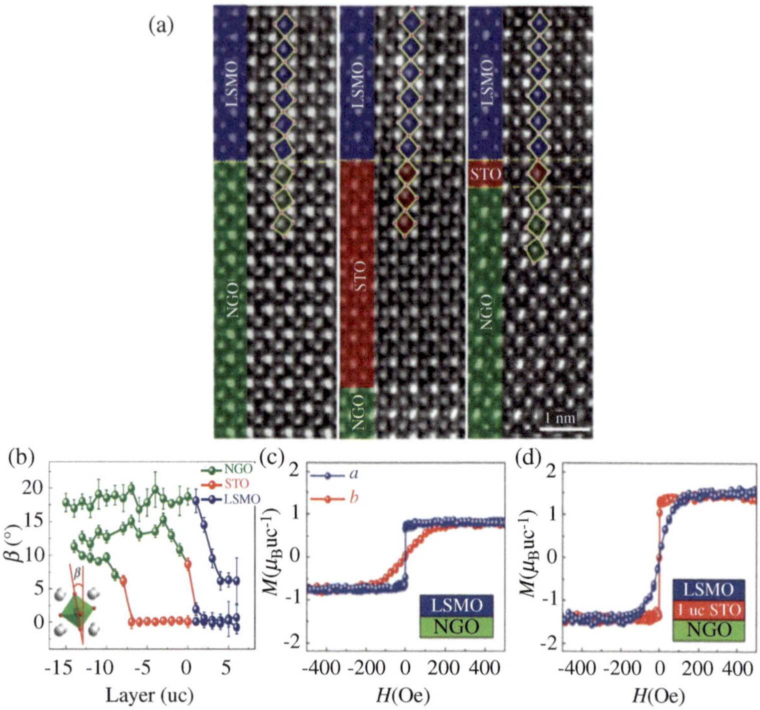

Figure 1.2 Interfacial octahedral rotations at the $La_{0.7}Sr_{0.3}MnO_3/NdGaO_3$ (001) interface controlled by a $SrTiO_3$ interlayer: (a) Inversed annular bright field scanning transmission electron microscopy image with marked octahedrons. (b) Layer-wise recorded tilt angle β (inset) for the three interface types shown in (a). (c, d) In-plane magnetization along two perpendicular axes *a*, *b* showing effective control of magnetic anisotropy. Adapted from Ref. [17], with permission from Springer Nature, Copyright 2016.

Until recently, experimental access to depth-resolved interfacial rotations was strongly limited. Volume-averaged rotations, both type of rotation pattern and rotation angles, have been derived from intensities of half-order X-ray diffraction peaks [13, 18, 19].

Collective rotations of oxygen octahedrons cause a doubling of the pseudocubic unit cell in one, two, or three directions. Recent advances in scanning transmission electron microscopy make it possible to image the octahedral rotations directly or derive even three-dimensional rotation patterns [1, 4, 15, 16].

1.1.3 Electrical Charge and Electron Orbitals

Any two fairly conducting materials forming an interface experience an electron exchange as a consequence because of the different chemical potentials for electrons on both sides. (The chemical potential is the energy required to add an electron.) Well-known from semiconductors, this mechanism induces band bending at interfaces. The resulting charge transfer can shift the valence of atoms near the interface.

Additionally, oxide interfaces may carry an electrical charge because of the charged terminating atomic planes of the oxides. A charged (infinite) plane has a constant electric field and thus, a diverging electrostatic potential with growing distance from the interface, the so-called *polar catastrophe* [3, 5]. An example is $LaAlO_3/SrTiO_3$: $La^{3+}O^{2-}$ planes carry the charge of $+e$ (with the elementary charge e), $Al^{3+}O_2^{2-}$ planes the charge of $-e$. Thus, the $LaO–Ti^{4+}O_2$ interface is charged with $+e$, the $AlO_2–Sr^{2+}O$ interface with $-e$. Transfer of ½e of the opposite sign to the interface removes the diverging potential. If no mobile electrons are available, however, ionic shifts may occur to compensate for the interface charge [8]. For instance, this could be an enrichment of (positive) oxygen vacancies at a negatively charged interface.

Magnetic B site ions have partially occupied electronic 3d, 4d, or 5d levels. In a cubic ABO_3 lattice, there are two groups of electronic states separated by the crystal field energy, t_{2g} and e_g (Fig. 1.1). Electron orbitals of t_{2g} states lie between B–O bond axes, while orbitals of e_g states lie along the bond axes giving them larger overlap with oxygen 2p orbitals and thus, larger electrostatic energy. Lattice distortions of the lower crystal symmetries (e.g., tetragonal) stabilize the orbitals along the elongated directions due to the lowered electrostatic energy. Very few ABO_3 oxides are truly cubic, but lattice distortions are often small. Hence, the orbital occupation at interfaces can deviate from that found in the bulk. Interfacial orbital reconstruction can be driven, for example, by the formation of B–O–B* hybrid orbitals bridging the interface (Fig. 1.1c) [20].

Active control of the interface charge has been demonstrated at interfaces of a ferroelectric oxide whose normal component of polarization can be switched [21, 22]. As shown by Preziosi et al., the charge redistribution can be associated with an altered orbital occupation [22].

1.1.4 Magnetic Order and Coupling across the Interface

Magnetic ordering of B site transition metal ions in ABO_3 oxides originates from superexchange interactions [23–26] of localized 3d, 4d, or 5d electrons along the B–O–B bonds. Additionally, some oxides show the double-exchange interaction that relies on delocalized electrons moving with conserved spin between B ions, which have a localized magnetic moment. Both types of exchange interactions require overlapping electronic orbitals of B ions with oxygen 2p orbitals. This makes orbital occupation a key parameter for the sign and strength of magnetic ordering. Magnetic coupling at ABO_3–$A^*B^*O_3$ interfaces works through the same exchange interactions as in bulk oxides. It has been reviewed recently by Bhattacharya and May [10], see their article for details. Here, the focus is on the elastic control of interface magnetism, which can take two distinct routes. The first is through the effect of coherent film strain on bulk magnetism of the interface partners addressed in Section 1.2. This can include changes of lattice symmetry, octahedral rotations and orbital occupation with regard to the unstrained bulk crystal. The second is through the interfacial pattern of oxygen octahedral rotations between the two oxides. Since exchange interactions are highly sensitive to B–O–B* bond angles, the interfacial rotation pattern is expected to directly control the magnetic coupling across the interface. Probably, the interfacial rotation pattern and the exchange interactions are interrelated rather than being in a causal relation. The interfacial rotation pattern has a characteristic decay length into the depth of the films, which typically is below 10 uc (Section 1.1.2). Interfaces with smaller distance from each other than the decay length are not independent and may show different rotations and magnetic ordering [10]. This leads to distinct properties of short-period superlattices.

1.2 Elastic Control of Bulk Magnetism in Coherently Strained Films

Every thin film grown epitaxially on a monocrystalline substrate of another material experiences elastic strain because lattice parameters and thermal expansion are different. Previously a nuisance, elastic strain has long been recognized as a design tool for tuning electronic and magnetic properties of complex oxides. Today, there is a long list of altered structural, magnetic, and electronic phases obtained in strained films [27–31].

In this chapter, effects of coherent elastic strain are outlined for $SrRuO_3$, manganites $AMnO_3$, and nickelates $ANiO_3$, since these oxides form the interfaces discussed as examples in Section 1.3. Coherent strain, even though transferred at the interface from one component to the other, is not considered an interface property, because it can extend over more than 100 uc (~40 nm) away from the interface.

1.2.1 $SrRuO_3$

$SrRuO_3$ is a conducting ferromagnet with rather wide electronic bands because of the large 4d orbitals of Ru. Therefore, $SrRuO_3$ is often applied as electrodes in oxide heterostructures. $SrRuO_3$ (SRO) bulk crystals have orthorhombic *Pbnm* symmetry with lattice parameters of a = 5.57 Å, b = 5.53 Å, and c = 7.85 Å. The pseudocubic lattice parameter is $a_{pc} = (abc)^{1/3}$ = 3.925 Å. The bulk rotation pattern in Glazer notation is $a^-a^-c^+$ [32]. Above 550 °C, bulk SRO is tetragonal with rotation pattern $a^0a^0c^+$ [33]. Ru^{4+} ions have an intermediate $t_{2g}^{\uparrow 3}t_{2g}^{\downarrow 1}$spin state (S = 1), the large crystal field energy prevents e_g levels from being occupied. Ru spins order in a collinear ferromagnetic structure at the Curie temperature of T_C = 150 K. Magnetocrystalline anisotropy is large; magnetic fields of several Tesla at temperatures $T<<T_C$ are required to saturate the magnetization in magnetically hard directions. $SrRuO_3$ has metallic character, (i.e., positive temperature derivative of the electrical resistance) at all temperatures.

On $SrTiO_3$ (001) with cubic lattice parameter of a = 3.905 Å, SRO is weakly compressively strained, remains orthorhombic like the bulk phase and grows with the long axis (c) along the cubic [100] and [010] in-plane directions of the substrate [33, 34]. Vicinal substrates allow one to select one of these orientations and grow a single-crystalline SRO film [34, 35]. The magnetic easy axis is perpendicular to the c-axis and canted out-of-plane with a temperature-dependent canting angle [35]. The biaxial tensile strain of $\varepsilon \sim 1\%$ in the film plane induces tetragonal lattice symmetry, which is likely to coincide with the high-temperature tetragonal phase of bulk SRO [12, 14, 29]. The changes of lattice symmetry and octahedral rotations are associated with altered magnetocrystalline anisotropy: the tetragonal structure has two in-plane easy axes instead of the uniaxial anisotropy of the bulk crystal [29]. The spin state of Ru^{4+} ions has been predicted to change at larger tensile or compressive strains from $t_{2g}{}^{\uparrow 3}t_{2g}{}^{\downarrow 1}$ (S = 1) to $t_{2g}{}^{\uparrow 2}t_{2g}{}^{\downarrow 2}$ (S = 0) since t_{2g} levels split as a consequence of bond length changes (distortions of the oxygen octahedrons) [36]. Application of reversible strain (Section 1.2.4) to films with 1% of tensile strain confirmed a reduction of the ferromagnetic Ru moment with larger tensile strain [28]. The Curie temperature of SRO depends on the coherent strain in a non-trivial way: large strains of both signs reduce T_C, but the maximum is reached at moderate tensile strain around 0.5%, close to the orthorhombic to tetragonal phase transition [27, 29]. The reversible strain has been applied to record $dT_C/d\varepsilon$ in different strain states [29]. It never exceeds 10 K $\%^{-1}$ in the measured range of strain, showing a moderate strain response in comparison to other oxides.

1.2.2 Manganites $AMnO_3$

Manganites $A''_{1-x}A'_xMnO_3$ (with A″ = trivalent metal, A′ = divalent metal used for doping) are the most studied ABO_3 materials today. They exist in a wide range of compositions and show an impressive multitude of magnetic and electronic phases, since charge, orbital, spin, and lattice degrees of freedom are strongly coupled. Electrical conduction spans a range of many orders of magnitude and is unusually sensitive to external parameters, including the famous phenomenon of colossal magnetoresistance [37]. $La_{1-x}Sr_xMnO_3$ is the system with the largest ferromagnetic Curie temperature of T_C = 370 K reached at the doping level of x = 0.3. Therefore, the focus is here on this compound as an often-used prototype.

$La_{0.7}Sr_{0.3}MnO_3$ (LSMO) has a rhombohedral ($R\bar{3}c$) lattice with the pseudocubic lattice parameter of 3.87 Å and an $a^-a^-a^-$ rotation pattern of oxygen octahedrons. Manganese (Mn) has a mixed valence of 0.7 Mn^{3+} ($3d^4$, S = 2) and 0.3 Mn^{4+} ($3d^3$, S = 3/2) with the excess e_g electron being mobile in a narrow conduction band. The 3d electrons occupy t_{2g} and e_g levels with parallel spins according to Hund's rule. Orbital degeneracy of Mn^{3+} is removed by a strong Jahn–Teller effect, i.e., a spontaneous distortion of the oxygen octahedron lowers the energy of one of the e_g orbitals. The large Curie temperature is related to the fact that $La_{0.7}Sr_{0.3}MnO_3$ is the closest, among all manganites, to the cubic lattice structure where the ferromagnetic double-exchange interaction is the strongest because of the 180° Mn–O–Mn bond angles.

The coherent strain of both signs controlled by appropriate monocrystalline substrates causes distortions of the LSMO lattice leading to tetragonal (monoclinic) symmetry [14]. Rotation angles of oxygen octahedrons change with coherent strain [14, 17]. Larger coherent strain alters also bond lengths, whereby the orbital degeneracy is removed statically. Tensile strain stabilizes the in-plane $e_g\, x^2-y^2$ orbitals [38, 39], compressive strain the out-of-plane z^2-r^2 orbitals. The conducting ferromagnetic state is suppressed by both, (a) octahedral rotations reducing the Mn–O–Mn bond angle and thus, the ferromagnetic double-exchange interaction and (b) the static distortions of octahedrons leading to orbital ordering with dominating antiferromagnetic superexchange interactions [39, 40]. Indeed, the ferromagnetic metal phase of LSMO is strongly suppressed by the coherent elastic strain of either sign in experiments. Even though the Mn high spin state is stable against the elastic strain, the Curie temperature shows a strong strain-induced reduction of $|dT_C/d\varepsilon| \leq 200$ K $\%^{-1}$ [41, 42]. The largest values have been found in reversible-strain measurements where T_C of a single film is recorded in an adjustable strain range of the order of 0.1% [42].

1.2.3 Nickelates $ANiO_3$

Nickelates receive strong interest because they show a metal–insulator transition driven by strong electron correlations, which is sensitive to dimensionality and elastic strain [31, 43, 44]. $LaNiO_3$ is metallic and paramagnetic down to low temperatures, whereas rare earth metals at the A site result in an insulating antiferromagnetic ground state and a metal–insulator transition at a temperature T_{MI}. The effect of the rare earth metal on B site magnetic order is essentially an elastic one due to its lower ionic radius. The Néel temperature T_N and T_{MI} can differ by more than 100 K with $T_N \leq T_{MI}$ [31, 32, 44]. Metallic phases of $ANiO_3$ seem to show no magnetic long-range order. $LaNiO_3$ is rhombohedral with the rotation pattern $a^-a^-a^-$ and pseudocubic lattice parameter of 3.84 Å. Ni^{3+} ions have a nominal electronic $3d^7$ configuration with one electron in the e_g level and fully occupied t_{2g} levels (S = ½). Ni can take an electron configuration near $3d^8$ if a ligand hole at an adjacent oxygen ion is formed. In the insulating state, Ni ions show charge disproportionation associated with charge ordering, i.e., two different valence states of Ni ions exist and are regularly ordered [44, 45].

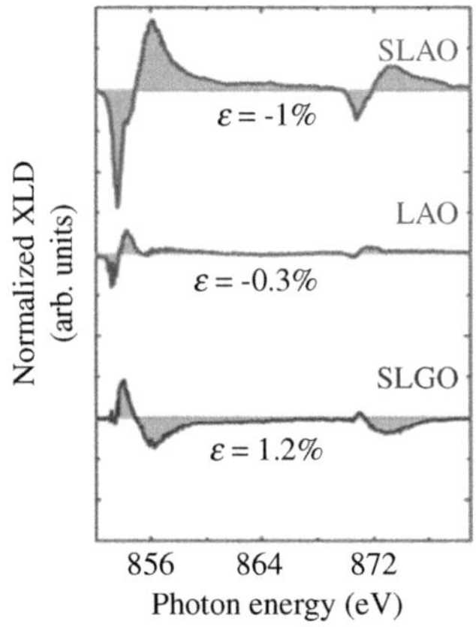

Figure 1.3 X-ray linear dichroism at the Ni $L_{2,3}$ edge of 10 nm thick $SmNiO_3$ films in different strain states ε as indicated. Measurements taken at 120 K. Substrates are $SrLaAlO_4$ (SLAO), $LaAlO_3$ (LAO), and $SrLaGaO_4$ (SLGO). Reprinted from Ref. [43], with permission from American Physical Society, Copyright 2013.

For nickelates, the effects of coherent elastic strain are not yet fully understood. Compressive strain stabilizes the metallic state in $SmNiO_3$ [43], $NdNiO_3$ [32], and $PrNiO_3$ [46]. Tensile strain can stabilize an antiferromagnetic insulating ground state [32]. X-ray linear dichroism of strained $SmNiO_3$ films confirms that tensile strain stabilizes the e_g in-plane (x^2–y^2) orbital, compressive strain the z^2–r^2 orbital (Fig. 1.3) [43].

1.2.4 Reversible Strain

Growing strained films on monocrystalline substrates with appropriate lattice parameters has been a very successful approach for measuring and controlling strain-dependent properties. In a few cases, however, strain-dependent or random defect formation hinders the evaluation of a true strain effect by comparing several differently strained films. An example is strain-dependent oxygen or metal vacancy formation [47, 48]. Reversible straining of a single film can overcome this problem and probe strain-dependent properties directly. Piezoelectric substrates of $0.72Pb(Mg_{1/3}Nb_{2/3})O_3$-$0.28PbTiO_3$ (PMN-PT) in (001) orientation have a sufficiently uniform in-plane lattice parameter to be used for quantitative studies of strain-dependent properties [29, 49, 50] (Fig. 1.4). The reversible strain of ≤ 0.15% at 300 K and ambient atmosphere [49], increased to 0.3% in ultrahigh vacuum [51, 52], can be linearly controlled by applying an electrical voltage across the substrate. The limited

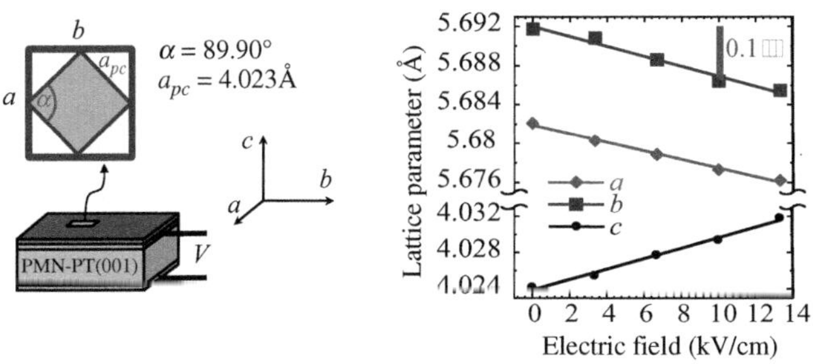

Figure 1.4 Piezoelectric substrate for reversible strain application. Lattice parameters at 300 K measured as a function of vertical electric field controlled by the voltage *V*. PMN-PT denotes the composition of $0.72PbMg_{1/3}Nb_{2/3}O_3$-0.28 $PbTiO_3$, which is on the rhombohedral side of the morphotropic phase boundary of PMN-PT. Adapted from Ref. [49], with permission from AIP Publishing, Copyright 2010.

strain range can be extended by growing films on appropriate buffer layers, for example, the compositionally tuned $LaSc_{1-x}Al_xO_3$ (LASO) [53]. Using this approach, strain-dependent Curie temperatures, ferromagnetic moments and electrical conduction of $SrRuO_3$ [29, 30], manganites [42, 54, 55], and cobaltites [56, 57] have been investigated (Fig. 1.5).

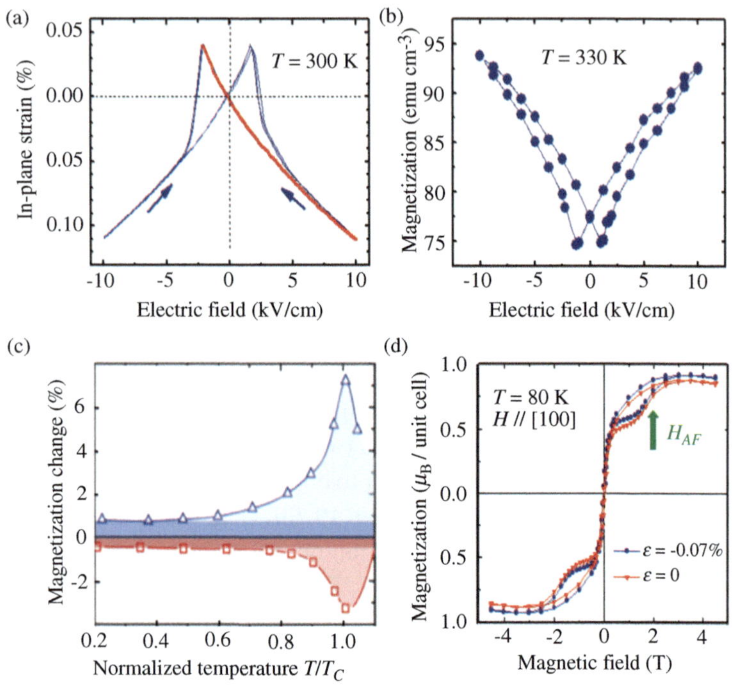

Figure 1.5 Strain-dependent magnetization measurements: (a) Biaxial in-plane strain butterfly loop for reversal of the polarization in the piezoelectric substrate (unipolar measurements along the red line). (b) Magnetization vs strain of a 50 nm thick $La_{0.7}Sr_{0.3}MnO_3$ film measured in 10 mT [42]. (c) Temperature-dependent magnetization change upon applying an electric field of 10 kV/cm to the piezoelectric substrate for $SrRuO_3$ films in different strain states [in-plane lattice parameter of 3.93 Å (red) and 3.96 Å (blue)] [29]. (d) Magnetic hysteresis loop of a $[La_{0.7}Sr_{0.3}MnO_3$ (22 Å)/$SrRuO_3$ (55 Å)$]_{15}$ superlattice in two reversibly controlled strain states [59].

Permanent, but removable lattice strain can be introduced into an oxide film by low-energy He atom implantation (Fig. 1.6) [19, 58]. The strain can reach a few percents. This type of strain is special in that it is purely along the film normal, while the in-plane lattice parameter stays constant. Thus, novel states of matter beyond the known elastic response of thin films or under hydrostatic pressure may be accessible. Annealing restores the original strain state without significant damage caused by He atoms [19, 58]. This approach seems promising also for strain patterning of nanoscale regions.

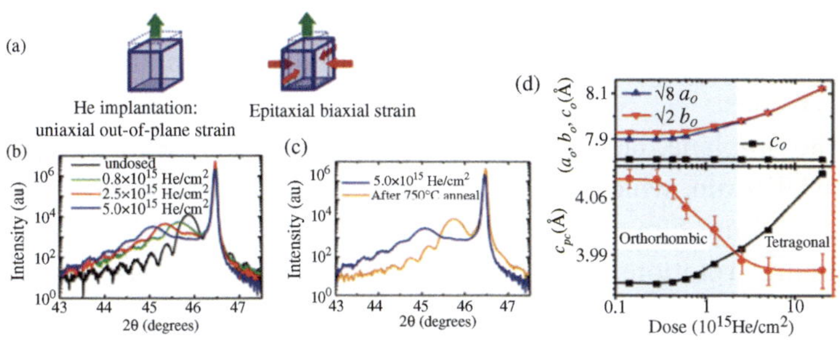

Figure 1.6 Uniaxial vertical strain controlled by implantation of low-energy He atoms: (a) Scheme highlighting the different types of epitaxial and implantation-induced strain. (b, c) X-ray diffraction patterns revealing the increasing out-of-plane lattice parameter of a $SrRuO_3$ film with the He dose and the recovery of initial strain state after annealing. (d) Lattice parameters of a $SrRuO_3$(20 nm)/$SrTiO_3$ (001) film vs the He dose. At a critical dose, the structure changes from orthorhombic to tetragonal. Note that the *c* parameter is in-plane. Adapted from Ref. [61], with permission from American Physical Society, Copyright 2011.

1.3 Elastic Control of Interface Magnetism

Here, examples of ABO_3–$A^*B^*O_3$ interfaces ordered in groups of B–B* ions are addressed, since these ions generate the magnetism. Mostly (001) oriented interfaces are considered. For a systematic discussion of other orientations, see Bhattacharya, A. et al. (2014) [10].

1.3.1 Magnetic Oxide–Non-magnetic Oxide Interfaces

In this section, interfaces are discussed where B is a magnetic ion (Ru, Mn, Ni) and B* a non-magnetic ion (Ti, Ga, Sc). At the interface, B* may become magnetic if a valence change occurs.

1.3.1.1 Mn–Ti, Ga

Interfaces between manganites and titanates are among the most studied interfaces because of their importance in magnetic tunnel junctions and magnetoelectric heterostructures [21, 22, 62, 63]. There are three inherent types of mismatch present at an ideal manganite–titanate interface caused by (i) the in-plane lattice parameter, (ii) the pattern of oxygen octahedral rotations, and (iii) the polar interface charge. For example, $Sr^{2+}O^{2-}$ and $Ti^{4+}O_2$ lattice planes of $SrTiO_3$ are neutral in contrast to (most) manganite A*O and MnO_2 planes.

The in-plane lattice parameter mismatch induces coherent epitaxial strain, which usually weakens ferromagnetic ordering of the manganite film as a bulk effect (Section 1.2.2). The interfacial octahedral rotations have been determined, for instance, at the $La_{0.7}Sr_{0.3}MnO_3/SrTiO_3$ (001) interface [17]. The first two unit cells of LSMO at the interface are found to be free of rotations ($a^0a^0a^0$); apparently, the cubic $SrTiO_3$ extinguishes the rotations in LSMO. At a larger distance from the interface, another rotation pattern is evident ($a^+a^-c^0$), which is different from the bulk pattern probably due to the effect of coherent strain. In superlattices of $[LaMnO_{3+d}]_n/[SrTiO_3]_n$, the $a^-a^-a^-$ rotations in the manganite decrease with increasing layer thickness (n = 2–16 uc, uc denotes unit cells) [64]. In agreement with enhanced double exchange for a larger bond angle, this is associated with a growing Curie temperature and ferromagnetic moment. The interfacial transfer of octahedral rotations works impressively on $NgGaO_3$ substrates where it has been applied to control the in-plane magnetic anisotropy of $La_{0.7}Sr_{0.3}MnO_3$ (Fig. 1.2) [15]. STEM annular bright field (ABF) images allow one to track the rotation angle about an in-plane axis across the interface. The manganite adopts a larger rotation angle for about 3 uc at the LSMO/$NdGaO_3$ interface. Interestingly, insertion of a single unit cell layer of $SrTiO_3$ at the interface suppresses the rotations in LSMO completely. As expected for an elastic response of two coupled oxides, the film can induce rotations into the substrate, too: a 9 uc thick $SrTiO_3$ interlayer effectively reduces the rotation angle in the $NdGaO_3$ substrate (Fig. 1.2). It remains an open question if an ideal interface can show several different (i.e., metastable) rotation patterns depending on growth conditions or if there is one specific interfacial rotation pattern for a pair of oxides.

An atomically sharp manganite–titanate interface of MnO_2-A*O termination has an excess negative charge of $(1 - x)e$, where the doping value x determines the fraction of Mn^{4+} and A* is a divalent metal. The AO–TiO_2 termination has a positive excess charge of the same magnitude. The 'polar catastrophe' (Section 1.1.3) can be avoided by transferring $(1 - x)/2$ electrons to the AO–TiO_2 interface. In agreement with this scenario, Bruno et al. detected Ti^{3+} at the LSMO/STO (001) interface using X-ray absorption spectroscopy. Thus, a magnetic moment has been induced to Ti at the interface. X-ray magnetic circular dichroism (XMCD) revealed the antiferromagnetic coupling of Ti spins to Mn spins in the LSMO/STO superlattices [62].

Other experiments on Mn–Ti interfaces detect the presence of interface roughness, which may be intrinsic or related to the film preparation process. $La_{1-x}Sr_xMnO_3$ ($x = 0.1 - 0.5$) films grown by pulsed laser deposition on TiO_2-terminated STO (001) substrates show a doping-dependent interface roughness of up to 4 uc, which is more pronounced when $La_{1-x}Sr_xMnO_3$ is an insulator (Fig. 1.7) [3]. This suggests an inherent interface reconstruction that involves moving ions to avoid the polar catastrophe when no mobile electrons are available [8]. Electronic charge distribution at the interface has been evaluated by electron energy loss spectroscopy (EELS) across the interface (Fig. 1.7). Ionic reconstruction in the sense of interdiffusion is often undesirable. Boschker et al. suggested a controlled compensation of the polar mismatch by growing a single unit cell layer of altered composition at the interface, which is $La_{0.33}Sr_{0.67}MnO_3$ for the TiO_2-terminated LSMO/STO interface [65].

Ferroelectric titanates such as $BaTiO_3$ and Pb(Zr, Ti)O_3 offer an active control of interface charge by reorienting the ferroelectric polarization. Thus, bound charge densities of up to ± 100 μC cm^{-2} {polarization of $PbZr_{0.2}Ti_{0.8}O_3$ (PZT) [66]} can be switched. Spurgeon et al. compare the Mn valence derived from EELS at both interfaces in an LSMO/PZT/LSMO tri-layer. A strong reduction occurs at the interface where the ferroelectric polarization is pointed to [63]. This indicates polarization screening by electrons, leading to a reduced concentration of hole-type charge carriers in LSMO. A strong drop of the ferromagnetic moment with respect to the bulk value is observed at this interface by polarized neutron reflectivity (PNR) measurements, whereas the other interface shows a more moderate reduction of the ferromagnetic moment. Realizing contacts for large-area switching of the polarization in $La_{0.825}Sr_{0.175}MnO_3$/PZT, Preziosi et al. could detect both valence and orbital control of Mn 3d e_g states by measuring X-ray linear dichroism (XLD) at the Mn $L_{2,3}$ edge [22]. The 3d x^2–y^2 in-plane orbital became more populated after switching the polarization to the direction pointing away from the interface, causing increased hole carrier (Mn^{4+}) density (Fig. 1.8). This is associated with a vertical shortening of the unit cell. Doping-dependent bulk phase diagrams of structural and magnetic properties often help to understand the effects of interface charge control but may fail if additional effects from different ionic size or reduced dimensionality dominate.

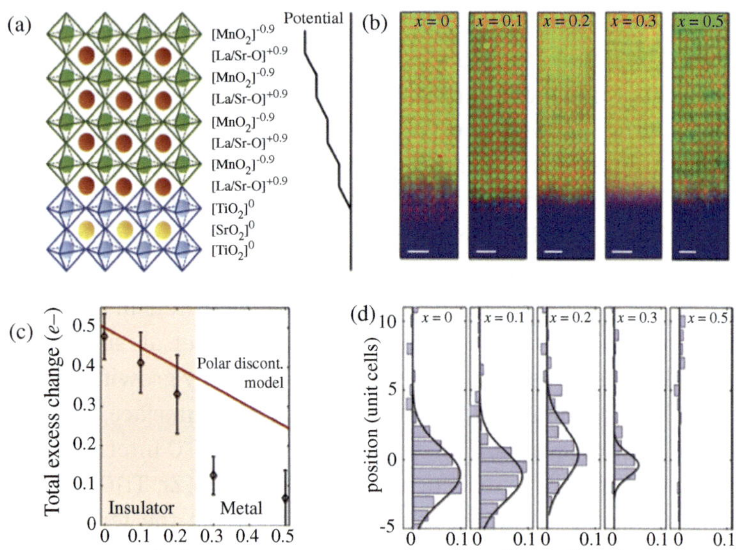

Figure 1.7 Charge and ionic distributions at $La_{1-x}Sr_xMnO_3/SrTiO_3$ (x = 0–0.5) interfaces as derived from scanning electron microscopy electron energy loss spectroscopy (STEM-EELS): (a) Lattice scheme with polar interface. (b) Elemental maps of Mn (green), La (red), and Ti (blue). (c) Total excess charge in the interface region vs Sr doping. (d) Depth-resolved excess charge of Mn ions. Reprinted from Ref. [3], with permission from Springer Nature, Copyright 2014.

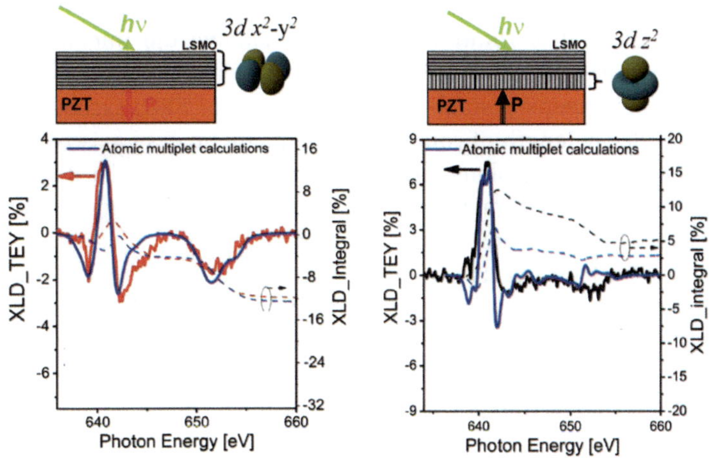

Figure 1.8 Control of orbital occupation by ferroelectric polarization reversal. X-ray linear dichroism at the Mn $L_{2,3}$ edge recorded at 300 K and simulated by atomic multiplet calculation. Adapted from Ref. [22], with permission from American Physical Society, Copyright 2015.

1.3.1.2 Ru–Ti, Ga, Sc

We again consider the types of mismatches at the interface, which are of structural or electric nature (charged lattice planes, oxygen octahedral rotations, and coherent strain). The effect of coherent strain on $SrRuO_3$ films has been addressed in Section 1.2.1. Octahedral rotation patterns do not match at any of the investigated interfaces regarding magnitude or type of the rotations, but there is growing experimental insight into the octahedral rotations of $SrRuO_3$ interfaces [16, 18, 67]. Kan et al. reported low interdiffusion and conserved B site termination, i.e., SrO–TiO_2 interfaces in films of $SrRuO_3/BaTiO_3$ [67] and $SrRuO_3/Ca_{0.5}Sr_{0.5}TiO_3$ [16] on $GdScO_3$ substrates. Interface properties, especially magnetic anisotropy, seem to be dominated by the transfer of octahedral tilts [16, 66]. $GdScO_3$ has a larger pseudocubic lattice parameter than SRO. Thus, it induces tensile strain and a tetragonal structure to the SRO layer (Section 1.2.1). At the interface to $GdScO_3$, interfacial rotations restore orthorhombic (monoclinic) symmetry of SRO [16]. A 1 uc thick $BaTiO_3$ layer at the interface suppresses the rotations [67], pointing to an immense stiffness of $BaTiO_3$ regarding octahedral rotations. This is in line with the similar effect of 1 uc of $SrTiO_3$ for

LSMO–$NdGaO_3$ [15] (Section 1.3.1.1). Quite striking is an apparent gradual tuning of rotation angles by changing the thickness of a $Ca_{0.5}Sr_{0.5}TiO_3$ interface layer [16]. Moreover, the tuned Ru–O–Ru bond angle keeps a stable value for 12 uc of SRO, exceeding the typical decay length of rotations. The magnetic easy axis reorients from canted out-of-plane into the film plane at 3 uc of $Ca_{0.5}Sr_{0.5}TiO_3$, demonstrating magnetic anisotropy control at another oxide interface based on the rotation pattern [16].

In $SrRuO_3$, SrO and RuO_2 lattice planes are charge-neutral. Hence, the nominal interface charge is zero for $SrTiO_3/SrRuO_3$, while it is $\pm e$ for SRO/$Nd^{3+}Ga^{3+}O_3$ and SRO/$GdScO_3$, changing the sign between the two terminations of the interfaces. Little data is yet available about a potential charge transfer to Ti, Ga, or Sc ions. The high conductivity of SRO serves to screen a polar interface, possibly preventing charge transfer to the non-magnetic ions of the other oxide.

On $SrTO_3$ (001) substrates, the coherent compressive strain of SRO stabilizes the $a^-a^-c^+$ rotation pattern of the orthorhombic lattice symmetry also found in bulk crystals [33]. Interestingly, a 4 nm $GdScO_3$ buffer layer seems to induce a tetragonal structure into a 10 nm thick SRO layer on STO (001) according to X-ray reciprocal space mapping [18]. On the other hand, measurements of half-order Bragg peaks indicate an orthorhombic rotation pattern [18]. Since interfacial rotations decay with growing distance from the interface, inhomogeneity is an inherent characteristic, which may lead to conflicting results from volume-averaging techniques.

1.3.1.3 Ni–Al, Ga, Sc

Nickelate interfaces with non-magnetic oxides have predominantly been studied in superlattices with ultrathin layers, because of the dimensionality effect on electron localization [68]. One should keep in mind that interfaces in such superlattices are not independent and single interfaces could have different properties.

Charge transfer due to polarity mismatch at these interfaces is experimentally yet little investigated. It may play a negligible role in superlattices with $RAlO_3$, $RScO_3$, or $RGaO_3$ (R = La or rare earth metal) because of identical charge of respective lattice planes if one assumes the nominal Ni^{3+} valence state. Charge disproportionation of Ni would, however, lead to a more complex situation at the interface. Terminating lattice planes are identical since A* = A in such superlattices.

Besides the coherent strain from substrates, the interfaces of nickelates with non-magnetic oxides are affected by the transfer of octahedral rotations, but single interfaces have not yet been studied. Qi et al. investigated the rotational pattern in a $LaNiO_3/LaGaO_3$ (4 uc/4 uc)$_8$ superlattice on $SrTiO_3$ (001) [4]. The rotation angle about an in-plane axis is zero at the substrate interface, consistent with the rotation-free substrate and continuously grows to 8° within 15 uc. This reveals correlated rotations in a superlattice of ultrathin layers, no fluctuations between layers could be resolved.

Contrary to the relative lack of structural data, systematic investigations of the orbital occupation have been conducted using linearly polarized X-ray absorption and X-ray reflectometry [32, 43, 60, 68, 69]. The depth-dependent orbital occupation has been derived with a precision of few percents from linearly polarized X-ray reflectometry data on a $LaNiO_3$ (4 uc)/$LaAlO_3$ (4 uc) superlattice [60] (Fig. 1.9). The Ni 3d x^2–y^2 orbital is more populated at the interface atomic layers than in the interior of the nickelate layer.

This behavior is similar in $PrNiO_3$ (4 uc)/$PrAlO_3$ (4 uc) superlattices [68]. Substrate-controlled strain states of the latter superlattices reveal a linear strain dependence of orbital occupation, with tensile strain stabilizing the in-plane x^2–y^2 orbitals (Fig. 1.9).

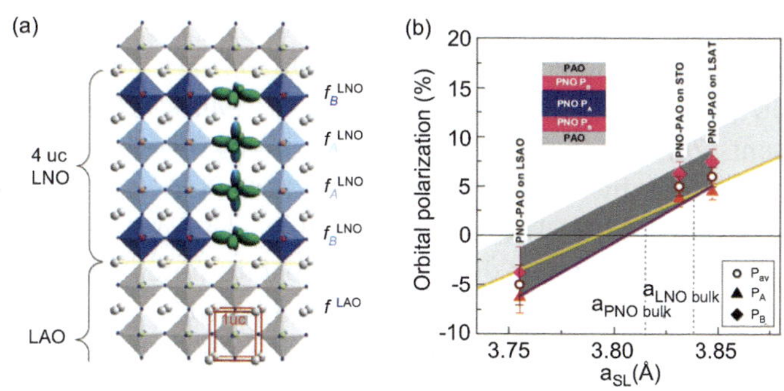

Figure 1.9 Layer-resolved occupation of Ni e_g orbitals derived from soft X-ray reflectometry: (a) Scheme of lattice and orbitals in 4 uc thick $LaNiO_3$ layers in a superlattice with $LaAlO_3$. (b) Strain-dependent orbital occupation in a similar $PrNiO_3/PrAlO_3$ superlattice grown coherently on different substrates with in-plane parameter a_{SL}. P_A and P_B denote the orbital polarization of the internal two atomic layers and the interface atomic layers, respectively. Adapted from Ref. [60], with permission from Springer Nature, Copyright 2011, and from Ref. [68], with permission from American Physical Society, Copyright 2015.

1.3.2 Interfaces between Two Magnetic Oxides

These interfaces naturally show coupling between magnetic transition metals at B sites via oxygen bonds in the ABO_3 lattice. Important aspects of spintronics are based on such interfaces, for instance, the magnetoelectric coupling of multiferroic $BiFeO_3$ to magnetic oxides [70], the exchange bias at oxide interfaces [71], and the spin filtering in magnetic tunnel barriers of tunnel junctions [72].

1.3.2.1 Mn–Fe

Ferrites like $BiFeO_3$ or $LaFeO_3$ are antiferromagnetic insulators. If they couple with ferromagnetic manganites, phenomena like exchange bias or induced ferromagnetic moments in the antiferromagnet may arise. In the beginning, various forms of mismatches at this type of interface are summarized: (i) polar mismatch between electric charges of the AO/BO_2 lattice planes and the difference of chemical potentials causing electron transfer, (ii) mismatch of oxygen octahedral rotations and (iii) coherent lattice strain due to different lattice parameters. Most interfaces can have two different terminations. Current research rarely addresses all of these aspects for a given interface, because it would still require extraordinary experimental efforts. In particular, there is yet little known about octahedral rotations or orbital occupation at Mn–Fe interfaces.

The most studied interface between ternary oxides, besides $LaAlO_3/SrTiO_3$, is probably that of $La_{0.7}Sr_{0.3}MnO_3/BiFeO_3$ (001) at present [70, 73]. Both interface terminations have been prepared and analyzed by high-angle annular dark field (HAADF) STEM and by EELS [70]. The polar mismatch changes sign with the termination, resulting in different electrostatic potentials at the two interfaces. This is reflected in opposite self-poled ferroelectric states and shifted ferroelectric hysteresis loops of BFO [70]. $BiFeO_3$ gains a ferromagnetic moment antiparallel to the in-plane Mn spins at the interface with LSMO according to XMCD measurements, while the antiferromagnetic Fe spin orientation is perpendicular to the interface [73]. In a BFO/LSMO (6 uc/5 uc) superlattice, BFO layers show significant in-plane magnetization observed by polarized neutron reflectivity [73]. Density functional theory in that work predicts ferromagnetic Mn–Fe coupling for the BiO-terminated interface and antiferromagnetic Mn–Fe coupling for the FeO_2

termination. This result suggests a case of heavy dependence of magnetic coupling on interface termination and a strong impact of interdiffusion at real interfaces for the achievable magnetic coupling.

An interface-induced ferromagnetic Fe moment has also been observed in antiferromagnetic $LaFeO_3$ sandwiched in a coherent trilayer with LSMO [72]. The two $LaFeO_3$ interfaces have distinctly different roughness, with the upper interface showing at least 1 uc of interdiffusion, whereas the lower interface is rather abrupt as derived from STEM-EELS elemental maps. XMCD photoemission electron microscopy (PEEM) was applied to image the in-plane magnetic domains at Mn and Fe L_3 edges for both interfaces (Fig. 1.10). BFO domains directly mirror the LSMO domains, with an antiparallel Mn–Fe alignment. As a consequence, the insulating $LaFeO_3$ acts as a tunnel barrier with the spin filter effect, i.e., a spin-dependent tunnel barrier height is imprinted by the domain pattern in the electrodes [72].

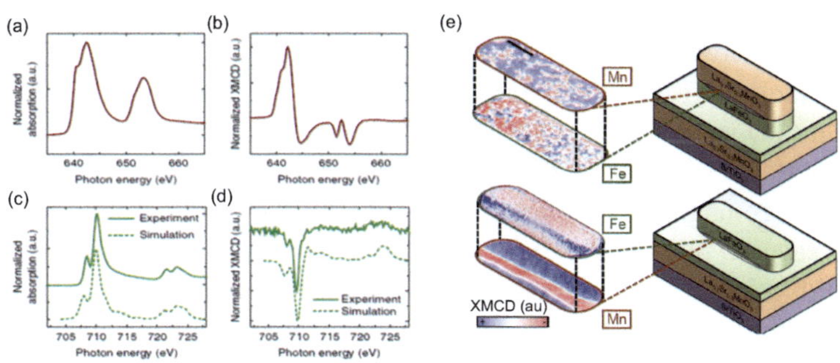

Figure 1.10 Induced ferromagnetic Fe moment at $LaFeO_3/La_{0.7}Sr_{0.3}MnO_3$ (001) interfaces. X-ray absorption (a, c) and X-ray magnetic circular dichroism (XMCD) (b, d) spectra showing the antiferromagnetic alignment of Mn and Fe. (e) XMCD photoemission electron microscopy images of the upper and the lower interface in a tri-layer magnetic tunnel junction. Reprinted from Ref. [72], with permission from American Physical Society, Copyright 2015.

$La_{0.7}Sr_{0.3}FeO_3$ (LSFO) is a related insulating G-type antiferromagnet with a Néel temperature of 360 K that has been investigated in superlattices with LSMO [74]. Nominally, this interface has no polar mismatch. Nevertheless, the mixed valence of Mn and Fe shifts according to the redox reaction of $Mn^{3+} + Fe^{4+} \rightarrow Mn^{4+} + Fe^{3+}$ [74], equilibrating the chemical potential. Jia et al. studied LSFO/

LSMO (111) superlattices and found the (111) plane offers parallel alignment of Fe spins with the interface. Interestingly, this interface does not show an induced Fe magnetic moment as concluded from Fe $L_{2,3}$ XMCD measurements. Magnetic Mn–Fe coupling is strong enough that the antiferromagnetic spin orientation of Fe switches together with the Mn magnetic moments [74]. X-ray magnetic linear dichroism (XMLD) at Fe $L_{2,3}$ edge has been used to determine the antiferromagnetic spin orientation of Fe [74, 75]. It changes from parallel to perpendicular to the interface with decreasing layer thickness of LSFO [74]. The perpendicular coupling (also termed *spin–flop coupling*) is probably responsible for the absence of an exchange bias effect. LSMO layers show an enhancement of magnetic coercive fields, but no shift of magnetic hysteresis along the field axis.

1.3.2.2 Mn–Ru

The interface between $SrRuO_3$ and a ferromagnetic manganite connects two conducting ferromagnets. Ke et al. reported on an exchange bias effect and antiferromagnetic coupling between LSMO and SRO in bilayers on $SrTiO_3$ (001) [76, 77]. Subsequent work on superlattices of these oxides revealed an immense strength of the antiferromagnetic coupling, with low-temperature coupling fields of several Tesla [78, 79]. (The coupling field H_{AF} is the magnetic field required to overcome the antiferromagnetic interface coupling and align both ferromagnetic layers in the field direction. H_{AF} approximately grows with the reciprocal layer thickness of the switching layer.) Therefore, the LSMO-SRO interface received profound interest in recent years. Mn XMCD has been applied to probe the orientation of Mn moments after aligning Ru moments in an external field [80], indicating a sign change of the Mn–Ru coupling to ferromagnetic for a smaller in-plane lattice parameter of 3.79 Å on $LaAlO_3$ (001) substrates. The strain-dependent sign change of magnetic coupling has been suggested to originate from altered orbital occupation at the interface [80]. In a thorough study of magnetic switching of an LSMO/SRO/STO (001) bilayer, Solignac et al. concluded on the presence of two distinct coupling constants at the interface [81] (of unknown origin). XMCD measurements at the L_3 edges of both magnetic elements made it possible to track the elemental magnetic hysteresis of Ru and Mn, helping to clarify the contributions of component layers to the peculiar field- and temperature-dependent magnetization curves. Figure 1.11 shows an example recorded on an LSMO (9 uc)/SRO (14 uc)/STO (001) bilayer.

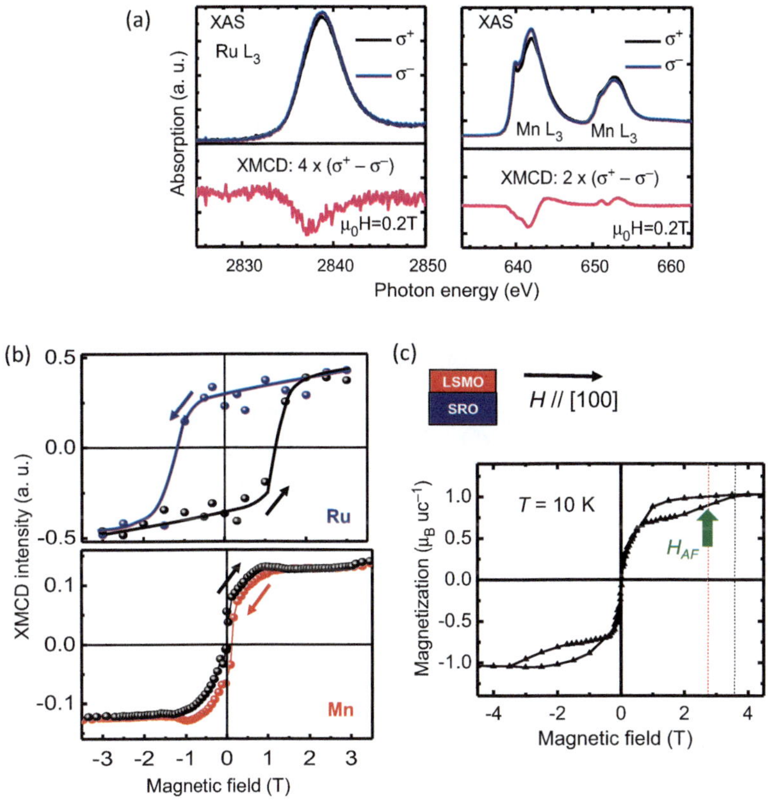

Figure 1.11 Element-resolved magnetic switching at the $La_{0.7}Sr_{0.3}MnO_3$/$SrRuO_3$ (RuO_2-terminated) interface: (a) X-ray absorption and magnetic circular dichroism spectra of a LSMO(9 uc)/SRO(14 uc)/$SrTiO_3$ (001) bilayer at 60 K recorded along the in-plane (100) direction in grazing incidence. (b) Elemental hysteresis loops of Ru and Mn derived from field-dependent spectra. The loop inversion of Mn results from antiferromagnetic interface coupling. Measurements taken at ALBA synchrotron light source. (c) Magnetization loop along the same in-plane direction at 10 K. Ru and Mn are aligned in field direction at H_{AF}.

The LSMO-SRO interface has two compositional variants depending on the terminating atomic planes of the layers (Section

1.1.1), MnO_2–SrO(–RuO_2) and (MnO_2–)$La_{0.7}Sr_{0.3}O$–RuO_2. Systematic STEM analysis is available [2, 7, 78, 79] showing that this interface (and related ones with other manganites) can be fabricated with very low interdiffusion of metal ions. Figure 1.12 gives an example of the two terminations observed in LSMO/SRO/STO (001) and SRO/LSMO/STO (001) bilayers grown by pulsed laser deposition. HAADF Z contrast imaging has been favorable for characterizing this interface because of the large mass differences of metals Mn and Ru at B sites as well as La and Sr at A sites. Bilayers with one or the other growth sequence on TiO_2-terminated STO substrates show opposite terminations with low interdiffusion (<10%).

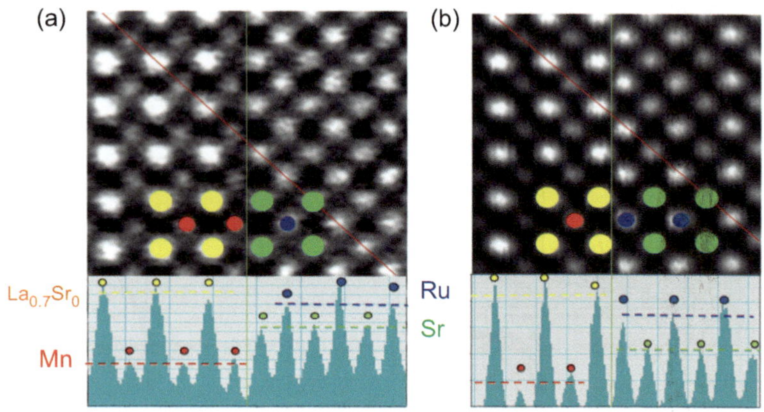

Figure 1.12 Scanning transmission electron microscopy high-angle annular dark field (STEM-HAADF) images of the two terminations of the $La_{0.7}Sr_{0.3}MnO_3$/$SrRuO_3$ interface. Lower panels are intensity profiles along the indicated red lines. The interface is marked by a green line. Average intensities for the four lattice sites ($La_{0.7}Sr_{0.3}$, Sr, Mn, Ru) indicated by dashed lines.

Polar lattice planes of LSMO cause a nominal charge of 0.7e at the interface, which has opposite sign for the two terminations. Screening by mobile electrons is efficient here since both oxides are good conductors. No systematic X-ray absorption studies of the valence and orbital states seem to be available for this interface type at present. On the structural side, octahedral rotation patterns are well-known for single layers of the components (Section 1.2), whereas the interfacial rotations are yet to be investigated. Different magnetic anisotropy for the two interface terminations suggests termination-dependent interfacial rotation pattern.

Density functional theory (DFT) calculations indicate that antiferromagnetic coupling is about three times stronger for the MnO_2–SrO termination than for the other one. Experimentally, the magnetic switching of Ru and Mn moments has been derived from field-dependent XMCD data at the Mn $L_{2,3}$ and RuL_3 edges. Figure 1.11 shows in-plane magnetization loops of Ru and Mn at 60 K. The Mn hysteresis is inverted because of antiferromagnetic coupling between Mn and Ru, which leads to a reversal of Mn to establish the antiferromagnetic configuration at remanence. For the other (i.e., the MnO_2–SrO) termination, the rigid coupling of Ru and Mn spins at the interface results in simultaneous switching of the LSMO layer and an interface layer in $SrRuO_3$, whereas the interface-far part of the SRO layer remains unchanged. Here, the magnetic interface coupling is apparently stronger than the internal ferromagnetic exchange in SRO. Thus, XMCD measurements confirm the different antiferromagnetic coupling strength for different interface terminations as predicted by DFT and reveal that magnetic interface coupling can induce a non-collinear spin texture in $SrRuO_3$.

The coupling strength at the interface can also be estimated from temperature- and field-dependent magnetization data. Cooling runs in small applied magnetic field show a distinct drop of magnetization at the Curie temperature of SRO. A sufficiently large cooling field can overcome the antiferromagnetic interface coupling and align the magnetization of both layers. Switching off the field at low temperatures leads to recovery of the antiferromagnetic alignment. Similarly, magnetization loops show a field-induced transition at the coupling field H_{AF} originating from breaking the antiferromagnetic coupling [78, 79] (Fig. 1.11). It must be considered, though, that SRO is not in a collinear single-domain state. Due to the large magnetocrystalline anisotropy of SRO, removal of domains does require a large field, too, and may in some cases be hard to distinguish from the first mechanism without insight into an element-specific switching. The strain dependence of H_{AF} of an LSMO/SRO superlattice has been probed by inducing reversible epitaxial strain from a piezoelectric substrate (Fig. 1.5d) [59]. The large strain response, however, may rather reflect the behavior of the entire superlattice with ultrathin layers (2.2 nm LSMO, 5.5 nm SRO) than that of the antiferromagnetic coupling at a single interface.

1.3.2.3 Mn–Ni

Interfaces of paramagnetic $LaNiO_3$ with manganites present a fascinating opportunity for achieving interface-induced magnetic order. Between parent compounds $La^{3+}Ni^{3+}O_3$ and $La^{3+}Mn^{3+}O_3$, there is no polar mismatch at interfaces. Nevertheless, there is strong charge transfer according to the redox reaction of $Mn^{3+} + Ni^{3+} \rightarrow Mn^{4+} + Ni^{2+}$ for equilibrating chemical potentials of electrons. This agrees with the fact that in double perovskites La_2NiMnO_6, valence states are Mn^{4+} and Ni^{2+} [61]. X-ray absorption at the Mn $L_{2,3}$ edge reveals charge transfer at the interface as a shift of the peaks towards higher energy [82]. On the structural side, octahedral rotations at Mn–Ni type interfaces are mismatching and yet unknown. For example, bulk $LaMnO_3$ is orthorhombic with rotation pattern $a^-a^-c^+$ and Mn–O–Mn bond angles of 155°, while $LaNiO_3$ is rhombohedral with $a^-a^-a^-$ pattern and Ni–O–Ni bond angles of 162°.

An impressive body of results on $LaMnO_3/LaNiO_3$ interfaces has been collected by Gibert and coworkers [71, 82]. In (001)-oriented bilayers on STO substrates, substantial intermixing occurs within 2–3 uc at the interface if $LaNiO_3$ is grown on top of $LaMnO_3$. On the contrary, the same interface is atomically sharp if films are grown in reversed sequence. This kind of interdiffusion asymmetry with respect to growth sequence is repeatedly observed for oxide interfaces. XMCD at the Mn $L_{2,3}$ edge reveals a ferromagnetic Mn moment, which is much larger in the bilayer with the rough interface. Further, XMCD at the Ni $L_{2,3}$ edge gives evidence for a ferromagnetic Ni moment, which is again larger in the rough-interface sample. These results indicate, in agreement with expectations, that charge transfer is more efficient at the interdiffused interface. Ni^{2+}–O–Mn^{4+} superexchange is positive, leading to the ferromagnetic ordering of Mn and Ni moments near the interface [82]. Consequently, the total magnetization of the rough-interface sample is higher than the sum of the magnetizations of single layers. Artificially created intermixing by deposition of submonolayers at the smooth interface confirmed this result.

In a bilayer system of $LaNiO_3$ (t)/LSMO (24 uc)/STO (001), t = 0–72 uc, Peng et al. investigated the Mn e_g orbital occupation based on X-ray linear dichroism (XLD) at the Mn $L_{2,3}$ edge [69]. The tensile strain of the manganite layer stabilizes the x^2–y^2 orbitals

(Section 1.2.2). Strikingly, with increasing thickness of $LaNiO_3$, the observed occupation of Mn z^2–r^2 out-of-plane orbitals grows and even dominates. This interface type agrees with the interdiffused one from Gibert′s studies and is, thus, likely to show strong charge transfer. The electron transferred from Mn to Ni is suggested by the authors to reside in the Ni z^2–r^2 orbital, which mediates ferromagnetic coupling by hybridization with the Mn z^2–r^2 orbital (Fig. 1.1c). Further enhancement of Mn z^2–r^2 orbital occupation with increasing thickness of the $LaNiO_3$ top layer beyond, say, $t = 20$ uc is unlikely as an interface effect. Peng et al. suggest a top layer–induced compressive strain that would favor the out-of-plane e_g orbitals [69], but break the coherent strain of the bilayer.

References

1. He, Q. et al. (2015). Towards 3d mapping of BO_6 octahedron rotations at perovskite heterointerfaces, unit cell by unit cell, *ACS Nano.*, **9**, pp. 8412–8419.
2. Hillebrand, R. et al. (2011). A study of intermixing in perovskite superlattices by simulation-supported Cs-corrected HAADF-STEM, *Phys. Status Solidi A*, **208**, pp. 2144–2149.
3. Mundy, J. A. et al. (2014). Visualizing the interfacial evolution from charge compensation to metallic screening across the manganite metal-insulator transition, *Nat. Commun.*, **5**, p. 3464.
4. Qi, H. Y. et al. (2015). Local octahedral rotations and octahedral connectivity in epitaxially strained $LaNiO_3$/$LaGaO_3$ superlattices, *J. Mater. Sci.*, **50**, pp. 5300–5306.
5. Ohtomo, A. et al. (2004). A high mobility electron gas at the $LaAlO_3$/$SrTiO_3$ heterointerface, *Nature*, **427**, pp. 423–426.
6. Park, J. et al. (2013). Oxygen-vacancy-induced orbital reconstruction of Ti ions at the interface of $LaAlO_3$/$SrTiO_3$ heterostructures: a resonant soft X-ray scattering study, *Phys. Rev. Lett.*, **110**, p. 017401.
7. Borisevich, A. Y. et al. (2012). Interface dipole between two metallic oxides caused by localized oxygen vacancies, *Phys. Rev. B*, **86**, p. 140102(R).
8. Nakagawa, N. et al. (2006). Why some interfaces cannot be sharp, *Nat. Mater.*, **5**, pp. 204–209.
9. Glazer, A. M. (1975). *Acta. Cryst. A*, **31**, pp. 756–762.

10. Bhattacharya, A. et al. (2014). Magnetic oxide heterostructures, *Annu. Rev. Mater. Res.*, **44**, pp. 65–90.

11. Biegalski, M. D. et al. (2014). Interrelation between structure and magnetic properties in $La_{0.5}Sr_{0.5}CoO_3$, *Adv. Mater. Interfaces*, **1**, p. 1400203.

12. Herklotz, A. et al. (2015). Characterization of the tetragonal phase of $SrRuO_3$ under epitaxial strain by density functional theory, *Eur. Phys. J. B*, **88**, p. 60.

13. May, S. et al. (2010). Quantifying octahedral rotations in strained perovskite oxide films, *Phys. Rev. B*, **82**, p. 014110.

14. Vaillionis, A. et al. (2011). Misfit strain accommodation in epitaxial ABO_3 perovskites: lattice rotations and lattice modulations, *Phys. Rev. B*, **83**, p. 064101.

15. Liao, Z. et al., (2016). Controlled lateral anisotropy in correlated manganite heterostructures by interface engineered oxygen octahedral coupling, *Nat. Mater.*, **15**, pp. 425–432.

16. Kan, D. et al. (2016). Tuning magnetic anisotropy by interfacial engineering the oxygen coordination environment in a transition metal oxide, *Nat. Mater.*, **15**, pp. 432–438.

17. Vaillionis, A. et al. (2014). Symmetry and lattice mismatch induced strain accommodation near and away from correlated perovskite interfaces, *Appl. Phys. Lett.*, **105**, p. 131906.

18. Gao, R. et al. (2016). Interfacial octahedral rotation mismatch control of the symmetry and properties of $SrRuO_3$, *ACS Appl. Mater. Interfaces*, **8**, pp. 14871–14878.

19. Herklotz, A. et al. (2016). Continuously controlled optical band gap in oxide semiconductor thin films, *Nano. Lett.*, **16**, pp. 1782–1786.

20. Okamoto, S. (2010). Magnetic interaction at an interface between manganite and other transition metal oxides, *Phys. Rev. B*, **82**, p. 024427.

21. Molegraaf, H. J. A. et al. (2008). Magnetoelectric effects in complex oxides with competing ground states, *Adv. Mater.*, **21**, pp. 3470–3474.

22. Preziosi, D. et al. (2015). Electric field control of the orbital occupancy and magnetic moment of a transition-metal oxide, *Phys. Rev. Lett.*, **115**, p. 157401.

23. Anderson, P. W. (1950). Antiferromagnetism. Theory of superexchange interaction, *Phys. Rev.*, **79**, pp. 350–356.

24. Goodenough, J. B. (1955). Theory of the role of covalence in the perovskite-type manganites [La,M(II)]MnO_3, *Phys. Rev.*, **100**, pp. 564–573.

25. Kanamori, J. (1959). Superexchange interaction and symmetry properties of electron orbitals, *J. Phys. Chem. Solids*, **10**, pp. 87–98.

26. Khomskii, D. (2001). *Spin Electronics*, Chapter 'Electronic structure, exchange and magnetism in oxides', Thornton, M. J. and Ziese, M. (eds.), Berlin/Heidelberg, Springer-Verlag, pp. 89–116.

27. Middey, S. et al. (2011). Dependence of magnetism on $GdFeO_3$ distortion in the t_{2g} system $ARuO_3$ (A = Sr, Ca), *Phys. Rev. B*, **83**, p. 014416.

28. Fuchs, D. et al. (2008). Tuning the magnetic properties of $LaCoO_3$ films by epitaxial strain, *Phys. Rev. B*, **77**, p. 014434.

29. Herklotz, A. et al. (2013). Magnetism of the tensile-strain-induced tetragonal state of $SrRuO_3$ films, *Phys. Rev. B*, **88**, p. 144412.

30. Herklotz, A. et al. (2914). Strain response of magnetic order in perovskite-type oxide films, *Phil. Trans. R. Soc. A*, **372**, p. 20120441.

31. Lee, J. H. (2010). A strong ferroelectric ferromagnet created by means of spin-lattice coupling, *Nature*, **466**, pp. 954–958.

32. Liu, J. et al. (2013). Heterointerface engineered electronic and magnetic phases of $NdNiO_3$ thin films, *Nat. Commun.*, **4**, p. 2714.

33. Choi, K. J. et al. (2010). Phase-transition temperatures of strained single-crystal $SrRuO_3$ thin films, *Adv. Mater.*, **22**, pp. 759–762.

34. Koster, G. et al. (2012). Structure, physical properties and applications of $SrRuO_3$ thin films, *Rev. Mod. Phys.*, **84**, pp. 253–298.

35. Ziese, M. et al. (2010). Structural symmetry and magnetocrystalline anisotropy of $SrRuO_3$ films on $SrTiO_3$, *Phys. Rev. B*, **81**, p. 184418.

36. Zayak, A. T. et al. (2006). Structural, electronic and magnetic properties of $SrRuO_3$ under epitaxial strain, *Phys. Rev. B*, **74**, p. 094104.

37. Dagotto, E. et al. (2008). Recent developments in the theoretical study of phase separation in manganites and underdoped cuprates, *J. Phys. Cond. Matter*, **20**, p. 434224.

38. Huijben, M. et al. (2008). *Phys. Rev. B*, **78**, p. 094413.

39. Tebano, A. et al. (2008). Evidence of orbital reconstruction at interfaces in ultrathin $La_{0.67}Sr_{0.33}MnO_3$ films, *Phys. Rev. Lett.*, **100**, p. 137401.

40. Tokura, Y. et al. (2000). Orbital physics in transition metal oxides, *Science*, **288**, pp. 462–468.

41. Adamo, C. et al. (2009). Effect of biaxial strain on the electrical and magnetic properties of (001) $La_{0.7}Sr_{0.3}MnO_3$ thin films, *Appl. Phys. Lett.*, **95**, p. 112504.
42. Thiele, C. et al. (2007). Influence of strain on the magnetization and magnetoelectric effect in $La_{0.7}A_{0.3}MnO_3$/PMN-PT (001) (A = Sr, Ca), *Phys. Rev. B*, p. 054408.
43. Bruno, F. Y. et al. (2013). Rationalizing strain engineering effects in rare earth nickelates, *Phys. Rev. B*, **88**, p. 195108.
44. Catalan, G. (2008). Progress in perovskite nickelate research, *Phase Transit.*, **81**, pp. 729–749.
45. Garcia-Munoz, J. L. et al. (2009). Structure and charge order in the antiferromagnetic band-insulating phase of $NdNiO_3$, *Phys. Rev. B*, **79**, p. 134432.
46. Kumar, A. et al. (2006). Substrate effect on electrical transport properties of $RNiO_3$ thin films prepared by pulsed laser deposition, *J. Phys. D*, **39**, p. 5310.
47. Aschauer, U. et al. (2013). Strain-controlled oxygen vacancy formation and ordering in $CaMnO_3$, *Phys. Rev. B*, **88**, p. 054111.
48. Kushima, A. et al. (2010). Competing strain effects in reactivity of $LaCoO_3$ with oxygen, *Phys. Rev. B*, **82**, p. 115435.
49. Biegalski, M. D. et al. (2010). Applying uniform reversible strain to epitaxial oxide films, *Appl. Phys. Lett.*, **96**, p. 151905.
50. Guo, E. J. et al. (2015). Ferroelectric 180° domain wall motion controlled biaxial strain, *Adv. Mater.*, **27**, pp. 1615–1618.
51. Ding, F. et al. (2010). Stretchable graphene: a close look at fundamental parameters through biaxial straining, *Nano. Lett.*, **10**, pp. 3454–3458.
52. Rastelli, A. et al. (2012). Controlling quantum dot emission by integration of semiconductor nanomembranes onto piezoelectric actuators, *Phys. Status Solidi B*, **249**, pp. 687–696.
53. Herklotz, A. et al. (2010). Wide-range strain tunability provided by epitaxial $LaSc_{1-x}Al_xO_3$ template films, *New J. Phys.*, **12**, p. 113053.
54. Das, S. et al. (2014). Static and reversible elastic strain effects on magnetic order of $La_{0.7}Ca_{0.3}MnO_3/SrTiO_3$ superlattices, *J. Appl. Phys.*, **115**, p. 143902.
55. Dekker, M. C. et al. (2011). Magnetoelastic response of $La_{0.7}Sr_{0.3}MnO_3/$ $SrTiO_3$ superlattices to reversible strain, *Phys. Rev. B*, **84**, p. 054463.
56. Herkotz, A. et al. (2010). Reversible strain effect on magnetization of $LaCoO_3$ films, *Phys. Rev. B*, **79**, p. 092409.

57. Rata, A. D. et al. (2008). Strain-induced insulator state and giant gauge factor of $La_{0.7}Sr_{0.3}CoO_3$ films, *Phys. Rev. Lett.*, **100**, p. 076401.

58. Herklotz, A. et al. (2016). Controlling octahedral rotations in a perovskite via strain doping, *Sci. Rep.*, **6**, p. 26491.

59. Das, S. et al. (2015). Strain dependence of antiferromagnetic interface coupling in $La_{0.7}Sr_{0.3}MnO_3/SrRuO_3$ superlattices, *Phys. Rev. B*, **91**, p. 134405.

60. Benckiser, E. et al. (2011). Orbital reflectometry of oxide heterostructures, *Nat. Mater.*, **10**, pp. 189–193.

61. Das, H. et al. (2008). Electronic structure, phonons and dielectric anomaly in ferromagnetic insulating double perovskite La_2NiMnO_6, *Phys. Rev. Lett.*, **100**, p. 186402.

62. Bruno, F. Y. et al. (2011). Electronic and magnetic reconstructions in $La_{0.7}Sr_{0.3}MnO_3/SrTiO_3$ heterostructures: a case of enhanced interlayer coupling controlled by the interface, *Phys. Rev. Lett.*, **106**, p. 147205.

63. Spurgeon, S. R. et al. (2014). Polarization-screening induced magnetic phase gradients at complex oxide interfaces, *Nat. Commun.*, **6**, p. 6735.

64. Zhai, X. et al. (2013). Correlating interfacial octahedral rotations with magnetism in $(LaMnO_{3+d})_N/(SrTiO_3)_N$ superlattices, *Nat. Commun.*, **5**, p. 4283.

65. Boschker, H. et al. (2012). Preventing the reconstruction of the polar discontinuity at oxide interfaces, *Adv. Funct. Mater.*, **22**, pp. 2235–2240.

66. Vrejoiu, I. et al. (2006). Intrinsic ferroelectric properties of strained tetragonal $PbZr_{0.2}Ti_{0.8}O_3$ obtained on layer-by-layer-grown, defect-free single-crystalline films, *Adv. Mater.*, **18**, pp. 1657–1661.

67. Kan, D. et al. (2014). Unit-cell thick $BaTiO_3$ blocks octahedral tilt propagation across oxide heterointerface, *J. Appl. Phys.*, **115**, p. 184304.

68. Wu, M. et al. (2015). Orbital reflectometry of $PrNiO_3/PrAlO_3$ superlattices, *Phys. Rev. B*, **91**, p. 195130.

69. Peng, J. et al. (2015). Charge transfer and orbital reconstruction in strain-engineered $(LaSr)MnO_3/LaNiO_3$ heterostructures. *ACS Appl. Mater. Interfaces*, **7**, pp. 17700–17706.

70. Yu, P. et al. (2012). Oxide interfaces: pathways to novel phenomena, *Mater. Today*, **15**, pp. 320–327.

71. Gibert, M. et al. (2012). Exchange bias in $LaNiO_3$–$LaMnO_3$ superlattices, *Nat. Mater.*, **11**, pp. 195–198.

72. Bruno, F. Y. et al. (2015). Insight into spin transport in oxide heterostructures from interface-resolved magnetic mapping, *Nat. Commun.*, **6**, p. 6306.

73. Singh, S. et al. (2014). Induced magnetization in $La_{0.7}Sr_{0.3}MnO_3/BiFeO_3$ superlattices, *Phys. Rev. Lett.*, **113**, p. 047204.

74. Jia, Y. et al. (2016). Thickness dependence of exchange coupling in (111)-oriented perovskite oxide superlattices, *Phys. Rev. B*, **93**, p. 1044403.

75. Lüning, J. et al. (2003). Determination of the antiferromagnetic spin axis in epitaxial $LaFeO_3$ films by X-ray magnetic linear dichroism spectroscopy, *Phys. Rev. B*, **67**, p. 214433.

76. Ke, X. et al. (2004). Positive exchange bias in ferromagnetic $La_{0.67}Sr_{0.33}MnO_3/SrRuO_3$ bilayers, *Appl. Phys. Lett.*, **84**, p. 5458.

77. Ke, X. et al. (2005). Antiferromagnetic exchange bias in epitaxial ferromagnetic $La_{0.67}Sr_{0.33}MnO_3/SrRuO_3$ bilayers, *J. Appl. Phys.*, **97**, p. 10K115.

78. Ziese, M. et al. (2010). Tailoring magnetic interlayer coupling in $La_{0.7}Sr_{0.3}MnO_3/SrRuO_3$ superlattices, *Phys. Rev. Lett.*, **104**, p. 167203.

79. Ziese, M. et al. (2010). Inverted hysteresis and giant exchange bias in $La_{0.7}Sr_{0.3}MnO_3/SrRuO_3$ superlattices, *Appl. Phys. Lett.*, **97**, p. 052504.

80. Seo, J. W. et al. (2010). Tunable magnetic interaction at the atomic scale in oxide heterostructures, *Phys. Rev. Lett.*, **105**, p. 167206.

81. Solignac, A. et al. (2012). Dual antiferromagnetic coupling at $La_{0.67}Sr_{0.33}MnO_3/SrRuO_3$ interfaces, *Phys. Rev. Lett.*, **109**, p. 027201.

82. Gibert, M. et al. (2015). Interfacial control of magnetic properties at $LaNiO_3/LaMnO_3$ interfaces, *Nano Lett.*, **15**, pp. 7355–7361.

Chapter 2

Interface Engineering in $La_{0.67}Sr_{0.33}MnO_3$–$SrTiO_3$ Heterostructures

Hans Boschker,[a,b] Zhaoliang Liao,[b] Mark Huijben,[b] Gertjan Koster,[b] and Guus Rijnders[b]
[a]*Max Planck Institute for Solid State Research, 70569 Stuttgart, Germany*
[b]*Mesaplus Institute for Nanotechnology and Faculty of Science and Technology, University of Twente, 7500AE Enschede, The Netherlands*
h.boschker@fkf.mpg.de

2.1 Introduction

Oxide epitaxy has made tremendous advances in the past 15 years. Using techniques such as pulsed laser deposition (PLD) and molecular beam epitaxy (MBE), oxide epitaxy has evolved from merely depositing material with the right crystal structure onto a substrate to growing the material in a layer-by-layer fashion. The latter allows interventions during the growth and therefore enables the creation of artificial materials. The progress in epitaxy is not only due to improvements in the growth systems and growth monitoring but also to a better understanding of the ways materials influence

Oxide Spintronics
Edited by Tamalika Banerjee

ISBN 978-981-4774-99-4 (Hardcover), 978-0-429-46819-3 (eBook)
www.jennystanford.com

each other during heteroepitaxy. Research has shown that perovskite oxide epitaxial layers are influenced by the boundary conditions applied by the epitaxy process. Properties are modified due to the presence of such phenomena as strain, polar discontinuities, and octahedral connectivity mismatch. This sensitivity offers the unique opportunity to engineer the materials structure. In this way, advanced epitaxy has become a research tool for understanding materials, because it allows different structural parameters to be disentangled. Some examples of the possibilities of advanced epitaxy are tricolor superlattices [1], compositional interface engineering [2–4], modulation doping [5], oxygen octahedra rotation engineering [6–11], interface dipole engineering [12], transfer of electron–phonon coupling [13], and defect mitigation [14].

$La_{0.67}Sr_{0.33}MnO_3$ is interesting because it is a ferromagnetic metal with 100% spin polarization [15]. It is therefore relevant to a variety of devices such as magnetic tunnel junctions [16, 17], diodes [18, 19], transistors [12], and ferroelectric tunnel junctions [20]. All of these applications require control of the properties of $La_{0.67}Sr_{0.33}MnO_3$ at the interface. The half-metallic state in $La_{0.67}Sr_{0.33}MnO_3$ arises from a complex balance between competing interactions. The double-exchange interaction favors conductivity, whereas the super-exchange interaction favors an insulating state. Furthermore, the material is susceptible to Jahn–Teller distortions that also favor electron localization. All these factors depend on the local crystal environment of the Mn ions. At interfaces, the balance between the competing interactions is modified and an electrical and magnetic dead-layer is present [21, 22]. Here we present a study of different interface engineering schemes for $La_{0.67}Sr_{0.33}MnO_3$ interfaces with a focus on the $La_{0.67}Sr_{0.33}MnO_3$–$SrTiO_3$ interface. The goal is to systematically investigate different mechanisms thought to be responsible for the dead-layer behavior. In the following sections, we discuss the polar discontinuity problem, the effect of disorder and the oxygen octahedra rotations.

2.2 Experiments

$SrTiO_3$ substrates were prepared using a standard BHF etching procedure [23]. The $La_{0.67}Sr_{0.33}MnO_3$ thin films were grown with pulsed laser deposition (PLD) (TSST system). The substrate temperature during growth was 750–800°C in an oxygen environment of 0.27 mbar. The laser beam was produced by a 248 nm wavelength KrF excimer laser (LPXPro from Coherent, Inc.) with a typical pulse duration of 20–30 ns. With a 4 mm by 15 mm rectangular mask, the most homogeneous part of the laser beam was selected. A sharp image of the mask was created on the stoichiometric target (Praxair electronics) with a lens, resulting in a spot size of 2.3 mm^2 (0.9 mm by 2.5 mm). The beam energy was controlled with a variable attenuator or with the laser voltage, yielding a fluence at the target of 2 J/cm^2. The repetition rate was 1 Hz and the substrate was placed at 5 cm distance directly opposite to the target. Before deposition, the target was pre-ablated for 2 minutes at 5 Hz to remove any possible surface contamination. After deposition, the PLD chamber was flooded with pure oxygen (typically 100 mbar) and the samples were cooled down by switching off the heater power. Typically, the cooldown required 2 hours. For the PLD growth of $SrTiO_3$ and $SrMnO_3$, identical settings as for the $La_{0.67}Sr_{0.33}MnO_3$ growth were used. For the $LaTiO_3$ growth and the growth of the $SrTiO_3$ layer on top of the $LaTiO_3$, the deposition pressure was lowered to $3{\times}10^{-5}$ mbar, in order to prevent the formation of $La_2Ti_2O_7$ [24]. Some of the thicker $La_{0.67}Sr_{0.33}MnO_3$ films were grown with a repetition rate of 5 Hz and an oxygen pressure of 0.16 mbar.

The magnetization of the samples was measured with a vibrating sample magnetometer (VSM) Physical Properties Measurement System (PPMS®) by Quantum Design]. In order to determine the saturation magnetization and Curie temperature unambiguously, magnetic hysteresis loops were obtained at all temperatures as discussed in detail in the supplementary information of [2]. The resistivity of the samples was measured in the van der Pauw configuration [25] (PPMS by Quantum Design). In order to obtain ohmic contacts between the aluminium bonding wires and the $La_{0.67}Sr_{0.33}MnO_3$ layer, gold contacts were deposited on the corners of the sample with the use of a shadow mask. The electric field gating experiments were performed with a back gate geometry, using the 0.5 mm substrate as a gate dielectric. In order to ensure a uniform electric field, the back of the substrate was glued to a copper plate using silver epoxy.

2.3 Thin Film Growth and Characterization

The growth of the films was studied in situ with reflection high-energy electron diffraction (RHEED) during the growth. The substrate RHEED pattern is shown in Fig. 2.1 on the top left. The main specular spot is very intense compared to the two side spots. This is the typical signature of TiO_2 terminated $SrTiO_3$ [26]. Kikuchi lines are visible as well, indicating the smoothness of the substrate. The side spots are doubled, which is due to the additional periodicity at the surface from the regularly spaced terrace steps. Figure 2.1, top right, shows the RHEED image of the $La_{0.67}Sr_{0.33}MnO_3$ film after the deposition of 5 unit cell (uc) layers. Clear two-dimensional spots are visible, but also 2D streaks are present. The streaks are attributed to the scattering of the RHEED beam off the unit cell high steps at the surface. Similar RHEED images were observed after the completion of films with thicknesses up to 40 nm.

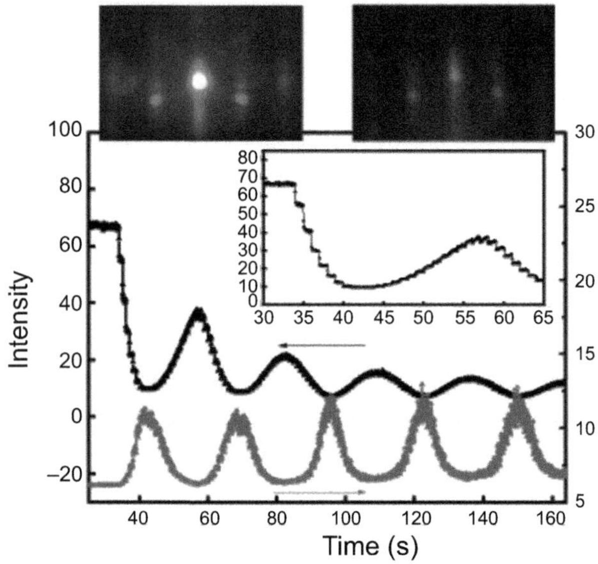

Figure 2.1 RHEED specular spot intensity and full width at half maximum (FWHM) as monitored during the initial growth of $La_{0.67}Sr_{0.33}MnO_3$ on $SrTiO_3$ $(001)_c$. The inset shows an expanded view of the growth of the first unit cell layer. Two RHEED images are presented, the left image was taken before deposition (at low pressure) and the right image was taken after deposition of 5 uc layers. Reprinted from Ref. [27], Copyright 2011, with permission from IOP Publishing.

The main graph in Fig. 2.1 shows the intensity of the specular reflection as measured during the initial growth of $La_{0.67}Sr_{0.33}MnO_3$. The intensity shows oscillations that correspond to the growth of the individual unit cell layers. Within the oscillations, recovery of the intensity after the sudden decrease during the laser burst is observed, as shown more clearly in the inset of the graph. The oscillation amplitude decreases with the amount of material deposited during the first part of the growth. This decrease of the intensity is due to the difference in reflectivity of the $SrTiO_3$ surface and the $La_{0.67}Sr_{0.33}MnO_3$ surface, due to the increased scattering from the heavy La ions in the lattice. Finally, the FWHM of the intensity of the specular spot, measured along the (10) direction, is presented as well. During the growth, the RHEED spots are periodically more streaky, because more step edge scattering is present at half unit cell layer coverage compared to full unit cell layer coverage. Therefore, the FWHM oscillates as well. The FWHM depends only on the shape of the intensity distribution and not on the total intensity and it is, therefore, a better indicator of the surface morphology than the main intensity of the reflection. As can be seen in the figure, the FWHM during the growth is almost equal to the FWHM of the substrate reflection indicating a smooth surface morphology. From these measurements, it is concluded that the initial stage of the $La_{0.67}Sr_{0.33}MnO_3$ growth proceeds in the ideal 2D layer-by-layer growth mode.

During the growth of the film, an increase of the RHEED oscillation maximum intensity was observed, as presented in Fig. 2.2. The oscillation intensity maximum typically peaked around 20–25 uc layers. The maximum in intensity was not observed during growth on an *A*-site terminated substrate surface, so it might indicate a termination conversion. After this peak, the oscillation intensity maximum decreased rapidly and the oscillation amplitude became comparable to the intensity variations in the laser pulse recovery cycles. It is concluded that the $La_{0.67}Sr_{0.33}MnO_3$ growth mode during the latter part of the growth is close to the steady-state growth mode, which is characterized by a relatively constant step density, which is large compared to the step density of the initial substrate surface.

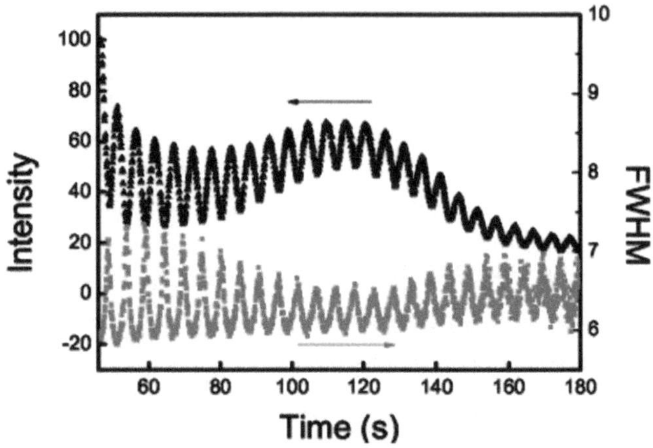

Figure 2.2 RHEED specular spot intensity and FWHM as monitored during the growth of a relatively thick film of $La_{0.67}Sr_{0.33}MnO_3$ on $SrTiO_3$ $(001)_c$. The graph shows the 7th to 32nd oscillation. A recovery of the RHEED oscillation intensity maximum, followed by a rapid decrease, is observed around the 20th unit cell layer. Reprinted from Ref. [27], with permission from IOP Publishing, Copyright 2011.

In order to check the hypothesis about a termination conversion during growth, three samples with respectively a 20, 26, and 30 uc thick $La_{0.67}Sr_{0.33}MnO_3$ layer were measured with angle-dependent time-of-flight mass spectroscopy. The relative intensity of the Mn ions with respect to Sr and La is shown in Fig. 2.3. The strong angular dependence in the Mn/La ratio of the 20 uc thick sample is consistent with a dominant MnO_2 crystal termination. With increasing film thickness, the angular dependence is reduced indicating a reduction of the MnO_2 termination. The data is more consistent with a mixed termination for the thicker films than with a complete surface termination conversion. The Mn/Sr ratio has a similar angular dependence, although the magnitude of the signal is reduced. The lower Mn/Sr ratio compared to Mn/La is also seen in experiments using controlled AO and BO_2 terminated $La_{0.67}Sr_{0.33}MnO_3$ films [28]. In conclusion, the surface termination changes around 26 uc towards mixed termination, in agreement with the deductions from the growth analysis.

A multilayer $La_{0.67}Sr_{0.33}MnO_3$–$SrTiO_3$ sample was characterized with scanning transmission electron microscopy (STEM) to study the interface atomic structure. The STEM data presented was measured with an FEI Titan microscope. The sample is comprised of several layers, from substrate to surface: 4 nm $La_{0.67}Sr_{0.33}MnO_3$, 2 nm $SrTiO_3$, 16 nm $La_{0.67}Sr_{0.33}MnO_3$, 2 nm $SrTiO_3$, and 8 nm $La_{0.67}Sr_{0.33}MnO_3$.

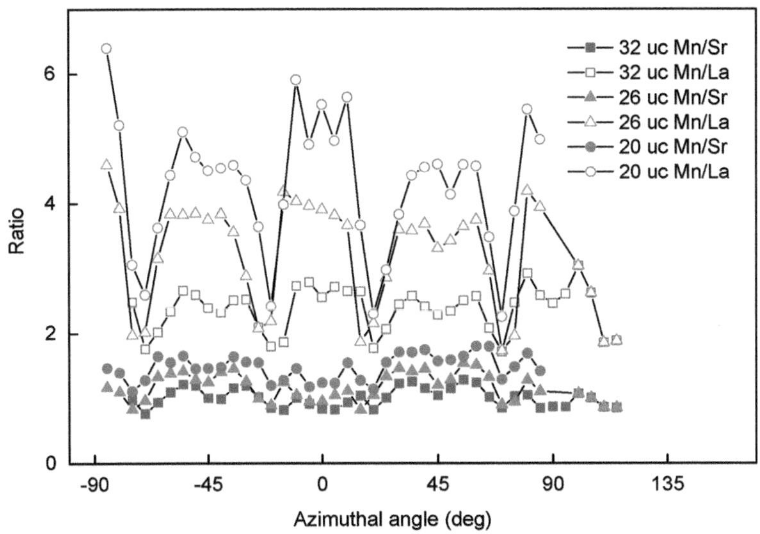

Figure 2.3 Angle-dependent time-of-flight mass spectroscopy. The angular dependence of the relative intensities is due to the atomic shadowing effect.

Figure 2.4a presents a low magnification high-angle annular dark field (HAADF) image of the different layers in the sample. In HAADF microscopy, the observed intensity of a column of atoms scales with the atomic weight of the elements within the columns. Therefore, the $La_{0.67}Sr_{0.33}MnO_3$ is brighter in the image than the $SrTiO_3$. Figure 2.4b presents a higher magnification image obtained from the layers close to the substrate. Both the *A*-site and *B*-site columns are observed. The multilayer structure is grown coherently and following a quantitative analysis by statistical parameter estimation [29], it can be concluded that the interfaces are well-defined with a chemical roughness of maximum 1 uc.

The $La_{0.67}Sr_{0.33}MnO_3$–$SrTiO_3$ heterostructures have not only excellent structural properties, but they also have excellent functional properties. Thicker films of 10 nm show a Curie temperature of 350 K, a low-temperature saturation magnetization of 4 μB/Mn and a residual resistivity of 60 μΩcm. The heterostructures compare favorably to the samples presented in the literature [27]. The combination of high-quality films with excellent functional properties and the two-dimensional layer-by-layer growth allows for the interface engineering described in the remainder of this chapter. However, because atomic control of the crystal termination during growth is only possible up to a finite film thickness, reliable interface engineering can only be performed for thin $La_{0.67}Sr_{0.33}MnO_3$ heterostructures.

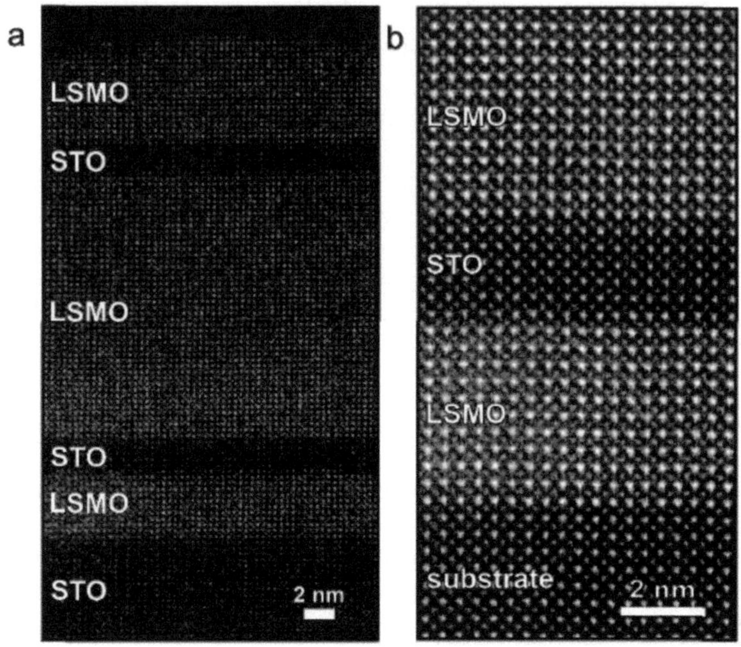

Figure 2.4 Scanning transmission electron microscopy, (a) HAADF-STEM image of a $La_{0.67}Sr_{0.33}MnO_3$–$SrTiO_3$ heterostructure. (b) High-magnification image of the area close to the substrate. Reprinted from Ref. [27], with permission from IOP Publishing, Copyright 2011.

2.4 Polar Discontinuities at the Interface

$La_{0.67}Sr_{0.33}MnO_3$–$SrTiO_3$ interfaces are polar and therefore ionic or electronic reconstructions are to be expected. Two different atomic stacking sequences are possible for the $La_{0.67}Sr_{0.33}MnO_3$–$SrTiO_3$ interfaces, due to the possible AO or BO_2 termination of each material.

One will be referred to as the $La_{0.67}Sr_{0.33}O$ terminated interface and the other as the MnO_2 terminated interface. The $La_{0.67}Sr_{0.33}O$ terminated interface is shown in Fig. 2.5a. At this interface, the atomic stacking sequence is $SrO–TiO_2–La_{0.67}Sr_{0.33}O–MnO_2$. The MnO_2 terminated interface is shown in Fig. 2.5b. Here the atomic stacking sequence is $TiO_2–SrO–MnO_2–La_{0.67}Sr_{0.33}O$. Both interface configurations are polar and without reconstruction would result in a diverging electrostatic potential, as indicated in the figures. As $La_{0.67}Sr_{0.33}MnO_3$ is conducting, the mobile charges can screen the diverging potential. This screening, however, is only partial because the charges are confined to the Mn sublattice. In Fig. 2.5(a, b), the electrostatic potential after the screening (effectively an electronic reconstruction of the polar discontinuity) is shown as well. Here it is assumed that the screening occurs completely in the first MnO_2 layer, which is consistent with the Thomas Fermi length of $La_{0.67}Sr_{0.33}MnO_3$ of ≈0.31 nm [19]. The diverging potential is eliminated, but a band offset, especially in the case of the $La_{0.67}Sr_{0.33}O$ terminated interface is present and will result in intermixing of the cations [30]. Note that the Ti^{4+} valence state is stable at all these interfaces, as the Ti t_{2g} levels are 1 eV higher in energy compared with the Mn e_g levels [31].

In order to remove the polar discontinuity, a subunit cell atomic layer with a net charge of $1/3\ e^-$/uc has to be inserted at the interface. This is shown in Figs. 2.5c and 2.5d, in which two methods are shown for interface engineering by inserting a single atomic layer of $La_{0.33}Sr_{0.67}O$ between the two materials through the deposition of either a $La_{0.33}Sr_{0.67}MnO_3$ layer (I) or a $La_{0.67}Sr_{0.33}$ TiO_3 layer (II). This gives an atomic stacking sequence at the interface of $SrO–TiO_2–La_{0.33}Sr_{0.67}O–MnO_2–La_{0.67}Sr_{0.33}O$. In this case, the diverging potential is absent and no driving force for reconstruction exists. A band offset is present at this interface as well, so some intermixing can still be expected.

Several heterostructures with and without interface engineering (denoted IE and non-IE, respectively) were grown. The structures consist of a TiO_2 terminated STO substrate, a variable thickness $La_{0.67}Sr_{0.33}MnO_3$ layer and a $SrTiO_3$ cap layer with a thickness of 5 uc. In order to realize the interface engineered heterostructure, first, a single unit cell layer of $La_{0.33}Sr_{0.67}MnO_3$ was grown. This was followed with the growth of n – 1 layers of $La_{0.67}Sr_{0.33}MnO_3$, then a single unit cell layer of $La_{0.33}Sr_{0.67}TiO_3$ and finally, 4 uc layers of

$SrTiO_3$. This recipe resulted in heterostructures with *n* layers of Mn ions, which can, therefore, be compared with the *n* unit cell layer $La_{0.67}Sr_{0.33}MnO_3$–$SrTiO_3$ structure without the interface engineering. Next to the single layer samples, also two superlattice samples (IE and non-IE) with five repetitions of $(La_{0.67}Sr_{0.33}MnO_3)_8(SrTiO_3)_5$ were grown.

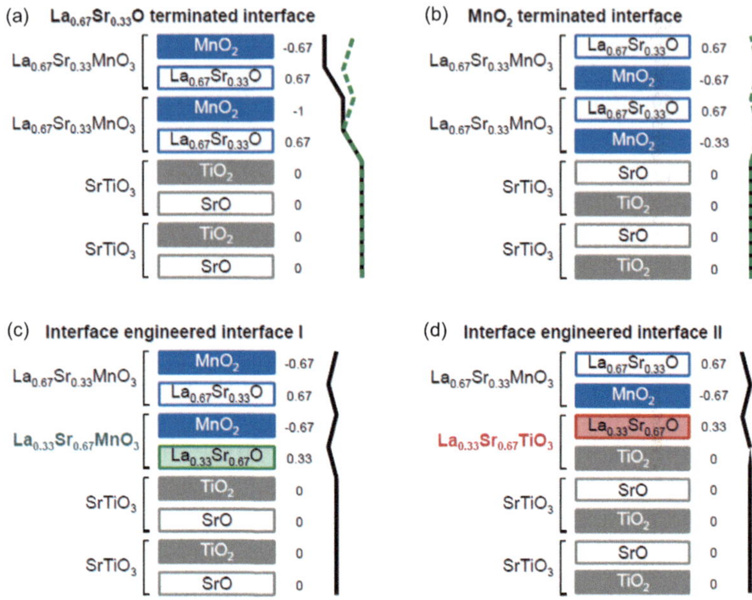

Figure 2.5 The $La_{0.67}Sr_{0.33}MnO_3$–$SrTiO_3$ interface configurations, (a) $La_{0.67}Sr_{0.33}O$ terminated interface, (b) MnO_2 terminated interface, (c) Interface engineered $TiO_2/La_{0.33}Sr_{0.67}O/MnO_2$ interface I. (d) Interface engineered $TiO_2/La_{0.33}Sr_{0.67}O/MnO_2$ interface II. The electrostatic potential due to the polar discontinuity is indicated with a black line (before reconstruction) and a green dotted line (after reconstruction). The numbers indicate the net charge at each layer, after reconstruction. Reprinted from Ref. [2], with permission from John Wiley and Sons, Copyright 2012.

The samples were analyzed extensively by STEM and electron energy loss spectroscopy (EELS) [2, 4]. It was found that in the non-IE samples, differences between the top and bottom interface exist, due to the difference in stacking sequence at the interface. At the bottom interface ($La_{0.67}Sr_{0.33}O$ terminated interface), a relatively large diffusion of La ions into the $SrTiO_3$ was found, which is probably the result of a reconstruction of the polar discontinuity. In contrast, the IE samples show similar top and bottom interfaces, in agreement with the design of the heterostructures.

The depth profiles of the magnetization in the superlattice samples were measured with polarized neutron reflectometry (PNR) [4]. The main result of the analysis is shown in Fig. 2.6. These profiles show that the topmost $SrTiO_3$ and $La_{0.67}Sr_{0.33}MnO_3$ layers exhibited different structural properties, most likely due to relaxation at the surface. However, all other four buried $La_{0.67}Sr_{0.33}MnO_3$–$SrTiO_3$ bilayers showed to be structurally equal as demonstrated by showing the first (LI) and fourth (L4) $La_{0.67}Sr_{0.33}MnO_3$ layer in Figs. 2.6(c, e) and 2.6(d, f), respectively. The chemical roughness in the IE heterostructure is typically 1.5 times larger than that in the non-IE heterostructure. This can to some degree be attributed to the more gradual La distribution in the design of the IE heterostructure. The magnetic roughness, in contrast, is similar in both heterostructures.

Most importantly, for the non-interface engineered (non-IE) heterostructure, a clear 5.5 Å ± 1.0 Å difference between the magnetic and chemical thickness of the $La_{0.67}Sr_{0.33}MnO_3$ layer was observed (Figs. 2.6c and 2.5e). The optimal model indicates that only the top part of each $La_{0.67}Sr_{0.33}MnO_3$ layer was strongly magnetically reduced. Because the La profile in the non-IE structures is shifted with respect to the Mn profile (STEM-EELS measurements), we also tested an alternative model with equally sized dead-layers at both interfaces. The model also yields a total 5 Å magnetically dead region (2.5 Å each at the top and bottom interfaces), similar to the optimal model, albeit with a slightly increased fitting error. Note that the difference between magnetic and chemical profiles at the bottom interface is largely due to La diffusion, as the first Mn layer is still magnetic. On the other hand, at the top interface, a reduction of the magnetization is found in a layer where Mn exists, therefore a true magnetic dead-layer $La_{0.67}Sr_{0.33}MnO_3$ only resides at the top interface. This is in good agreement with a previous observation that a surplus of La does not deteriorate the magnetism [21]. Overall, in the non-interface engineered heterostructure, the magnetic thickness is ~5 Å less than the chemical thickness and there is at least a magnetically reduced region at the top and possibly at the bottom, the combination of which is 5.5 Å ± 1.0 Å thick.

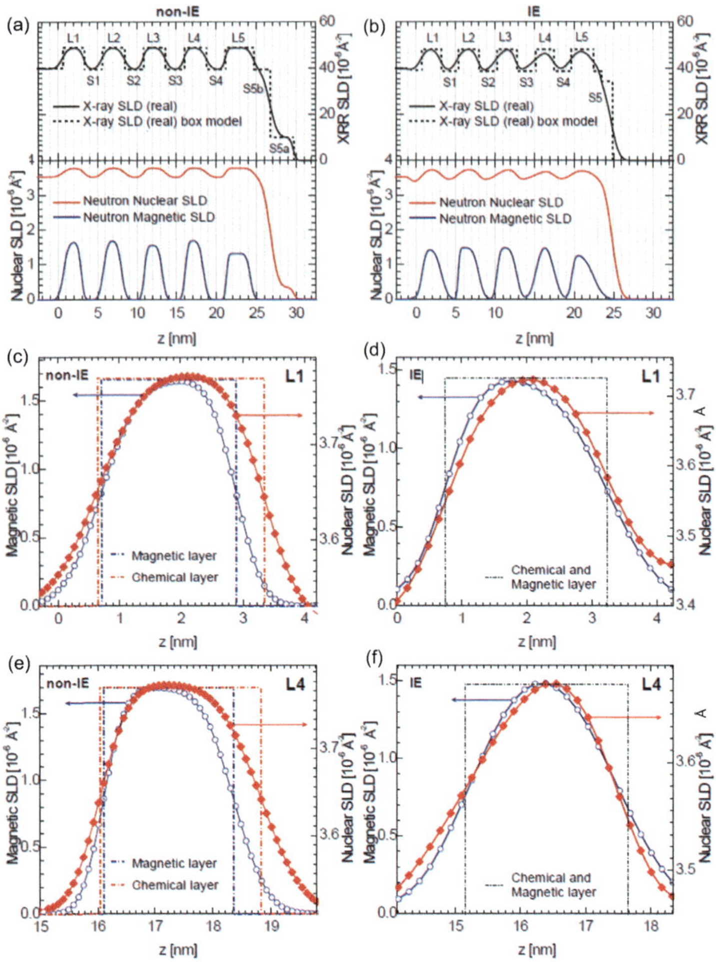

Figure 2.6 The magnetic and chemical scattering length density (SLD) profiles at various film thickness, *z* obtained from fitting neutron and X-ray reflectivity data for $La_{0.67}Sr_{0.33}MnO_3$–$SrTiO_3$ heterostructures (a) without and (b) with interface engineering. The dashed lines represent the SLD profiles with sharp interfaces (a so-called box model), (c–f) overlay of the chemical and magnetic profile without (c, e) and with (d, f) interface engineering for respectively the first (LI) and fourth (L4) $La_{0.67}Sr_{0.33}MnO_3$ layer. Reprinted from Ref. [4], with permission from John Wiley and Sons, Copyright 2015.

In contrast, the interface engineered (IE) heterostructure demonstrated an equal magnetic and chemical thickness of the $La_{0.67}Sr_{0.33}MnO_3$ layer [Fig. 2.6(d and f)]. The integrated magnetization at 120 K over 8 uc thick $La_{0.67}Sr_{0.33}MnO_3$ layer, extracted from the PNR data, is ~3.3 μB/Mn for the heterostructure with interface engineering, while a heterostructure without interface engineering exhibited a magnetization of only ~2.8 μB/Mn. The absolute magnetization values extracted from PNR are confirmed by VSM measurements and the observed magnetization enhancement of 18% is in very good agreement with the observed increase in magnetization of a single $La_{0.67}Sr_{0.33}MnO_3$–$SrTiO_3$ bilayer when interface engineering was applied.

Figure 2.7 presents temperature-dependent magnetization and resistivity measurements of the bilayer samples. The magnetic field was applied in-plane along the $[100]_c$ $SrTiO_3$ crystal direction during the magnetization measurements. Figure 2.7a shows the temperature dependence of the saturation magnetization for the n = 5, 6, 8, and 13 uc samples. A clear difference between the IE and non-IE samples is observed. The IE samples have significantly higher saturation magnetization and Curie temperature (T_C). Figure 2.7b presents the temperature dependence of the resistivity of the samples. For all thicknesses, the interface engineering results in a lower resistivity. The n = 5 samples and the non-IE n = 6 sample are insulating at low temperature. The n = 6 IE sample has an upturn in the resistivity at low temperature, but conductivity was observed down to the lowest temperature (10 K) in the measurement. An enhancement of the magnetization and electrical conductivity can still be observed for thicker $La_{0.67}Sr_{0.33}MnO_3$ layers in the 13 uc thick samples, however, the effect is smaller as the enhanced interface properties provide a smaller contribution to the properties of the complete $La_{0.67}Sr_{0.33}MnO_3$ layer when thicker layers are studied. The obtained properties of the non-interface engineered 13 uc sample are in good agreement with previously obtained results in ultrathin $La_{0.67}Sr_{0.33}MnO_3$ films [22].

These results provide strong evidence of the elimination of a magnetic dead-layer at the $La_{0.67}Sr_{0.33}MnO_3$–$SrTiO_3$ interface when applying interface engineering by incorporating of a single $La_{0.67}Sr_{0.33}O$ layer, which leads to the embedding of the interfacial MnO_2 atomic layer in between two (La, Sr)O layers. Furthermore, both the interface asymmetry of the magnetically reduced layer in the non-IE heterostructure and the absence of the dead-layer in the IE heterostructure confirm the reconstructions induced by the polar discontinuity at the interface as the origin of the former magnetic dead-layer at $La_{0.67}Sr_{0.33}MnO_3$–$SrTiO_3$ interfaces. However, even in the IE samples, the electrical dead-layer is still present and the question remains whether we can engineer even better interfaces.

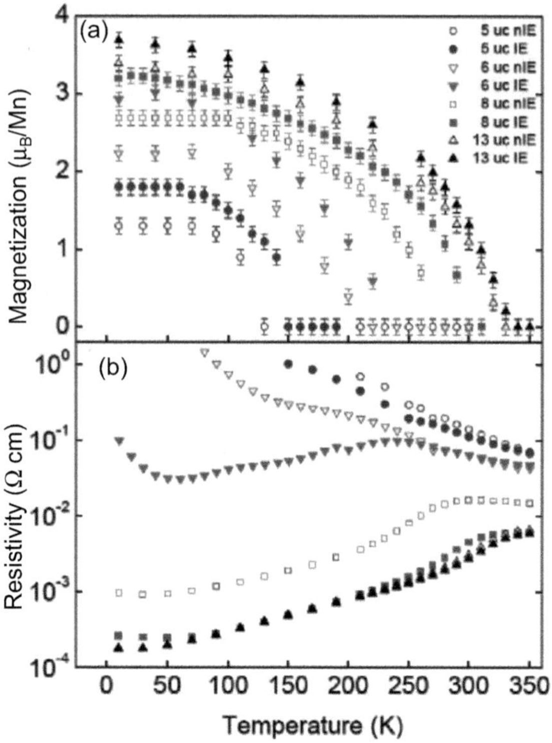

Figure 2.7 Temperature dependent (a) magnetization and (b) resistivity measurements for $La_{0.67}Sr_{0.33}MnO_3$–$SrTiO_3$ heterostructures of various *n* thicknesses in unit cells (uc). Results of samples with interface engineering (IE) and without interface engineering (non-IE) are shown by closed and open symbols, respectively. Reprinted from Ref. [2], with permission from John Wiley and Sons, Copyright 2012.

2.5 Order and Disorder

$La_{0.67}Sr_{0.33}MnO_3$ is a solid solution of $LaMnO_3$ and $SrMnO_3$. This implies $La_{0.67}Sr_{0.33}MnO_3$ is an intrinsically disordered material, where the *A*-site ion on a certain lattice site can be occupied with either a La or a Sr ion. It is an interesting question whether the disorder affects the properties, e.g., electrical transport. Several studies have presented results on the properties of $(LaMnO_3)_{2n}/(SrMnO_3)_n$ superlattices, which have an identical La/Sr ratio with respect to $La_{0.67}Sr_{0.33}MnO_3$ [32–34]. A decrease of the residual resistivity, with respect to the solid solution, was observed for the $n = 1$ superlattice [34].

Here, the effect of the disorder on the properties at the interface between $La_{0.67}Sr_{0.33}MnO_3$ and $SrTiO_3$ is studied. It is proposed that the $SrTiO_3$–$SrMnO_3$ interface has no intrinsic disorder, as the A-site ions of both materials are equal. To obtain the $La_{0.67}Sr_{0.33}MnO_3$ functional properties at the interface, charge has to be supplied to the $SrMnO_3$ layer. This is accomplished by chemically doping the $SrTiO_3$ away from the interface. This results in modulation doping [35], in which charge is transferred from the doped $SrTiO_3$ layer to the $SrMnO_3$ layer. The amount of charge that can be transferred is limited by the potential drop due to the electric dipole between the transferred electrons and the positively charged dopants. Eventually, the potential drop increases the energy of the $SrMnO_3$ conduction band to the energy of the $SrTiO_3$ conduction band and then the charge transfer stops. The maximum charge transfer is then given by:

$$x = (\Delta \varepsilon_0 K)/d$$

in which x is the maximum charge transfer, Δ is the energy difference between the $SrTiO_3$ and $SrMnO_3$ conduction bands, ε_0 is the permittivity of free space, K is the dielectric constant, and d is the distance over which the charge transfer occurs. Therefore, the use of $SrTiO_3$ has a great advantage: its huge dielectric constant screens the potential drop. It is possible to dope 2 e^-/uc over a distance of 2 nm, which is an order of magnitude larger as the charge transfer by modulation doping in semiconductor structures. And at low temperatures, significant charge transfer can be present without significant band bending, as the dielectric constant of $SrTiO_3$ increases massively with decreasing temperature.

We tested the modulation-doping scheme by growing two samples with 10 uc thick manganite layers. A schematic image of the two samples is shown in Fig. 2.8. Sample A is a reference sample, where a $SrMnO_3$ layer is inserted between the $La_{0.67}Sr_{0.33}MnO_3$ and the $SrTiO_3$ to obtain the $SrTiO_3$–$SrMnO_3$ interface. The thickness of the $SrMnO_3$ layer is 3 uc, which corresponds to the amount of electrically dead-layers at the $La_{0.67}Sr_{0.33}MnO_3$–$SrTiO_3$ interface. Next, a 1 uc thick layer of $La_{0.67}Sr_{0.33}MnO_3$ is present between the $SrMnO_3$ and the $La_{0.67}Sr_{0.33}MnO_3$, as this results in a higher quality growth of the $La_{0.67}Sr_{0.33}MnO_3$. The $La_{0.67}Sr_{0.33}MnO_3$ layer thickness is 6 uc, so the sample can be compared with the $La_{0.67}Sr_{0.33}MnO_3$ sample with a thickness of 10 uc. For this sample, no dopants are present in the $SrTiO_3$, so the $La_{0.67}Sr_{0.33}MnO_3$ layer is expected to be over doped. Sample *B* is similar to sample *A*, but here a 2 uc thick layer of $LaTiO_3$ is inserted in the $SrTiO_3$ 2 uc away from the interface. The $LaTiO_3$ layer dopes 2 e^-/uc to the structure, which is precisely the amount required in the $SrMnO_3$ layer to obtain $La_{0.67}Sr_{0.33}MnO_3$'s $Mn^{3.33+}$ valence state. Note that only the bottom interface has been engineered to remove the disorder, while the top interface is a disordered $La_{0.67}Sr_{0.33}MnO_3$–$SrTiO_3$ interface.

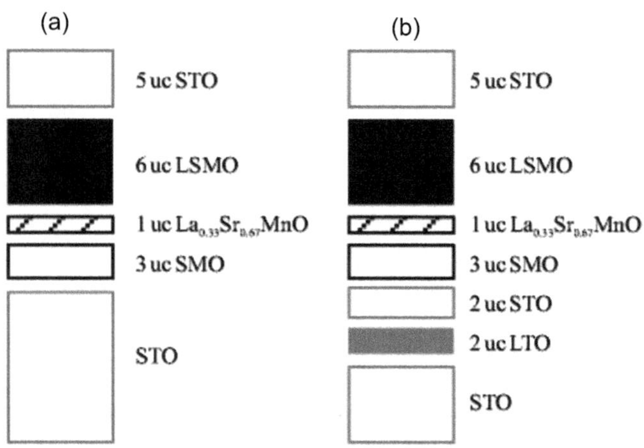

Figure 2.8 A schematic of the sample structure: (a) reference sample *A* and (b) modulation doped sample *B*.

The intensity of the specular spot, as monitored with RHEED during the growth of sample *B*, is shown in Fig. 2.9. Clear oscillations are observed during the growth of $LaTiO_3$ and $SrTiO_3$. The initial growth of $SrMnO_3$ on the $LaTiO_3$–$SrTiO_3$ stack started with an increase of the RHEED intensity to a maximum, followed by a prolonged oscillation. This second maximum occurred at the expected time for the deposition of 1.5 uc layers. The period of the second oscillation corresponds to the time required to deposit a single unit cell layer. As the $SrMnO_3$ deposition was stopped at the third RHEED intensity maximum, only 2.5 uc of $SrMnO_3$ was grown instead of 3 uc. This initial growth of $SrMnO_3$ was not observed for the $SrMnO_3$ growth directly on the $SrTiO_3$ substrate. The inset of Fig. 2.9a shows the RHEED specular spot intensity observed during the $SrMnO_3$ growth of sample *A*. Here three clear oscillations with equal periods are observed. After the growth of the $SrMnO_3$, the growth of the $La_{0.33}Sr_{0.67}MnO_3$, $La_{0.67}Sr_{0.33}MnO_3$, and $SrTiO_3$ layers showed RHEED oscillations.

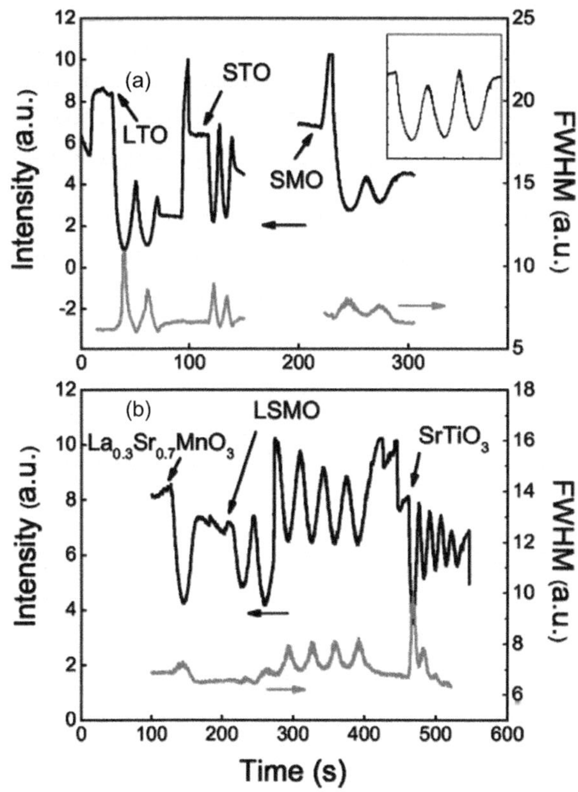

Figure 2.9 RHEED specular spot intensity oscillations and FWHM during the growth of sample *B*. The starting point of each layer is indicated. The inset in (a) shows the RHEED specular spot intensity observed during the growth of $SrMnO_3$ directly on the $SrTiO_3$ substrate, sample *A*.

Therefore, it is concluded that the entire heterostructure was grown in 2D layer-by-layer fashion.

Electrical transport measurements of the $n = 10$ samples are presented in Fig. 2.10. For comparison, also two other samples are shown. Sample C is a 10 uc layer $La_{0.67}Sr_{0.33}MnO_3$ sample capped with 5 uc layers of $SrTiO_3$. Sample D is similar to sample C, but with engineered interfaces to remove the polar discontinuities, as discussed in Section 1.4. The resistivity of sample B is similar to the values observed for samples C and D, indicating that sample B has the properties of a 10 uc layer $La_{0.67}Sr_{0.33}MnO_3$ sample. Sample A, in contrast, has a higher resistivity, lower T_C, and negative magnetoresistance present at all temperatures. The transport behavior of sample A is actually similar to that of 7 uc thick $La_{0.67}Sr_{0.33}MnO_3$ samples, consistent with the presence of 7 uc of $La_{0.67}Sr_{0.33}MnO_3$ in the sample structure. This result indicates that charge is transferred from the $LaTiO_3$ layer to the $SrMnO_3$ layer and that the charge transfer significantly changes the conductivity.

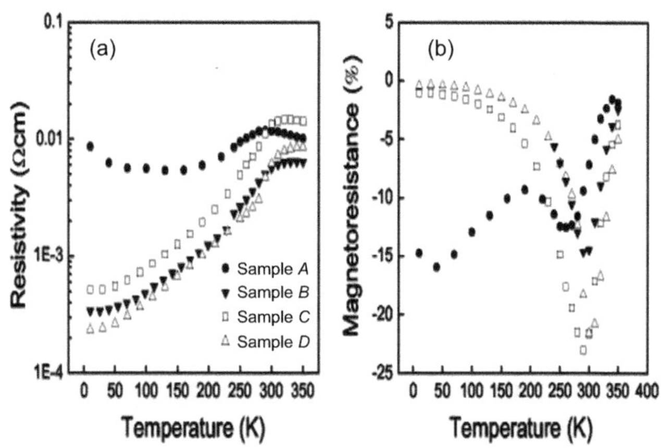

Figure 2.10 Temperature dependent resistivity measurements of four 10 uc thick $La_{0.67}Sr_{0.33}MnO_3$ samples. Sample A contains a $SrMnO_3$ layer but has no modulation doping. Sample B contains the $SrMnO_3$ layer and has modulation doping. Sample *C* is a standard $La_{0.67}Sr_{0.33}MnO_3$ sample, capped with $SrTiO_3$ and sample D is similar to sample C, but with engineered interfaces to remove the polar discontinuity.

In order to verify that the optimum amount of charge is present in the $SrMnO_3$ layer, electrical field effect measurements were performed. The charge carrier density of the sample can be controlled with the gate voltage applied to the back of the sample. Figure 2.11 presents the gate voltage dependence of the resistivity, measured at 10 K. Sample A shows some hysteretic behavior, probably due to charge impurities in the $SrTiO_3$ substrate, but the general dependence is monotonic with gate voltage. A decrease of the resistivity is observed for positive gate voltage (electron accumulation in the $SrMnO_3$). This corresponds well with the expected over doping of sample A. For sample B, a minimum in the resistivity is observed at a gate voltage of 50 V. This indicates that the optimum charge carrier density for the conductivity of the $La_{0.67}Sr_{0.33}MnO_3$ is very close to the doping achieved with the modulation doping. With a $SrTiO_3$ dielectric constant of 20000, at most 0.015 e^-/uc charge carriers are accumulated by the electric field. From a practical point of view, the control over the amount of dopants is 5% (approximately one laser pulse during deposition) of a unit cell layer of $LaTiO_3$, which corresponds to 0.05 e^-/uc. Therefore, it is concluded that sample B is as good as can presently be realized and little room for improvement exists in optimizing the inserted amount of dopants. The experiments indicate that modulation doping with a significant charge transfer was achieved. However, $La_{0.67}Sr_{0.33}MnO_3$'s conductivity at the interface was not much improved, indicating that disorder is not the key parameter governing the properties at the interface.

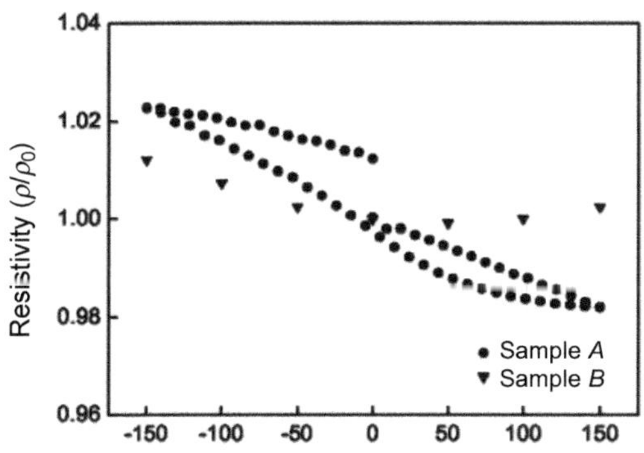

Figure 2.11 The gate voltage dependence of the resistivity.

2.6 Octahedra Rotations at the Interface

In the proceeding sections, we presented improvements of the $La_{0.67}Sr_{0.33}MnO_3$ interface properties by removing the polar discontinuities and the intrinsic disorder at the interfaces. However, the electrical dead-layer still remains. The likely origin of the dead-layer lies in the structural reconstructions at the interface [36, 37]. In this section, we study the structure of the oxygen octahedra at the interface in detail.

The changes in the unit cell symmetry resulting from different octahedral rotations have been systematized by Glazer [38] and later expanded by Woodward [39, 40]. Octahedral rotations in the perovskite-type unit cell can be described as a combination of rotations about three symmetry axes of the pseudocubic unit cell. The relative magnitudes of the tilts are denoted by letters a, b, and c, e.g., aab means equal rotations around the [100] and [010] axes and a different tilt around the [001] axis. Two adjacent octahedra around one of the <100> axes can rotate either in-phase or out-of-phase, which is indicated by the + or – sign, respectively. No rotation is indicated by the 0 sign.

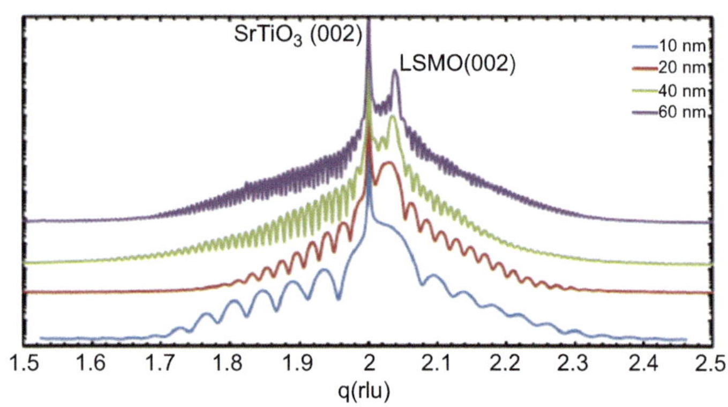

Figure 2.12 Synchrotron XRD data for $La_{0.67}Sr_{0.33}MnO_3$ films of different thicknesses.

$SrTiO_3$ is a cubic crystal (above 105 K) with Glazer tilt system $a^0a^0a^0$. $La_{0.67}Sr_{0.33}MnO_3$ is rhombohedral with tilt system $a^-a^-a^-$. $La_{0.67}Sr_{0.33}MnO_3$ films grown on $SrTiO_3$, however, have tilt system $a^+a^-a^0$ in order to accommodate the tensile strain applied to the films [41]. Clearly, a tilt system mismatch exists at the interface and because the oxygen octahedra network has to be continuous across the interface, structural distortions of the octahedra have to be present. These distortions have been predicted to occur on the $La_{0.67}Sr_{0.33}MnO_3$ side of the interface [37].

Four $La_{0.67}Sr_{0.33}MnO_3$ samples with different thicknesses (10, 20, 40, and 60 nm) were studied with synchrotron X-ray diffraction (XRD). Figure 2.12 shows out-of-plane momentum scans around the (002) Bragg reflection. Next to the Bragg peaks, fringes are present due to the finite thickness of the samples. Due to the low surface and interface roughnesses in the samples, a large number of fringes is seen and quantitative analysis is possible. The data contains two features that are inconsistent with the $La_{0.67}Sr_{0.33}MnO_3$ having a uniform thickness throughout the sample: the fringe intensity at small momentum transfer (with respect to the Bragg peak) is larger than that at high momentum transfer and the fringe intensities exhibit a modulation with a larger period. Dynamical XRD calculations can only reproduce these features when several layers with different out-of-plane lattice parameters are included in the model of the sample structure [42]. Three distinct layers are present in the samples: an interface layer, an intermediate layer and the main layer, see Fig. 2.13. Here we focus on the interface layer. Surprisingly we find that the interface layer is 2 uc thick with an out-of-plane lattice parameter of 3.93–3.94 Å, independent of the total film thickness. This implies that the first atomic layers of the films are under compressive strain as opposed to the tensile strain in the rest of the films. The result can easily be understood by considering the interface mismatch in the oxygen octahedra rotation patterns between $La_{0.67}Sr_{0.33}MnO_3$ and $SrTiO_3$. The Mn–O bond length in $La_{0.67}Sr_{0.33}MnO_3$ is 1.957 Å, so with the $a^0a^0a^0$ tilt pattern of $SrTiO_3$, the $La_{0.67}Sr_{0.33}MnO_3$ would be cubic with a lattice constant of 3.914 Å. This is larger than the lattice constant of $SrTiO_3$ and the $La_{0.67}Sr_{0.33}MnO_3$ interfacial layers are under compressive strain. The strain distorts the Mn–O bond length and with a Poisson ratio of 0.4, the out-of-plane lattice parameter would be 3.93 Å, consistent with the experiment. In conclusion, the first 2 uc of $La_{0.67}Sr_{0.33}MnO_3$ at the $La_{0.67}Sr_{0.33}MnO_3$–$SrTiO_3$ interface are severely distorted due to structural coupling to the $SrTiO_3$. This interfacial layer has an $a^0a^0a^0$ tilt pattern and the oxygen octahedra are elongated along the out-of-plane direction. We note that structural distortions are also found in photoemission experiments at this interface [43, 44]. The elongation of the octahedra at the interface breaks the orbital symmetry and thereby the double exchange interactions are weakened at the interfaces.

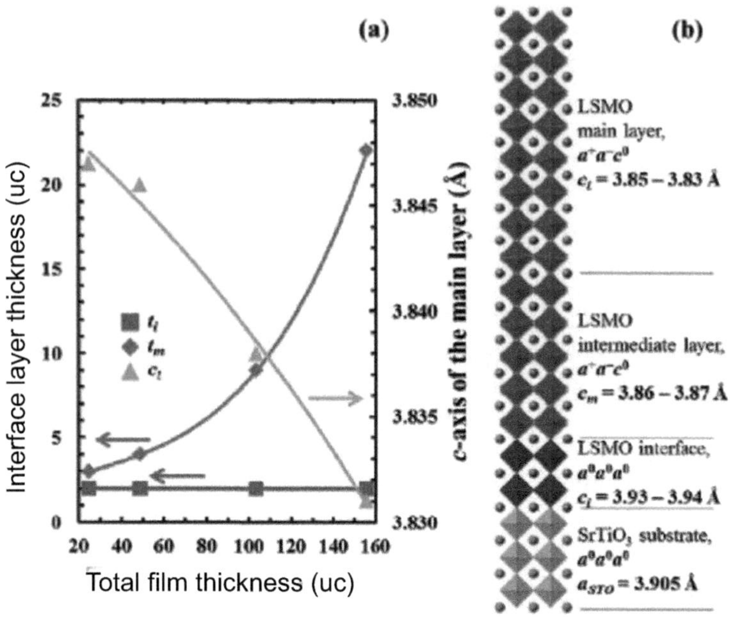

Figure 2.13 Structural evolution of the $La_{0.67}Sr_{0.33}MnO_3$ thin film along the out-of-plane direction. (a) Variation of the interface and intermediate layer thickness and *c*-axis lattice parameter of the main layer as a function of the total film thickness. (b) Schematic diagram of the $La_{0.67}Sr_{0.33}MnO_3$ layer structural evolution near and away from the $La_{0.67}Sr_{0.33}MnO_3$–$SrTiO_3$ interface. Reprinted from Ref. [42], with the permission from AIP Publishing, Copyright 2014.

The situation is different in $La_{0.67}Sr_{0.33}MnO_3$–$NdGaO_3$ heterostructures. $NdGaO_3$ is orthorhombic with the $a^-a^-c^+$ tilt pattern, whereas $La_{0.67}Sr_{0.33}MnO_3$ grown on $NdGaO_3$ (110) has the $a^+b^-c^-$ tilt pattern [41]. These structural characteristics give rise to an in-phase versus out-of-phase rotation type mismatch occurring along the [001]-axis that is absent along the [1$\bar{1}$0]-axis where both materials' rotations are out-of-phase. The magnitude of the bond angle θ also has a certain degree of mismatch: ~154° in $NdGaO_3$

versus 166.3° in $La_{0.67}Sr_{0.33}MnO_3$. As a result, both the anisotropic rotation type mismatch and the large difference (~12°) in bond angle will cause a strong discontinuity of the octahedra. In order to retain the connectivity of the corner shared oxygen octahedra across the interface, the large octahedral tilt present in the $NdGaO_3$ substrate will propagate into $La_{0.67}Sr_{0.33}MnO_3$ near the interface region, similar to what occurs in $La_{0.67}Sr_{0.33}MnO_3$–$SrTiO_3$ interfaces as mentioned above. Figure 2.14 presents STEM images of three $La_{0.67}Sr_{0.33}MnO_3$–$NdGaO_3$ heterostructures together with schematics of the bulk structures of the materials. A direct visualization of the octahedra rotations is possible with Cs-corrected STEM via which the oxygen atoms can be imaged through the angular bright-field technique. Figure 2.14b clearly shows that $NdGaO_3$ induces large rotations of octahedra in the $La_{0.67}Sr_{0.33}MnO_3$ layers close to the interface. The octahedral tilt angle can be quantified by statistical parameter estimation theory [45] and is presented in Fig. 2.14e. The tilt angle continuously changes from the $NdGaO_3$ substrate value to that of bulk $La_{0.67}Sr_{0.33}MnO_3$ (far from the interface). Interestingly, the first 2 uc layers of the $La_{0.67}Sr_{0.33}MnO_3$ have almost the same tilt angle as the $NdGaO_3$. The impact of the octahedral coupling decays rapidly away from the interface and disappears above 4 uc layers.

Because of the short impact length scale of oxygen octahedra coupling, the rotations of the $La_{0.67}Sr_{0.33}MnO_3$ can be significantly altered by inserting a non-tilted $SrTiO_3$ buffer layer (Fig. 2.14c). Within the $SrTiO_3$ layer, the oxygen octahedra rotations are also coupled to those of the $NdGaO_3$, but the tilt angle relaxes quickly, i.e., the tilt of the TiO_3 octahedra disappears above 2 uc layers. Consequently, the $La_{0.67}Sr_{0.33}MnO_3$ connects to a non-tilted octahedra structure and therefore, within the STEM spatial resolution, there is no tilting of the MnO_3 octahedra (see Fig. 2.14e), neither at the interface nor away from the interface. Due to the steep decay of the tilts in the $SrTiO_3$, a 1 uc thick $SrTiO_3$ buffer layer suffices to significantly reduce the tilts in the $La_{0.67}Sr_{0.33}MnO_3$ [see Fig. 1.14(d, e)]. These results indicate that the local oxygen octahedra rotations at the substrate surface act as a controllable template for the structure of the epitaxial $La_{0.67}Sr_{0.33}MnO_3$ film.

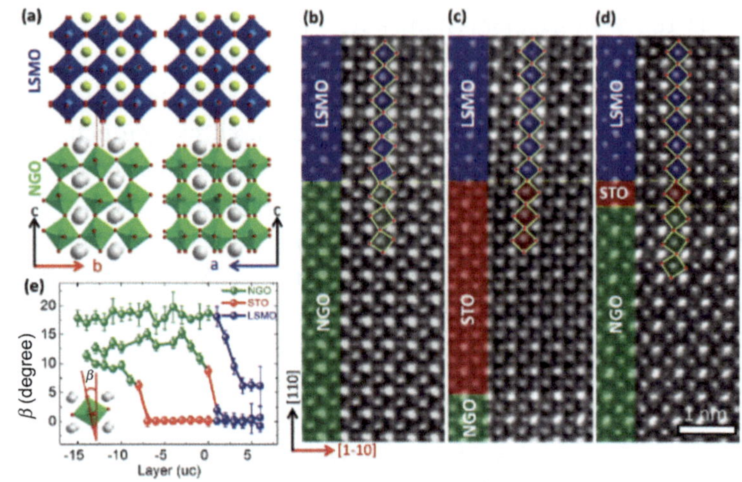

Figure 2.14 Oxygen octahedral coupling at interfaces in manganite heterostructures. (a) Schematic models of atomic ordering in $La_{0.67}Sr_{0.33}MnO_3$ and $NdGaO_3$ crystal structures. (b–d) Inversed annular bright-field STEM images of $La_{0.67}Sr_{0.33}MnO_3$–$NdGaO_3$ (left) and $La_{0.67}Sr_{0.33}MnO_3$–$SrTiO_3$–$NdGaO_3$ (right) heterostructures. The oxygen atoms are clearly visible and the connectivity of oxygen octahedra across the interfaces is indicated. The MnO_3 octahedra show a clear in-phase rotation following the $NdGaO_3$. The $La_{0.67}Sr_{0.33}MnO_3$ films are 6 uc thick while the $SrTiO_3$ buffer layer has a thickness of 9 uc or 1 uc. (e) Layer-position dependent octahedron tilt angle (β) in $La_{0.67}Sr_{0.33}MnO_3$–$NdGaO_3$ heterostructures with and without a $SrTiO_3$ buffer layer. Reprinted from Ref. [45] with permission from Springer Nature, Copyright 2016.

The strong octahedral tilts in $La_{0.67}Sr_{0.33}MnO_3$ near interface in $La_{0.67}Sr_{0.33}MnO_3$–$NdGaO_3$ heterostructures affects the magnetic properties when the films are thin enough for the interfacial structure to dominate. In strong contrast to the normally observed easy axis along the long $[1\bar{1}0]$ axis [46], the easy axis of ultrathin $La_{0.67}Sr_{0.33}MnO_3$ with $t < 8$ uc is along the short [001] axis due to the interfacial magnetic anisotropy (IMA), as shown in Fig. 2.15a. The contribution of anisotropic strain to the magnetic anisotropy will increase with increasing thickness, hence thicker films ($t > 8$ uc) exhibit bulk magnetic anisotropy (BMA) with the easy axis along the $[1\bar{1}0]$ axis. At 8 uc, the competition between the IMA and the BMA results in biaxial anisotropy with the easy axes along the $[1\bar{1}1]$ and $[1\bar{1}\bar{1}]$ axes. The engineering of octahedral tilting by introducing an ultrathin $SrTiO_3$ buffer layer [see Fig. 2.15(b, d)] also provides us with capabilities to switch the IMA to BMA in ultrathin LSMO films, e.g., an 8 uc thick $SrTiO_3$ buffered 6 uc thick $La_{0.67}Sr_{0.33}MnO_3$ film exhibits an easy axis along the $[1\bar{1}0]$ axis (see Fig. 2.15c). Therefore, the short impact length scale of octahedra coupling becomes a unique tool that allows us to atomically engineer octahedra rotations at the interface and thus anisotropic properties at the atomic scale.

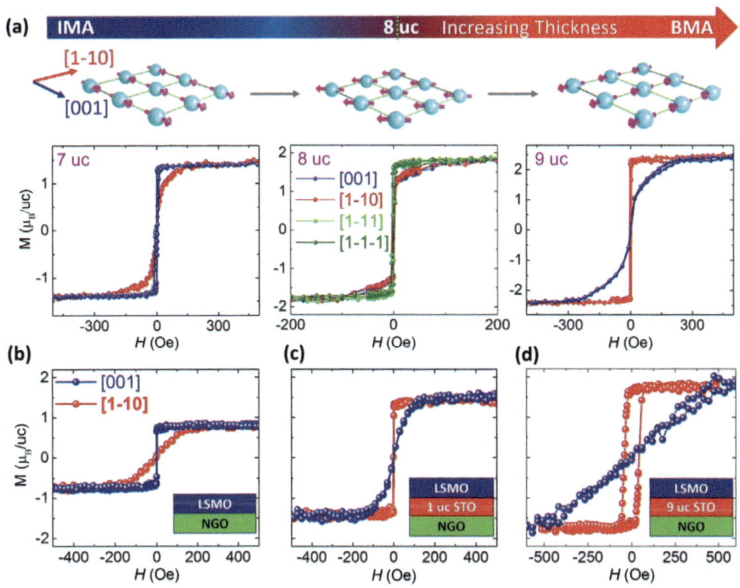

Figure 2.15 Atomic scale control of magnetic anisotropy. (a) The *M(H)* curves at 100 K along the [001] and $[1\bar{1}0]$ axis of $La_{0.67}Sr_{0.33}MnO_3$ films with thicknesses of 7, 8, and 9 uc on $NdGaO_3$ substrates. The schematics at the top show the corresponding ground state of the Mn spin orientation. The *M(H)* curves at 100 K along the [001] and $[1\bar{1}0]$ axes of the 6 uc $La_{0.67}Sr_{0.33}MnO_3$ films on $NdGaO_3$ substrates without (b) and with a 1 uc (c) and 9 uc (d) $SrTiO_3$ buffer layer. Reprinted from Ref. [45] with permission from Springer Nature, Copyright 2016.

We achieved the largest reduction of the electrical dead-layer by growing $La_{0.67}Sr_{0.33}MnO_3$ on $SrTiO_3$-buffered and $SrTiO_3$-capped $NdGaO_3$ [48], see Fig. 2.16. In this case, a 4 uc thick film is still metallic and localization sets in at 3 uc. A few ingredients contribute to the reduction of the dead layer. First of all, the films are under compressive strain as opposed to films grown on $SrTiO_3$ that are under tensile strain. Compressive strain induces octahedra rotations around the out-of-plane axis, whereas the tensile strain induces rotations around the in-plane axes [41]. The double exchange mechanism that drives the conductivity of the manganites depends on the Mn–O–Mn bond angles; strong octahedra rotation reduces the conductivity. Therefore, compressive strain promotes double exchange in the critical out-of-plane direction. Second, the adjacent $SrTiO_3$ layers prevent the rotations from the $NdGaO_3$ from inducing large rotations in the $La_{0.67}Sr_{0.33}MnO_3$, which would be detrimental for the electrical transport. However, it could be expected that the $SrTiO_3$ layers induce octahedra deformation just as is the case of $La_{0.67}Sr_{0.33}MnO_3$ grown on $SrTiO_3$ crystals. Judging from the reduction of the dead-layer, this is not so much the case for these heterostructures. Some structural differences are present: the $SrTiO_3$ is strained and probably has an $a^0a^0c^-$ tilt pattern instead of the standard cubic $a^0a^0a^0$ tilt pattern and the $La_{0.67}Sr_{0.33}MnO_3$ is orthorhombic ($a^+b^-c^-$ tilt pattern) under compressive strain instead of tetragonal ($a^+a^-c^0$ tilt pattern) under tensile strain. Presumably, one of these structural factors together with the limited thickness of the $SrTiO_3$ layers prevents the $SrTiO_3$ from inducing the Jahn–Teller distorted interfacial layer that occurs in $La_{0.67}Sr_{0.33}MnO_3$ grown on $SrTiO_3$ crystals.

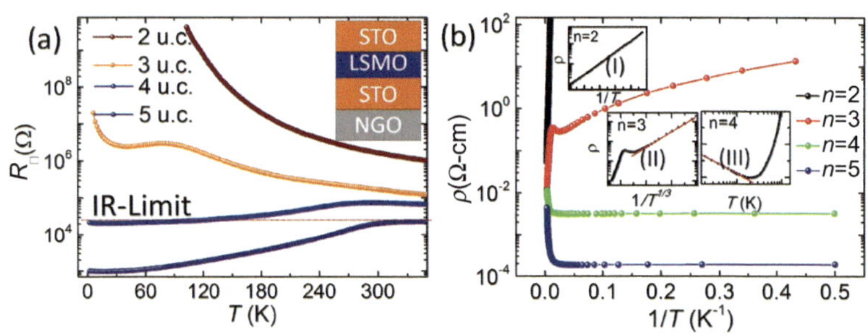

Figure 2.16 (a) Temperature dependence of the sheet resistance for different thicknesses of $La_{0.67}Sr_{0.33}MnO_3$ films grown on 9 uc $SrTiO_3$ buffered $NdGaO_3$ with 2 uc $SrTiO_3$ capping layers. The red dotted line indicates the IR-limit for the onset of strong localization [47]. (b) The corresponding log-plot of resistivity as a function of $1/T$ respectively. The insets show the low-T zoom-in plot of (I) $\log(\rho)$ versus $1/T$ for 2 uc, (II) $\log(\rho)$ versus $1/T^{1/3}$ for 3 uc and (III) ρ versus $\log(T)$ for the 4 uc film, respectively. Reprinted from Ref. [48], with permission from American Physical Society, Copyright 2015.

2.7 Conclusions

The magnetic dead-layer at the $La_{0.67}Sr_{0.33}MnO_3$–$SrTiO_3$ interface can be eliminated with the use of compositional interface engineering. The electrical dead-layer, however, remains present in all the heterostructures. The structural distortions induced by oxygen octahedra connectivity are likely the dominant cause for the electrical dead-layer, consistent with the conclusions of a variety of other experiments and theoretical models [36, 37, 43, 49]. Also, experiments performed on (011) oriented $SrTiO_3$ clearly indicated the importance of the oxygen octahedra rotation pattern on the functional properties [7].

The experiments demonstrate that interface engineering is a powerful tool for studying the properties of materials at interfaces systematically. Due to the importance of the local crystal structure, the main challenge for materials engineers is the control of the amount and direction of the oxygen octahedra rotations at the interface. Next to the results shown in Section 1.6, other possible approaches to optimize $La_{0.67}Sr_{0.33}MnO_3$ interfaces are using buffer layers with the right rotation patterns, changing the A-cation size at the interface, similar to bulk experiments in the manganites [50] and using infinite layer compounds to eliminate the oxygen octahedra connectivity.

2.8 Acknowledgments

The authors gratefully acknowledge key contributions to the research made by Arturas Vailionis (XRD); Sandra van Aert, Sara Bals, Ricardo Egoavil, Nicolas Gauquelin, Staf van Tendeloo, and Jo Verbeeck (STEM); Valeria Lauter, Yaohua Liu, and Suzanne te Velthuis (PNR); and Ramamoorthy Ramesh and Pim Rossen (TOF-MS). The transport measurements discussed in Section 1.5 were performed during a stay of H. B. at the University of Tokyo and would not have been possible without the help of Chris Bell, Yasuyuki Hikita, Harold Hwang and Takeaki Yajima.

References

1. Lee, H. N., Christen, H. M., Chisholm, M. P., Rouleau, C. M., and Lowndes, D. H. (2005). Strong polarization enhancement in asymmetric three-component ferroelectric superlattices, *Nature,* **433(7024)**, pp. 395–399.
2. Boschker, H., Verbeeck, J., Egoavil, R., Bals, S., van Tendeloo, G., Huijben, M., et al. (2012). Preventing the reconstruction of the polar discontinuity at oxide heterointerfaces, *Adv. Funct. Mater.,* **22(11)**, pp. 2235–2240.
3. Peng, R., Xu, H. C., Xia, M., Xia, M., Zhao, J. F., Xie, X., et al. (2014). Tuning the dead-layer behavior of $La_{0.67}Sr_{0.33}MnO_3/SrTiO_3$ via interfacial engineering, *Appl. Phys. Lett.,* **104(8)**, p. 081606.
4. Huijben, M., Liu, Y., Boschker, H., Lauter, V., Egoavil, R., Verbeeck, J., et al. (2015). Enhanced local magnetization by interface engineering in perovskite-type correlated oxide heterostructures, *Adv. Mater. Interfaces,* **2(3)**, p. 1400416.
5. Chen, Y. Z., Trier, P., Wijnands, T., Green, R. J., Gauquelin, N., Egoavil, R., et al. (2015). Extreme mobility enhancement of two-dimensional electron gases at oxide interfaces by charge-transfer-induced modulation doping, *Nat. Mater.,* **14(8)**, p. 801.
6. Bousquet, E., Dawber, M., Stucki, N., Lichtensteiger, C., Hermet, P., Gariglio, S., et al. (2008). Improper ferroelectricity in perovskite oxide artificial superlattices, *Nature,* **452(7188)**, pp. 732–736.
7. Boschker, H., Kautz, J., Houwman, E. P., Siemons, W., Blank, D. H. A., Huijben, et al. (2012). High-temperature magnetic insulating phase in ultra-thin $La_{0.67}Sr_{0.33}MnO_3$ films, *Phys. Rev. Lett.,* **109(15)**, p. 157207.
8. Rondinelli, J. M., May, S. J., and Freeland, J. W. (2012). Control of octahedral connectivity in perovskite oxide heterostructures: An emerging route to multifunctional materials discovery, *MRS Bulletin,* **37(3)**, pp. 261–270.
9. Aso, R., Kan, D., Shimakawa, Y., and Kurata, H. (2014). Octahedral tilt propagation controlled by *A*-site cation size at perovskite oxide heterointerfaces, *Cryst. Growth Des.,* **14(5)**, pp. 2128–2132.
10. Moon, E. J., Balachandran, V., Kirby, B. J., Keavney, D. J., Sichel-Tissot, R. J., Schlepuetz, C. M., et al. (2014). Effect of interfacial octahedral behavior in ultrathin manganite films, *Nano Lett.,* **14(5)**, pp. 2509–2514.

11. Zhai, X., Cheng, L., Liu, Y., Schlepuetz, C. M., Dong, S., Li, H., et al. (2014). Correlating interfacial octahedral rotations with magnetism in ($LaMnO_3$+delta)(N)/($SrTiO_3$)(N) superlattices, *Nat. Commun.,* **5**, p. 4283.

12. Yajima, T., Hikita, Y., and Hwang, H. Y. (2011). A heteroepitaxial perovskite metal-base transistor, *Nat. Mater.,* **10(3)**, pp. 198–201.

13. Driza, N., Blanco-Canosa, S., Bakr, M., Soltan, S., Khalid, M., Mustafa, L., et al. (2012). Long-range transfer of electron-phonon coupling in oxide superlattices, *Nat. Mater.,* **11(8)**, pp. 675–681.

14. Lee, C.-H., Orloff, N. D., Birol, T., Zhu, Y., Goian, V., Rocas, E., et al. (2013). Exploiting dimensionality and defect mitigation to create tunable microwave dielectrics, *Nature,* **502(7472)**, p. 532.

15. Park, J. H., Vescovo, E., Kim, H. J., Kwon, C., Ramesh, R., and Venkate-san, T. (1998). Direct evidence for a half-metallic ferromagnet, *Nature,* **392**, pp. 794–796.

16. Sun, J. Z., Abraham, D. W., Rao, R. A., and Eom, C. B. (1999). Thickness-dependent magnetotransport in ultrathin manganite films, *Appl. Phys. Lett.,* **74**, pp. 3017–3019.

17. Bowen, M., Bibes, M., Barthelemy, A., Contour, J. P., Anane, A., Lemaitre, Y., et al. (2003). Nearly total spin polarization in $La_{2/3}Sr_{1/3}MnO_3$ from tunneling experiments, *Appl. Phys. Lett.,* **82**, pp. 233–235.

18. Postma, F. M., Ramaneti, R., Banerjee, T., Gokcan, H., Haq, E., Blank, D. H. A., et al. (2004). Epitaxial diodes of a half-metallic ferromagnet on an oxide semiconductor, *J. Appl. Phys.,* **95**, pp. 7324–7326.

19. Hikita, Y., Nishikawa, M., Yajima, T., and Hwang, H. Y. (2009). Termination control of the interface dipole in $La_{0.7}Sr_{0.3}MnO_3$/Nb:$SrTiO_3$ (001) Schottky junctions, *Phys. Rev. B,* **79**, p. 073101.

20. Garcia, V., Bibes, M., Bocher, L., Valencia, S., Kronast, P., Crassous, A., et al. (2010). Ferroelectric control of spin polarization, *Science,* **327**, pp. 1106–1110.

21. Kavich, J. J., Warusawithana, M. P., Freeland, J. W., Ryan, P., Zhai, X., Kodama, R. H., and Eckstein, J. N. (2007). Nanoscale suppression of magnetization at atomically assembled manganite interfaces: XMCD and XRMS measurements, *Phys. Rev. B,* **76**, p. 014410.

22. Huijben, M., Martin, L. W., Chu, Y. H., Holcomb, M. B., Yu, P., Rijnders, G., Blank, D. H. A., and Ramesh, R. (2008). Critical thickness and orbital ordering in ultrathin $La_{0.7}Sr_{0.3}MnO_3$ films, *Phys. Rev. B,* **78**, p. 094413.

23. Koster, G., Kropman, B. L., Rijnders, G. J. H. M., Blank, D. H. A., and Rogalla, H. (1998). Quasi-ideal strontium titanate crystal surfaces

through formation of strontium hydroxide, *Appl. Phys. Lett.,* **73**, pp. 2920–2922.

24. Ohtomo, A., Muller, D. A., Grazul, J. L., and Hwang, H. Y. (2002). Epitaxial growth and electronic structure of $LaTiO_x$ films, *Appl. Phys. Lett.,* **80**, pp. 3922–3924.
25. van der Pauw, L. J. (1958). A method of measuring the resistivity and Hall coefficient on lamellae of arbitrary shape, *Philips Tech. Rev.,* **20**, pp. 220–224.
26. Koster, G., Rijnders, G., Blank, D. H. A., and Rogalla, H. (2000). Surface morphology determined by (001) single-crystal $SrTiO_3$ termination, *Physica C Supercond.,* **339**, pp. 215–230.
27. Boschker, H., Huijben, M., Vailionis, A., Verbeeck, J., van Aert, S., Luys-berg, M., et al. (2011). Optimized fabrication of high-quality $La_{0.67}Sr_{0.33}MnO_3$ thin films considering all essential characteristics, *J. Phys. D-Appl. Phys.,* **44(20)**, p. 205001.
28. Yu, P., Luo, W., Yi, D., Zhang, J. X., Rossell, M. D., Yang, C. H., You, L., et al. (2012). Interface control of bulk ferroelectric polarization, *Proc. Natl. Acad. Sci. U.S.A.,* **109(25)**, pp. 9710–9715.
29. Van Aert, S., Verbeeck, J., Erni, R., Bals, S., Luysberg, M., Van Dyck, D., and Van Tendeloo, G. (2009). Quantitative atomic resolution mapping using high-angle annular dark field scanning transmission electron microscopy, *Ultramicroscopy,* **109**, pp. 1236–1244.
30. Nakagawa, N., Hwang, H. Y., and Muller, D. A. (2006). Why some interfaces cannot be sharp, *Nat. Mater.,* **5**, pp. 204–209.
31. Kumigashira, H., Chikamatsu, A., Hashimoto, R., Oshima, M., Ohnishi, T., Lippmaa, M., et al. (2006). Robust Ti^{4+} states in $SrTiO_3$ layers of $La_{0.6}Sr_{0.4}MnO_3$/ $SrTiO_3$/ $La_{0.6}Sr_{0.4}MnO_3$ junctions, *Appl. Phys. Lett.,* **88**, p. 192504.
32. Yamada, H., Kawasaki, M., Lottermoser, T., Arima, T., and Tokura, Y. (2006). $LaMnO_3/SrMnO_3$ interfaces with coupled charge-spin-orbital modulation, *Appl. Phys. Lett.,* **89**, p. 052506.
33. Adamo, C., Ke, X., Schiffer, P., Soukiassian, A., Warusawithana, M., Maritato, L., and Schlom, D. G. (2008). Electrical and magnetic properties of $(SrMnO_3)_{(n)}/(LaMnO_3)_{(2n)}$ superlattices, *Appl. Phys. Lett.,* **92**, p. 112508.
34. Bhattacharya, A., May, S. J., Velthuis, S. G. E. t., Warusawithana, M., Zhai, X., Jiang, B., et al. (2008). Metal-insulator transition and its relation to magnetic structure in $(LaMnO_3)_{(2n)}/(SrMnO_3)_{(n)}$ superlattices, *Phys. Rev. Lett.,* **100**, 257203.

35. Dingle, R., Stormer, H. L., Gossard, A. C., and Wiegmann, W. (1978). Electron mobilities in modulation-doped semiconductor hetero-junction super-lattices, *Appl. Phys. Lett.*, **33**, pp. 665–667.

36. Pruneda, J. M., Ferrari, V., Rurali, R., Littlewood, B., Spaldin, N. A., and Artacho, E. (2007). Ferrodistortive instability at the (001) surface of half-metallic manganites, *Phys. Rev. Lett.*, **99**, p. 226101.

37. He, J., Borisevich, A., Kalinin, S. V., Pennycook, S. J., and Pantelides, S. T. (2010). Control of octahedral tilts and magnetic properties of perovskite oxide heterostructures by substrate symmetry, *Phys. Rev. Lett.*, **105**, p. 227203.

38. Glazer, A. M. (1972). The classification of tilted octahedra in perovskites, *Acta Crystallogr. Sec. B*, **28**, pp. 3384–3392.

39. Woodward, M. (1997a). Octahedral tilting in perovskites I geometrical considerations, *Acta Crystallogr. B*, **53**, pp. 32–43.

40. Woodward, M. (1997b). Octahedral tilting in perovskites II Structure stabilizing forces, *Acta Crystallogr. B*, **53**, pp. 44–66.

41. Vailionis, A., Boschker, H., Siemons, W., Houwman, E. P., Blank, D. H. A., Rijnders, G., and Koster, G. (2011). Misfit strain accommodation in epitaxial ABO(3) perovskites: Lattice rotations and lattice modulations, *Phys. Rev. B*, **83(6)**, p. 064101.

42. Vailionis, A., Boschker, H., Liao, Z., Smit, J. R. A., Rijnders, G., Hui-jben, M., and Koster, G. (2014). Symmetry and lattice mismatch induced strain accommodation near and away from correlated perovskite interfaces, *Appl. Phys. Lett.*, **105(13)**, p. 131906.

43. Gray, A. X., Papp, C., Balke, B., Yang, S.-H., Huijben, M., Rotenberg, E., et al. (2010). Interface properties of magnetic tunnel junction $La_{0.7}Sr_{0.3}MnO_3/SrTiO_3$ superlattices studied by standing-wave excited photoemission spectroscopy, *Phys. Rev. B*, **82**, p. 205116.

44. Gray, A. X., Minar, J., Plucinski, L., Huijben, M., Bostwick, A., Roten-berg, E., et al. (2013). Momentum-resolved electronic structure at a buried interface from soft X-ray standing-wave angle-resolved photoemission, *Europhysics letters*, **104(1)**, p. 17004.

45. Liao, Z., Huijben, M., Zhong, Z., Gauquelin, N., Macke, S., Green, R. J., et al. (2016). Controlled lateral anisotropy in correlated manganite heterostructures by interface-engineered oxygen octahedral coupling, *Nat. Mater.*, **15(4)**, p. 425.

46. Boschker, H., Mathews, M., Houwman, E. P., Nishikawa, H., Vailionis, A., Koster, G., Rijnders, G., and Blank, D. H. A. (2009). Strong

uniaxial in-plane magnetic anisotropy of (001)- and (011)-oriented $La_{0.67}Sr_{0.33}MnO_3$ thin films on $NdGaO_3$ substrates, *Phys. Rev. B,* **79(21)**, p. 214425.

47. Lee, A., and Ramakrishnan, T. V. (1985). Disordered electronic systems, *Rev. Mod. Phys.,* **57(2)**, pp. 287–337.

48. Liao, Z., Li, F., Gao, P., Li, L., Guo, J., Pan, X., et al. (2015). Origin of the metal-insulator transition in ultrathin films of $La_{2/3}Sr_{2/3}MnO_3$, *Phys. Rev. B,* **92(12)**, p. 125123.

49. Petrov, A. Y., Torrelles, X., Verna, A., Xu, H., Cossaro, A., Pedio, M., et al. (2013). Surface octahedral distortions and atomic design of perovskite interfaces, *Adv. Mater.,* **25(29)**, pp. 4043–4048.

50. Radaelli, G., lannone, G., Marezio, M., Hwang, H. Y., Cheong, S. W., Jorgensen, J. D., and Argyriou, D. N. (1997). Structural effects on the magnetic and transport properties of perovskite $A_{1-x}A'_x$ MNO_3 (x = 0.25, 0.30), *Phys. Rev. B,* **56**, pp. 8265–8276.

Chapter 3

Electron Transport across Oxide Interfaces on the Nanoscale

Kumari Gaurav Rana, Saurabh Roy, and Tamalika Banerjee
Zernike Institute for Advanced Materials, University of Groningen, Nijenborgh 4, 9747 AG Groningen, The Netherlands
t.banerjee@rug.nl

3.1 Introduction

In recent decades, one of the most active and promising areas of research in condensed matter physics has been employing complex oxides and their heterostructures. With the advent of new growth and measurement techniques, a wealth of novel magnetic and electronic ground states has been found in complex oxide heterointerfaces. The interest in these materials stems from their multifunctional properties, such as the coexistence of different types of magnetic ordering, ferroelectricity, charge and orbital ordering, all promoted by a strong interplay between spin, charge and orbital degrees of freedom [1–6]. These materials represent a clear advancement over conventional Si-based technology as their physical properties can

Oxide Spintronics
Edited by Tamalika Banerjee

ISBN 978-981-4774-99-4 (Hardcover), 978-0-429-46819-3 (eBook)
www.jennystanford.com

be tuned; and when used in devices, they lead to functionalities not realized in conventional electronics. The tunability of this provides a diverse playground of competing ground states and is greatly multiplied in thin films and their heterostructures by the addition of parameters such as substrate-induced strain and interfacial electronic reconstruction.

In the family of complex oxides, thin films of metallic $La_{0.67}Sr_{0.33}MnO_3$ (LSMO) and $SrRuO_3$ (SRO) have attracted a lot of research interest. LSMO is a strongly correlated oxide, which is predicted to be a half metal and has been investigated as a source of spin polarized electrons for oxide-based spintronics. In LSMO, the spin up and the spin down bands are separated by 1 eV or more [7, 8]. The parent compound of LSMO is $LaMnO_3$ (LMO), which is a Mott insulator made conducting and magnetic by doping with strontium (Sr). Owing to the double exchange [9] mechanism in LSMO, the double valence states of manganese (Mn) ions couple indirectly with each other through the itinerant e_g electrons, thereby yielding a coupled magnetic and conducting ground state. Various experiments like spin-resolved photoemission spectroscopy, Andreev reflection point contact and tunnel junctions with superconductors have confirmed the high-spin polarization or half-metallicity in LSMO [10, 11]. Consequently, there is an endeavor to obtain the predicted properties and to understand the correlations between spin, charge, and lattice degrees of freedom, which are intrinsically nanoscopic in nature. The exciting multifunctional device prospects with LSMO are manganite-based transistors, photovoltaic effects in multiferroics, ferroelectric tunnel junctions, spin valve transistors, etc.

Similarly, $SrRuO_3$ is another promising complex oxide, where external stimuli like strain, temperature, and structural distortions control the stability of electronic and magnetic states across its magnetic phase transition, and are also useful for spintronics. $SrRuO_3$ has been studied extensively for its growth and physical properties and is widely used as an electrode in complex oxide devices. An initial study of $SrRuO_3$ thin films was driven by its good metallic characteristics (for oxide systems) and high-quality growth in heteroepitaxial structures, which make it ideal for device applications. In addition, interfaces coupling $SrRuO_3$ to other complex oxides exhibit novel functionalities, not present in their parent compounds [12–15]. These primarily stem from broken symmetries at the interface, which when carefully tailored, can lead to unique device functionalities. The microscopic origin of ferromagnetism is supported by the distortions of the oxygen octahedra that are coupled with the degree of hybridization between O 2p and the Ru 4d orbitals [12]. At the interface, an additional distortion originating from the substrate can cause geometrical changes, which strongly influences its magnetic and electronic properties on either side of the ferromagnetic transition, T_C [15].

These novel properties predominantly arise at the nanoscale and often a clear understanding of the origin and analysis of these features are typically governed by the size and properties of the probe. As a result, many scanning probes employed in such an investigation have undergone significant modifications enabling more detailed investigation of such novel functional properties. In this context, an important technique that has only been recently introduced to study correlation effects to electron transport in complex oxide heterointerfaces is a valuable addition. Based on the well-known principle of the scanning tunneling microscope (STM), this technique is known as the *ballistic electron emission microscopy* (BEEM). It reveals new features in electron transport at the nanoscale, across heterointerfaces in devices based on complex oxides; thereby providing insights into the correlation between local electronic transport and reconstructions at the buried interface due to changes in the perovskite structure. For designing and realization of oxide devices with novel attractive properties, this technique allows us to study any material system, whose electronic properties are susceptible to a temperature-driven phase transition, while simultaneously allowing us to investigate its homogeneity across such buried interfaces with a semiconductor, at the nanoscale. This is an important approach for material systems such as complex oxides, where small changes in the geometrical properties with thickness or temperature can produce large changes in the electric and magnetic properties.

3.2 Experimental Techniques

3.2.1 Growth of Thin Films

The functional properties of these materials emanate from the control and utilization of the charge, spin, and lattice degrees of freedom, which has largely been facilitated by advances in growing well-defined interfaces between complex oxide materials with atomic-scale precision, allowing for the engineering and cross-coupling of their unique magnetic and conducting properties. All of these necessitate a precise control of thin films growth for device fabrication aided by analysis techniques that investigate their electronic and magnetic properties. For the growth of high-quality crystalline LSMO and SRO films in our study, we use single crystalline $SrTiO_3$ (STO) or Nb:$SrTiO_3$ (Nb:STO) substrates. $SrTiO_3$ is the substrate of choice due to the close lattice matching to LSMO and SRO crystals and the availability of the semiconductor counterpart with a 0.01 wt.%. of Nb doping at local Ti sites. The substrates are chemically treated to ensure a uniform TiO_2 termination and characterized for their surface quality with atomic force microscopy (AFM). This process assumes critical importance because a clean and defect-free starting surface is highly desirable for the growth of crystalline epitaxial films that determines a uniform electronic transport across such a film-substrate (Nb:STO) interface. We used pulsed laser deposition (PLD) to grow epitaxial films and with the help of high-pressure reflection high-energy electron diffraction (RHEED), the growth speed and crystallinity were ensured.

3.2.2 Growth of $SrRuO_3$ Thin Films

For the growth of high-quality crystalline SRO films in our study, we use single crystalline $SrTiO_3$ or Nb:$SrTiO_3$ substrates. We used pulsed laser deposition (PLD) to grow epitaxial films of different thicknesses. The films were grown at 600 °C in 0.13 mbar of oxygen pressure with an optimum energy density of 2 Jcm^{-2} on the SRO target. The target-to-substrate separation (d_{TS}) was kept at 58 mm and with a pulse repetition rate of 1 Hz, an average deposition rate of 30 nm/h was obtained. High-pressure RHEED was used to monitor the growth speed and crystallinity of the SRO film in situ. Along the growth process, a transition of growth mode from an initial layer-by-layer to step-flow-like takes place implying a change in surface diffusivity during its initial growth. It has been reported that with the growth of the first monolayer of SRO, a termination conversion from the B-site (TiO_2-terminated $SrTiO_3$ or T–Nb:STO) to A-site (SrO-terminated $SrRuO_3$ or S–Nb:STO) occurs owing to the high volatility of the RuO_2 phase. This causes an enhanced diffusivity of the species leading to a change in the growth mode [16–18]. After deposition of approximately 3 uc, the intensity oscillations are damped and the average RHEED intensity remains constant during the rest of the deposition, indicating a step-flow like growth mode. Figure 3.1 shows the atomic force microscopic (AFM) image of an SRO film after it is cooled down to room temperature in an oxygen pressure of 100 mbar at the rate of 10 °C/min.

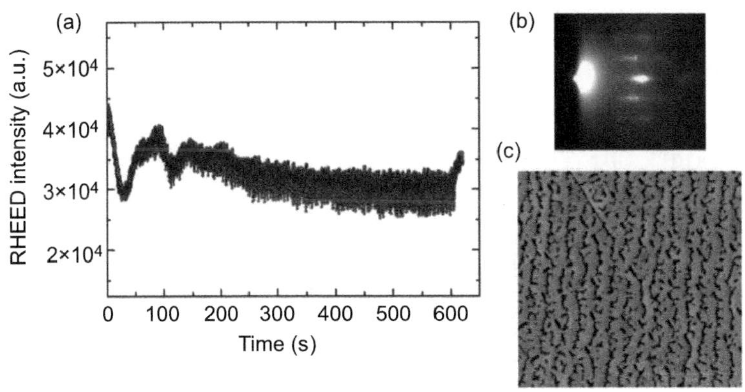

Figure 3.1 (a) RHEED oscillation during SRO growth. RHEED intensity oscillations saturate indicating the step-flow growth mode. (b) Typical RHEED pattern obtained after depositing 10 uc of SRO. (c) The AFM topography image of a 10 uc SRO film grown on Nb:STO substrate after cooling down to room temperature at the rate of 10 °C/min in the presence of 100 mbar of oxygen. The rms roughness of the surface is 0.853 nm.

3.2.3 Growth of LSMO Thin Films

Thin films of LSMO films are also grown epitaxially by using pulsed laser deposition. A layer-by-layer growth is monitored to a unit cell precision, using an in situ RHEED, as shown in Fig. 3.2a. Laser with a spot size of 2.2 mm^2 and a laser fluence of 2.2 J cm^{-2} is incident on optimum stoichiometry LSMO target. A singly terminated (TiO_2) Nb:STO substrate is held at a d_{TS} of 53 mm at 750 °C in 0.35 mbar oxygen pressure and epitaxial LSMO films are grown at the rate of 100 nm/h. A laser pulse repetition rate of 1 Hz is used. An AFM image obtained after cooling down the LSMO film to room temperature at 100 mbar oxygen pressure at the rate of 10 °C/min is shown in Fig. 3.2c. The root-mean-square roughness of the LSMO film surface is found to be 0.3 nm. The X-ray diffraction and transmission electron microscopy show that the LSMO films are grown in epitaxy with the underlying singly terminated (TiO_2) Nb:STO substrate with a tensile strain [19]. A small value of residual resistivity (50 μΩ-cm) for a 60 uc LSMO film also confirms the high quality of the thin films of LSMO [14].

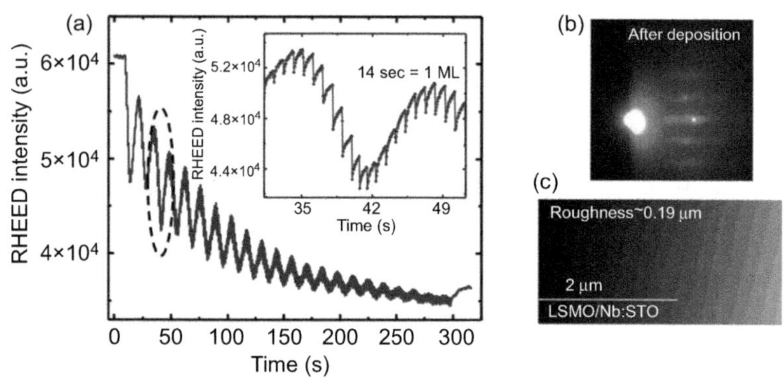

Figure 3.2 RHEED oscillations as observed during LSMO growth. The inset shows the zoomed-in plot of the third RHEED oscillation showing the laser pulses. (b) Typical RHEED pattern obtained after 10 nm of LSMO growth. (c) The AFM image of the film obtained after cooling down the LSMO film to room temperature at 100 mbar at the rate of 10 °C/min.

3.2.4 Ballistic Electron Emission Microscopy

BEEM is a unique technique [20] to study the nanoscale electronic transport across a wide variety of heterostructures. It is a three-terminal modified version of a scanning tunneling microscope (STM). The electrons tunnel from the negatively biased STM tip with a tunnel voltage V_T and tunnel current of I_T into a thin metallic film deposited on an n-type semiconductor as shown in Fig. 3.3. The injected electron distribution is termed as hot as they possess kinetic energies of a few electron volts (eV) above the thermal energy. These electrons tunnel into the unoccupied states of the metal overlayer (such as LSMO or SRO in this case). The metallic film and the semiconductor (Nb:STO) have distinct electrical contacts. With a distribution of hot electrons injected into the metallic film, they undergo scattering as they travel through the thickness of the film. Upon reaching the semiconducting interface, these electrons will thermalize in the conductance band of the semiconductor and will be collected as the BEEM current (I_B) provided they have energies above the Schottky barrier height (SBH, φ_B) at the metal-semiconductor interface. Usually, the tunnel current (I_T) is the range of one to tens of nanoamperes (nA) and the extracted BEEM current is of the order of a few picoamperes (pA). The energies of the injected hot electrons can be varied over a range and a spectroscopic study of the BEEM current for a range of voltage gives an accurate measure of the pristine energy band line-ups at the metal–semiconductor (M–S) interface. The lateral resolution, determined by the propagation of the transmitted electrons is strongly governed by the conservation of energy and transverse momentum at the Schottky interface. The BEEM spectroscopy thus provides information on the electronic band structure at the interface and when the tip is moved across the device gives useful information on the (in) homogeneity of the interface and its correlation with structural features of the buried interfaces. Further, the BEEM current is strongly attenuated by the inelastic scattering and elastic scattering occurring at the metal layers and at the M–S interface. A detailed description of the factors that influence the BEEM current can be found in Ref. [21]. A comprehensive study of hot electron transport and electron scattering mechanisms across epitaxially grown thin films, atomically engineered and well-characterized interfaces are the key to advancing oxide-based electronics. However, in order to ascertain the electronic transport across such an interface with Nb:STO, it is pertinent to first characterize the transport in the semiconducting substrate itself.

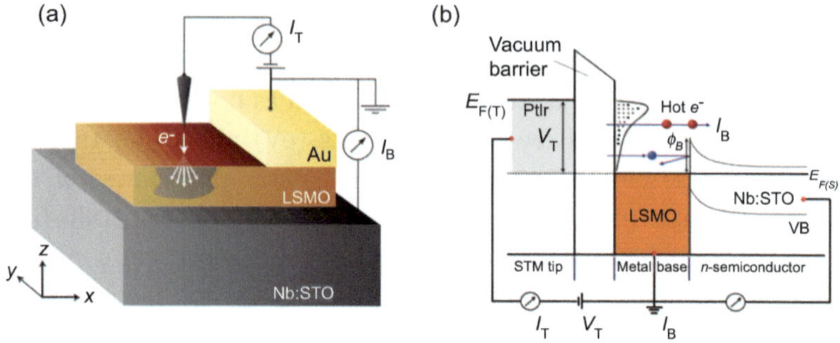

Figure 3.3 (a) Schematics of the BEEM experiment. The sample is a thin epitaxial film of LSMO of variable thickness deposited on 0.01 wt.% Nb:STO (001) substrate. A PtIr STM tip is used to locally inject electrons into the sample by tunneling at a sample-tip bias, V_T, between the tip and the LSMO surface. The electrons transmitted perpendicularly through the LSMO layer are collected as I_B in the Nb:STO with a third (rear) electrical contact. (b) Energy schematics of the BEEM technique. Hot electrons are emitted from the STM tip across the vacuum tunnel barrier and injected locally into the LSMO base. After transmission in the base, they are then collected in the conduction band of the Nb:STO, provided the energy and momentum criteria needed to overcome φ_B at the Nb:STO interface are satisfied (red circles; blue circle denotes those electrons which do not satisfy the criteria). All measurements were performed at 300 K [34].

3.3 Nb-Doped $SrTiO_3$ as a Semiconductor

For probing electron transport across an all oxide metal–semiconductor (M–S) Schottky interface, a well-characterized oxide semiconductor is needed. STO is a prototype perovskite, which in its d^0 state is a bandgap insulator. Pentavalent Nb is used as a donor in $SrTiO_3$ substituting at the tetravalent Ti sites and making it an n-type semiconductor. Electrical transport across conventional semiconductor (as Si)/metal Schottky is described by thermionic emission theory for moderately doped semiconductors, whereas for heavily doped semiconductors, quantum mechanical tunneling is the dominant transport process. Although such physics of charge transport across a Schottky interface between metal and Si is well understood, the underlying mechanism of electrical transport across an interface with oxide semiconductor as doped STO is not. Here we show that the temperature dependence of resistivity and the Hall coefficient were measured in the van der Pauw configuration, using a physical property measuring system (PPMS by Quantum Design) from 100–300 K. Figure 3.4 shows temperature-dependent ρ (resistivity) in Nb:STO for 0.01 and 0.1 wt.% Nb doping.

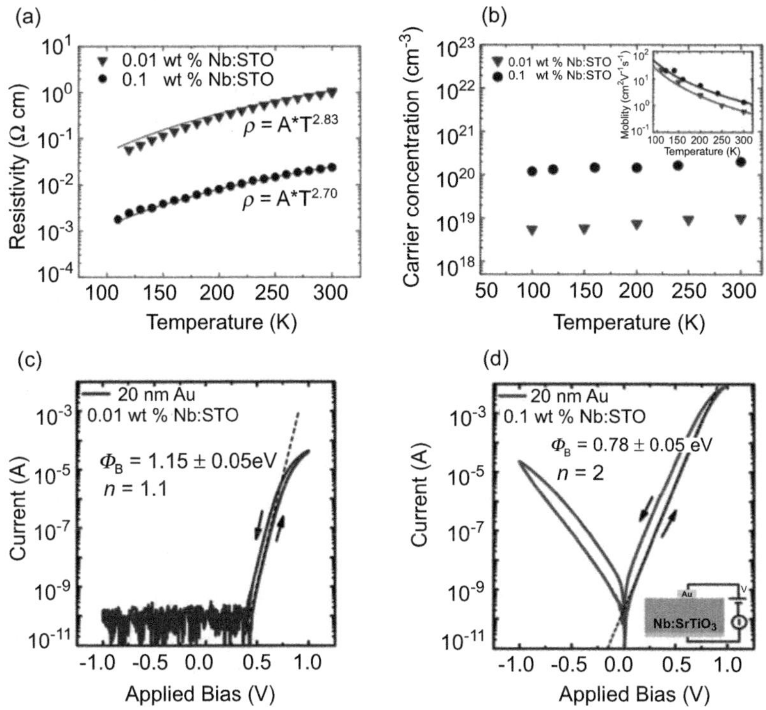

Figure 3.4 (a) Temperature dependence of resistivity. (b) Carrier concentration in Nb:STO for 0.1 and 0.1 wt.% Nb doping for the temperature range 100–300 K. Inset shows the temperature dependence of mobility, calculated from obtained resistivity and carrier concentration. The solid lines represent the power law fits. (c) Current (I)–voltage (V) characteristics for Au/Nb:STO junction for 0.1 and 0.01 wt.% Nb:STO. The up and down arrows indicate the direction of the voltage sweep. The dotted line represents the thermionic equation fit. (d) The inset shows the schematic layout of the device. Reprinted from Ref. [26], with permission from AIP publishing, Copyright 2012.

Figure 3.4a shows the temperature dependence of resistivity and Hall coefficient for 0.01 and 0.1 wt.% Nb doping. ρ decreases significantly with decreasing temperature strongly following a power law, as $\rho = AT^{\alpha}$, where A is a temperature independent parameter. Such a metal-like behavior of electrical resistivity is typical of unconventional heavily doped semiconductors. Temperature dependence of the carrier concentration (N_D), obtained using $R_H = \frac{1}{qN_D}$ from Hall measurements, consistent with electrons being the charge carriers is shown in Fig. 3.4b. Unlike conventional semiconductors like Si, a temperature invariant N_D is obtained. This is rationalized by using the hydrogenic theory of shallow donors, where the donor binding energy, E_D, is related to the dielectric constant (ε), the effective mass (m_e*), and free-electron mass (m_e) as $E_D = -13.6\ \text{eV}\left(\frac{m_e{}^*}{\varepsilon_r{}^2 m_e}\right)$ [21–24]. Due to the large temperature dependence of (ε in STO assumed to be the same in Nb:STO), the activation energy is small, even at 100 K. Thus, E_D at 300 K is 1.5 meV decreasing to only 0.2 μeV at 4.2 K. Unlike what is observed in conventional semiconductors, this thus indicates that the shallow donors remain ionized at all temperatures measured, implying the absence of carrier freeze out and explaining an almost temperature independence of the carrier concentration. However, a slight increase in the carrier concentration with increasing temperature, for both dopant densities (a factor of 1.4 in 0.01 wt.% and 2 in 0.1 wt.% of Nb doping) is observed [25]. This could be ascribed to additional conduction processes originating from charge carriers at energy levels below the conduction band whose temperature dependence could be different from the carriers at the bottom of the conduction band in Nb:STO. The resistivity and carrier density thus obtained are used to extract the mobility of the charge carriers in Nb:STO using $\rho = \frac{1}{\mu(T)\mathrm{q}\, N_D(T)\, e}$. The mobility is observed to decrease with increasing temperature and is fitted with a power law as $\mu(T) = \mu T^{-m}$ (Fig. 3.4), where μ_0 is a temperature independent coefficient and the exponent m is related to the different scattering mechanisms: electron–electron, electron-ionized impurity and electron–phonon scattering. Simple fitting of the obtained data gives m = 3.2 and 3 for Nb doping of 0.01 and of 0.1 wt.%, respectively. The temperature dependence of $\mu(T)$ suggests the role of electron–phonon scattering due to the strong ionic nature of the lattice. It could be related to the temperature dependence of ε, which implies effective screening of the ionized impurity scattering centers resulting in a metal-like behavior. These peculiar transport properties mark Nb:STO as unconventional semiconductor [21, 25]. Further, we have studied the electrical transport across an interface between a noble metal Au and well-characterized Nb:STO substrates. We use TiO_2 terminated Nb:STO to study electrical transport across the interface. Figure 3.4(c and d) show the current–voltage (I–V) characteristics, at RT, of Au/Nb:STO Schottky diodes with 0.01 and 0.1 wt. % Nb doping, which show a metal-semiconductor rectifying behavior. Equation 3.1 (as discussed in Section 3.4) is used to estimate the SBH of each Schottky diode. Although a large deviation from ideality factor ($n = 2$) is observed in case of heavier doped Nb:STO yet large rectifying (3 and 4 orders and $n = 1.1$) behavior is observed in low-doped Nb:STO in spite of degenerate kind of transport behavior as discussed earlier. $W_{\text{dep}} = \sqrt{2\varepsilon_s\left(\frac{V_{\text{in}}}{qN_D}\right)}$, where ε_s, N_D, V_{in} are semiconductor permittivity, carrier concentration and built-in potential, respectively, is used to estimate the depletion width and is found to be 70 nm and 13 nm for 0.01 wt.% and 0.1 wt.% Nb:STO, respectively. This supports the dominance of transport processes other than thermionic emission for Au on 0.1 wt. % Nb:STO interface [25–28].

3.4 Macroscopic Electronic Transport: LSMO/Nb:STO

Most reliable methods applied for understanding the electronic transport across a metal-semiconductor oxide interface are current–voltage (I–V) and capacitance–voltage (C–V) measurements and more recently, the internal photoemission (IPE) technique [27–29]. These studies have yielded insights that necessitate consideration of transport mechanisms beyond those commonly employed to describe electron transport in conventional (non-oxide based) Schottky interfaces. The influence of interface states and interface dipoles, temperature, and electric field dependence of the dielectric permittivity in complex oxides has to be incorporated for a complete description of the transport characteristics across such Schottky interfaces. However, studies related to the homogeneity of the transport properties at such complex oxide interfaces, where competing electronic phases might coexist, are absent, primarily due to its nanoscopic nature and due to the limitations of the techniques used thus far. In case of (*I*–*V*) measurements, the bias is applied at the metal-semiconductor (M–S) Schottky interface and the voltage is varied to record the diode characteristics. LSMO/Nb:STO Schottky interface shows a clear rectifying behavior with a linear dependence of the forward current at all temperatures and a shift in the onset of the current at lower temperatures (Fig. 3.5a). Forward characteristics tend to be less linear beyond a certain bias and are associated with the dominance of the forward series resistance, which has been calculated to be around 50 Ω at RT in these devices.

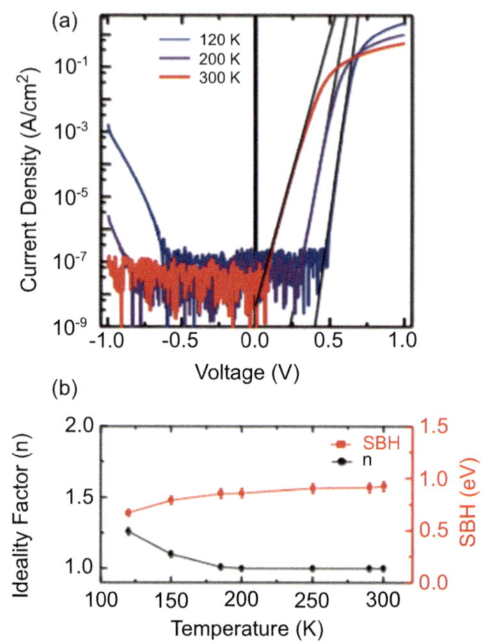

Figure 3.5 (a) Current–voltage (*I*–*V*) characteristics for LMSO/Nb:STO Schottky diode at room temperature, at 200 K and at 120 K. A clear rectification is observed in all cases. (b) The Schottky barrier heights and the ideality factors are extracted from (a) using Eq. 3.1 at different measurement temperatures. Reprinted from Ref. [33], with permission from the American Physical Society, Copyright 2013.

The shift of onset of the current decreases to a lower bias with decreasing temperature. These observations point to a higher probability of electrons to flow through the interface at low temperatures and suggest the role of transport mechanisms other than the commonly used thermionic emission model. Theoretical models that best describe the I–V characteristics of these Schottky junctions are based on thermionic emission, which is given by

$$I = A^* AT^2 \exp\left(q\frac{\varphi_B}{k_B T}\right)\left[\exp\frac{qV}{nk_B T} - 1\right] \tag{3.1}$$

where q is the charge of the electron, k_B is the Boltzmann constant, A is the area of the diode, T is the temperature, A^* is the Richardson constant (here assumed to be 156 $Acm^{-2}K^{-2}$), φ_B is the barrier height at zero bias and n is the ideality factor (which is unity for purely thermionic emission dominated transport) [30]. The Schottky barrier height at zero bias and the ideality factor n can be obtained from the log I–V plots by fitting the forward bias characteristics using the thermionic emission model. The linear part of the forward characteristics of the diode obtained from the I–V plot is used to obtain φ_B and n. The SBH of 0.92 eV gradually decreases till 180 K, below which the decrease is rather abrupt (at 120 K the SBH is 0.67 eV) as shown in Fig. 3.5b. The ideality factor concomitantly increases with decreasing temperature and is found to be 1.26 at the lowest temperature measured. Similar observations of a decrease in SBH and an increase in the ideality factor at low temperatures have also been reported for different combinations of manganite/Nb:STO interfaces [22, 29, 31]. However, there is not a very clear understanding at the nanoscale about the reasons behind the deviation from ideal thermionic emission transport. To address this, the nanoscale ballistic electron emission microscopy is employed to study unbiased oxide heterointerfaces.

The BEEM utilizes hot electrons (a few electron volts above E_F) to probe electron transport across thin metal layers on a semiconducting substrate. The energy of the injected electrons is varied by the voltage, V_T, applied between the scanning tunneling microscope (STM) tip and the metal base. Figure 3.3a shows the BEEM schematics and Fig. 3.3b shows the energy band diagram. The electrons are injected over the vacuum tunnel barrier into the metal base, where they undergo inelastic (electron–electron) and elastic (structural defects, grain boundaries) scattering, which reduce the number of electrons that have sufficient energy and momentum to overcome the Schottky barrier at the M–S interface. In BEEM, no bias is applied at the M–S interface (unbiased junctions) and electrons thus enter the conduction band of the semiconductor with their own kinetic energy. This is in contrast to the studies performed using standard I–V, where an external bias is applied at the M–S interface. The BEEM current, I_B, is recorded at a local area of a few nanometers at a constant tunnel current, I_T, while varying, V_T. I_B depends on the thickness of the metal layer and decreases exponentially with increasing thickness. Several such spectra are collected from one location and are then averaged to obtain a representative BEEM spectrum. By placing the STM tip at different locations of the device, similar BEEM spectra are recorded at different regions from which local Schottky barrier heights can be extracted and nanoscale homogeneity can be addressed. The onset of the BEEM current with V_T gives the SBH from the Bell–Kaiser (B–K) model [20]. According to this model, the BEEM transmission at the interface is given by

$$\frac{I_B}{I_T} \propto (V_T - \varphi_B)^2 \tag{3.2}$$

Hot electron transmission recorded using BEEM, across an LSMO (13 uc)/Nb:STO Schottky interface is shown in Fig. 3.6a at 300 K (red circles) and 120 K (green squares). Beyond a certain tip bias, the BEEM transmission is found to increase with increasing V_T while I_T is held constant. The transmission shown is an average of more than 50 individual spectra recorded at a particular location in the device. The hot electrons injected across the vacuum tunnel barrier propagate through the thin LSMO film and are collected in the conduction band of Nb:STO, provided they have the necessary energy and momentum to overcome the SBH at this interface, which is left unbiased. Electron–electron scattering processes at these energies can reduce I_B, which can additionally be influenced by other elastic scattering events during transmission in the LSMO film and at the M–S interface. The BEEM transmission is higher at 120 K, suggesting that scattering events such as inelastic scattering or scattering due to magnons and/or phonons are reduced and also in accord with the temperature dependence of resistivity observed in LSMO films. The onset in the BEEM transmission beyond a certain threshold voltage corresponds to the local SBH. Using the B–K model, SBH of the LSMO/Nb:STO interface is obtained by extrapolating the straight line so as to intersect the voltage axis in the plot of the square root of the BEEM transmission versus applied voltage, as shown in Fig. 3.6b. The SBH at RT is found to be 0.87 ± 0.02 eV (the error bar represents the error in the B–K fitting) and it is 0.83 ± 0.02 eV at 120 K. The SBH has been similarly extracted at different locations and is represented in the histograms shown in Figs. 3.6(b and c). We observe that on cooling, the SBH obtained from BEEM across unbiased interface does not match with that of macroscopic (*I*–*V*) precisely. The unchanged SBH values measured by BEEM technique cannot explain the macroscopic, as shown in Fig. 3.5. Additionally, as shown in Fig. 3.7, at a temperature of 120 K, there is a shift in the onset at which the forward current in the diode sets in. In order to understand this, the depletion width is calculated at 0.5 V, using Eq. 3.3:

$$W = \frac{b\varepsilon_0}{qN}\cosh^{-1}\left[1 + \frac{qN}{\sqrt{a}b\varepsilon_o}V_{bi}\right] \tag{3.3}$$

where N and V_{bi} are the carrier concentration in Nb:STO and the built-in potential, respectively and a and b are temperature dependent constants in the expression that describe the temperature dependence of the dielectric permittivity ε_r at zero electric fields [32]. The obtained values of depletion width as shown in the blue dashed curve in Fig. 3.7 are found to increase on cooling down at fixed bias. However, the electric-field-induced reduction in the W causes the electrons to tunnel through the interface at energies much lower than the actual SBH and leads to an apparent decrease in the SBH and increase in the ideality factor with temperature, as measured in macroscopic *I*–*V*. This additional current transport process at lower temperatures also enhances and leads to a non-saturating current for reverse bias in the diode (Fig. 3.5a). Electron transport and the extracted transport parameters, such as the SBH, from BEEM measurements are relatively unaffected at the unbiased interface (the bias applied here is in between the STM tip and the top surface of the LSMO film only), in contrast to the electric-field-dominated transport at the narrowed depletion width at the Schottky interface in *I*–*V* measurements. The local SBH as extracted from BEEM studies is reasonably similar at different locations of the device, both at RT (FWHM = 0.2 eV) and at 120 K (FWHM = 0.1 eV), as shown in the insets in Fig. 3.6 (b and c).

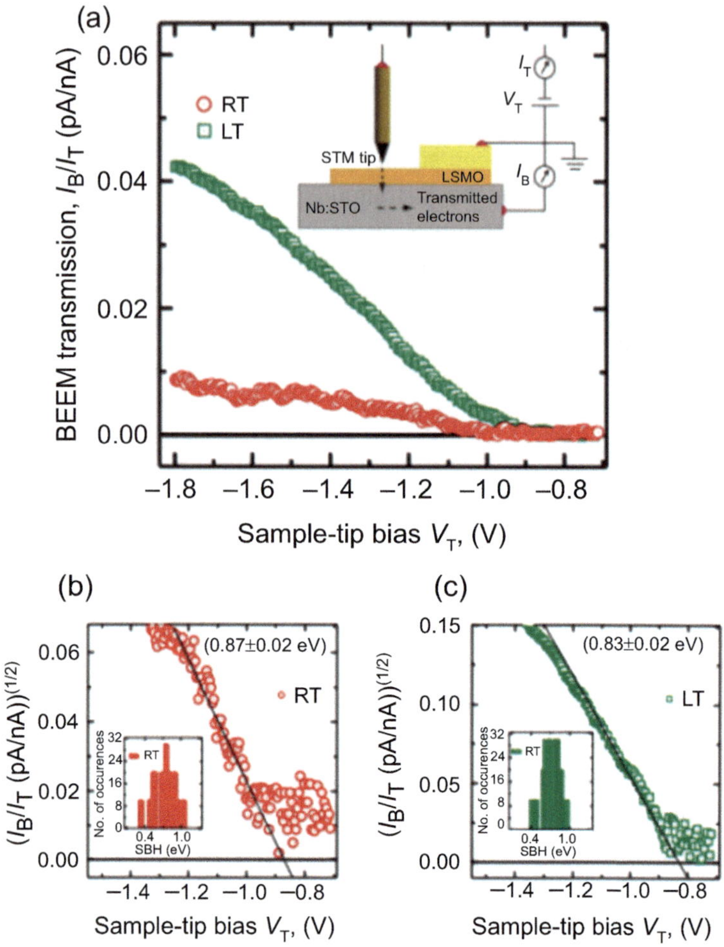

Figure 3.6 (a) BEEM spectra for the LSMO (13 uc)/Nb:STO Schottky interface at RT (red) and at 120 K (green). Each spectrum is a representative of several spectra taken at the same location. (b) The extracted SBH at the LSMO (13 uc)/ Nb:STO Schottky interface, obtained by plotting the square root of the normalized I_B with V_T and fitting it to the B–K model at RT. (c) The same plot at 120 K. The insets in (b) and (c) show the distribution of Schottky barrier heights obtained at different locations in the device at RT and at 120 K, respectively. Reprinted from Ref. [33], with permission from the American Physical Society, Copyright 2013.

Thus, by combining two independent probes, that are, current–voltage measurement and the technique of BEEM, we gain new insights into the transport characteristics in complex oxide heterostructures [32]. In contrast to the studies performed using standard *I–V*, where an external bias is applied at the M–S interface, BEEM can be performed for unbiased interfaces, revealing the intrinsic electronic properties of the interfaces which are not influenced by electric field effects on the non-linear dielectric permittivity of the semiconducting substrate. It is stressed that the decrease in SBH obtained from *I–V* measurements, in the temperature range studied, is primarily associated with the reduction in the depletion width (which otherwise broadens with decreasing temperatures in these oxides) with applied bias, which facilitates other transport processes, such as tunneling. In BEEM, where transport is by hot electrons, no such decrease in the local SBH with temperature is observed at the unbiased Schottky interface. Such studies of the local Schottky interface of complex oxides at the nanoscale can be extended to the study of complex oxides, where competing phases are believed to exist.

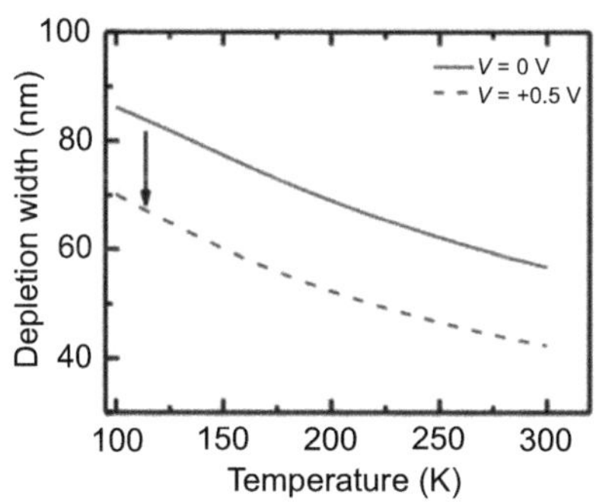

Figure 3.7 Depletion width in Nb:STO, calculated from Eq. 3.3 with varying temperature. The red solid curve is at zero bias, i.e., $V = 0$ V, whereas the blue dashed curve is at an applied bias of 0.5 V. The blue plot shows that the depletion width in Nb:STO reduces with the externally applied field at all temperatures.

3.5 Hot Electron Transport in a Strongly Correlated Transition-Metal Oxide: $La_{0.67}Sr_{0.33}MnO_3$

Engineered epitaxial LMSO/Nb:STO interfaces and their thickness dependence are investigated using BEEM. LSMO/Nb:STO interface is artificially engineered to obtain a relatively homogeneous interface with a macroscopic SBH of 1 eV. Interface dipoles have been tuned by an insertion of 1 uc of $SrMnO_3$ between LSMO and Nb:STO [22]. The BEEM spectra are obtained by recording the BEEM current I_B at a constant tunnel current I_T while varying V_T for at various locations. Thickness-dependent investigations (7–11 uc) are performed and an average BEEM spectrum is obtained for each thickness of LSMO, as shown in Fig. 3.8a. This gives an SBH of 1.06 ± 0.02 eV, obtained using the B–K model for all the thicknesses. I_B is observed to decrease with increasing thickness of LSMO. Figure 3.8b shows the I_B scaling with injected I_T for a 7 uc of LSMO thin film. The collected BEEM current in each BEEM spectra depends not only on the tunneling current injected into LSMO but also on the energy and momentum distribution of the carriers reaching the interface and the transmission probability at the LSMO/Nb:STO interface. The epitaxial LSMO/Nb:STO interfaces studied here have been optimized to be fully strained, atomically abrupt and with high crystalline perfection [33, 34]. Thus, they have fewer elastic scattering sites than the typical polycrystalline metal Schottky interfaces studied by BEEM, implying the conservation of transverse momentum of the transmitted electrons across the interface and with minimal influence on I_B. I_B also depends on the acceptance angle for electron collection at the Nb:STO interface, which is determined by the ratio of the effective masses of the Nb:STO and LSMO. This is calculated to be larger here than found for most standard M–S interfaces (such as Au on Si). Availability of allowed states in the conduction band minima (CBM) in k-space in Nb:STO is another criterion that further governs collection. Electronic band structure calculations show that the projected conduction band minima in doped STO are at the zone center, thus electrons with small parallel momenta should be easily collected in the available phase space [19, 35]. By way of comparison, I_B/I_T is almost two orders of magnitude higher for a similarly thick polycrystalline Ni film on an n–Si/Au M–S interface [16]. Thus, intrinsic strong correlation effects are concluded to have a considerable contribution in the measurement.

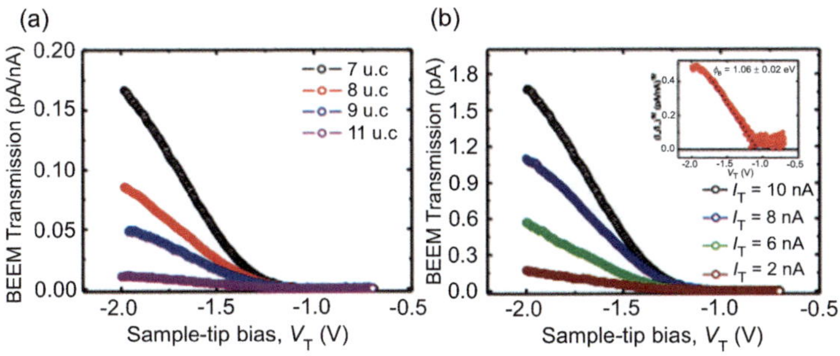

Figure 3.8 Energy dependence of hot electron transmission in LSMO at the nanoscale. (a) Electron current per nanoampere (nA) of injected tunnel current versus sample-tip bias across LSMO/Nb:STO (001), for LSMO thickness, *t*, of 7 uc, 8 uc, 9 uc and 11 uc. BEEM transmission decreases with increasing thickness of LSMO. With an increase in sample-tip bias an onset of I_B at around 1.06 ± 0.02 eV, corresponding to ϕ_B is seen. (b) I_B is observed to increase with I_T. The collected current is shown for tunnel currents of 2 nA (brown), 6 nA (green), 8 nA (blue) and 10 nA (black). The inset shows the extracted φ_B for LSMO (7 uc)/ Nb:STO(001), obtained by fitting to the B–K model, typically used for fitting BEEM spectra. The local ϕ_B is extracted by plotting the square root of the normalized I_B with V_T [34].

λ in LSMO is found to be 1.48 ± 0.10 uc at −1.9 V and increases to 2.02 ± 0.16 uc at −1.3 V at room temperature. Using Matthiessen's rule, the hot electron attenuation length, $\lambda(E)$, can be written as the sum of elastic λ_{elastic} and inelastic $\lambda_{\text{inelastic}}(E)$ scattering lengths as:

$$\frac{1}{\lambda(E)} = \frac{1}{\lambda_{\text{elastic}}} + \frac{1}{\lambda_{\text{inelastic}}(E)} \tag{3.4}$$

elastic scattering is minimal at such epitaxial heterostructures and λ(E) is thus dominated by $\lambda_{\text{inelastic}}$(E). The energy dependence of λ in LSMO is shown in Fig. 3.9b. Optimally doped LSMO is a transport half-metal with an insulating gap for minority spins at the Fermi level, E_F and conducting for majority spins. However, at energies higher than E_F, which are relevant for these devices, the density of states for both the majority and minority spin electrons increases. Conduction and ferromagnetism in LSMO are governed by the interaction of localized electrons from the incomplete *d* shell in Mn, by a process commonly referred to as the Zener double-exchange mechanism [9]. Furthermore, as suggested by neutron scattering experiments, LSMO at room temperature consists of dynamic nanoscale polarons,

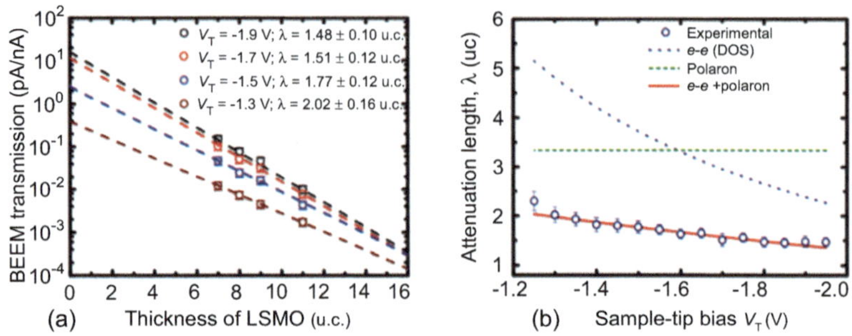

Figure 3.9 (a) BEEM transmission normalized per nanoampere (nA) of injected tunnel current versus the thickness of LSMO at different tunnel voltages. Dotted lines at each energy represent the exponential decay of I_B. These lines are extrapolated to zero LSMO thickness, which corresponds to attenuation due to the interface. (b) Attenuation length, λ, extracted from Fig. 3.8a for different energy values. The solid red curve is a theoretical estimate taking into account electron–electron scattering within Fermi liquid theory based on the DOS for the materials and an energy-independent polaron scattering term. The dotted curves depict the two contributions independently, where blue (green) shows the electron–electron (polaron) contribution to the full estimate [34].

which do not freeze out below the Curie temperature (as opposed to the situation in $La_{0.7}Ca_{0.3}MnO_3$) [17]. LSMO separates into regions, where the electrons are trapped in a local Jahn–Teller distortion and a conductive network without distortions [18]. The transport then occurs as the hot electrons move through the conductive network while being scattered by the local Jahn–Teller distortions, in addition to the electron–electron scattering of the injected charges themselves. Here, the polaron scattering can be regarded as static because the time scale of lattice motion (picoseconds) is much larger than that of hot electron transport through the LSMO film (femtoseconds). A Fermi liquid theory fit (blue dashed curve) is obtained based on the density of states (DOS) previously reported. However, the experimentally determined scattering rate decreases much slower than that predicted by this model. Incorporating an energy-independent fitting parameter that accounts for the polaronic scattering processes (~ 3.3 uc), the experimental data matches well as shown by the solid red curve in Fig. 3.9b. The obtained fitting constants are well within the ranges expected, based on estimates of the band masses and DOS from density functional theory [36]. The mean free path due to just the polaronic scattering is expected to be an order of the polaron separation. A previous study with neutron measurements estimated this value to be 2 uc (0.77 nm) [18], which is roughly consistent with the constant value obtained here (~3.3 uc). Thus, the hot electron energy dependence enables us to quantitatively isolate two dominant scattering mechanisms: electron–electron and polaronic scattering in such highly correlated electron complex oxides.

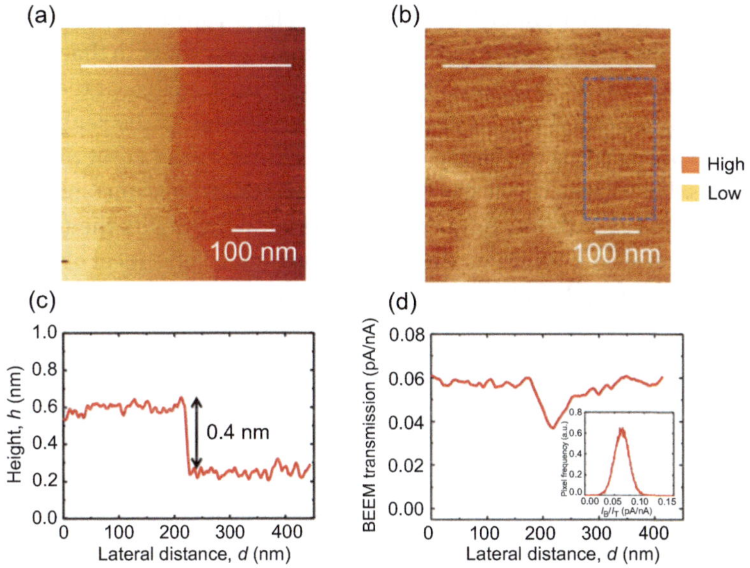

Figure 3.10 Spatially resolved electron transmission through LSMO (9 uc)/ Nb:STO (001). (a, c) Topographic STM image taken at $V_T = -2.5$ V and $I_T = 8$ nA over a 0.55 × 0.55 μm^2 area. A 0.4 nm step-height, is seen in the cross-section profile in (c). (b, d) The corresponding map of the BEEM current, I_B , acquired simultaneously. The BEEM transmission is seen to be reduced by a factor of two while approaching a step-edge. Quantitative analysis of I_B using a histogram of current distribution in the area under the blue box in (b) is shown in the inset of (d). The mean value of the distribution matches well with that of the BEEM transmission of this device as shown in Fig. 3.8a [34].

3.6 BEEM across the Steps and Edges

BEEM is also used to investigate the vertical transport through steps and edges of the surface. STM topography of the LSMO (9 uc)/

Nb:STO interface along with a simultaneously recorded spatial map of the transmitted current at $V_T = -2.5$ V and $I_T = 8$ nA is shown in Fig. 3.10. Since, Nb:STO surface can be terminated by either by SrO or TiO_2 unit cells in (001) direction, an atomically flat singly terminated TiO_2 surface with a step-height of 0.4 nm is observed from the STM topography, whereas a cross-section profile in the same location reveals different transmission at the terrace and at the step-edge with I_T being constant. A histogram of the transmitted current at the terrace (area under the blue box in Fig. 3.10b) is shown in the inset in Fig. 3.10d. The mean value of I_B matches well with that of the BEEM spectra for this film (Fig. 3.8a). The reduction in I_B at the step-edge as compared to that at the terrace arises due to the sensitivity of the propagating hot electrons to momentum scattering at the step edges. This broadens the hot electron distribution and concomitantly reduces I_B at such locations. This observation highlights the unique capability of the BEEM to study and directly visualize local electron transport in oxide heterointerfaces at the nanoscale.

3.7 Control of Film-Substrate Interface in Electronic Transport across Magnetic Phase Transition

Interfaces coupling complex oxides exhibit novel functionalities, not present in their parent compounds. These primarily stem from broken symmetries at the interface, which when are fully tailored, can lead to unique device functionalities. In this regard, SRO is a canonical metallic system that has been studied extensively for its growth and physical properties and is widely used as an electrode in complex oxide devices [37–41]. Despite this, not much has been studied to understand how external stimuli like strain, temperature, and structural distortions control the stability of electronic and magnetic states, across its phase transition. Interestingly, SRO undergoes a ferromagnetic transition at 140 K (T_C) and presents a distorted orthorhombic structure *Pbnm* below 820 K [42]. This strongly influences its magnetic and electronic properties on either side of the T_C. Here employing BEEM, we study such complex correlation across an interface between SRO on semiconducting Nb:$SrTiO_3$ (Nb:STO) and reveal that the geometrical reconstructions at the interface for a ferromagnetic transition have a strong influence on electronic transmission.

In this study, high-quality films of SRO of varying thicknesses 6, 8, 9, and 10 unit cells (uc) were grown on Nb:STO (as mentioned in details earlier). The films were characterized for their respective smoothness and the root-mean-square surface roughness of the 6 uc and 10 uc films are found to be 0.21 nm and 0.15 nm. The grown SRO films were characterized for their electronic transport and magnetic properties using standard van der Pauw method and superconducting quantum interference device (SQUID) magnetometry, respectively. The electrical resistivity studies on the films show a typical metallic behavior for all thicknesses. Resistivities of 6 uc and 10 uc SRO films, measured from 300 K down to 15 K, are shown in Fig. 3.11c. Their electrical properties are strongly affected by the film thickness, evidenced by a higher resistivity observed for the 6 uc film that decreases for the 10 uc film. A study of the temperature variation of resistivity shows a kink indicating the paramagnetic to ferromagnetic transition, with the corresponding T_C (marked by black arrows) derived from the temperature dependence of the derivative of resistivity plots (not shown). The Curie temperature (T_C) determined for the 6 uc film is 135 K and for 10 uc is 142 K. From SQUID magnetometry studies, the magnetic properties of the films were studied in an applied field of 0.1 T on field cooling for a temperature range between 5 K and 300 K (Fig. 3.11d). The T_C for the 6 uc and 10 uc films determined from these plots (derivative of magnetic moment with temperature, not shown) are found to be 135 K and 142 K, respectively and demonstrate the coupled magnetic and electric properties in SRO. The high value of the saturation magnetic moment obtained from experiments arises due to the strained films on (001) $SrTiO_3$, which decreases the orbital overlap via increased bond angle in Ru–O bonds, in such ultrathin films of SRO. This has been reported earlier in SRO films [43, 44].

In order to probe electronic transport of the deposited SRO films, they are patterned with standard UV lithography and further ion beam etched to be defined as devices as shown in Fig. 3.11e. As mentioned before, during the propagation of hot electrons through the thin films of SRO, the distribution encounters elastic and inelastic scattering while transmitting through the SRO metallic base and a concomitant dependence on the thickness dependence over the metallic SRO layer is established. The normalized BEEM transmission follows an exponential behavior (as described below):

$$\frac{I_B}{I_T} = C\exp\left[-\frac{d_{SRO}}{\lambda_{SRO}(E)}\right] \tag{3.5}$$

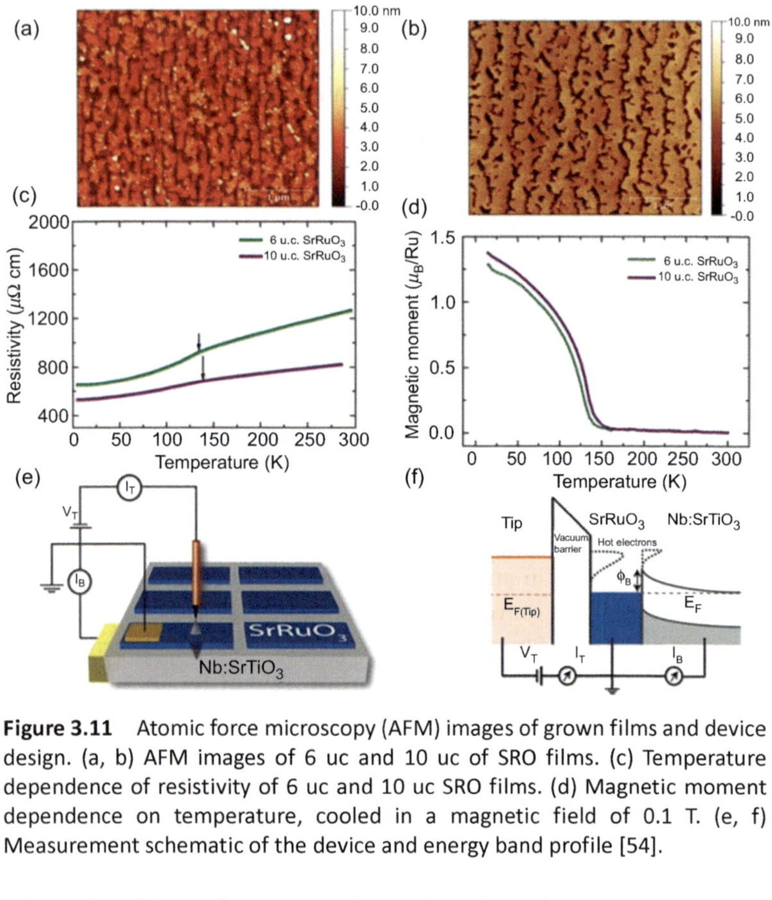

Figure 3.11 Atomic force microscopy (AFM) images of grown films and device design. (a, b) AFM images of 6 uc and 10 uc of SRO films. (c) Temperature dependence of resistivity of 6 uc and 10 uc SRO films. (d) Magnetic moment dependence on temperature, cooled in a magnetic field of 0.1 T. (e, f) Measurement schematic of the device and energy band profile [54].

where $\lambda_{SRO}(E)$ is the energy-dependent hot electron attenuation length, I_B is the collected BEEM current, I_T is the injected tunnel current, and C is a constant that is the kinematic transmission factor representative of electron transport across the M–S interface. The BEEM transmission data for various thicknesses of SRO thin films, in their paramagnetic (300 K) and ferromagnetic (120 K) phases is shown in Fig. 3.12. First, we discuss the observation for the paramagnetic phase (Fig. 3.12a). The transmission for all film thicknesses is low up to a certain bias voltage beyond which the transmission increases rapidly. This onset in bias voltage corresponds to the local Schottky barrier height (SBH) at the M–S interface. In accordance to the B–K model [20], a plot of the square root of BEEM current versus the bias voltage yields the value of SBH, which in this case is 1.13 ± 0.03 eV (inset, Fig. 3.12a). As the thickness of SRO films is increased, a decrease in transmission is observed. However, the local SBH extracted from measurements on films of different thicknesses are found to be the same, thus providing yet another proof of a high-quality growth of atomically sharp M–S interfaces. The representative BEEM spectra shown for each thickness are averages of several spectra taken at different devices and locations.

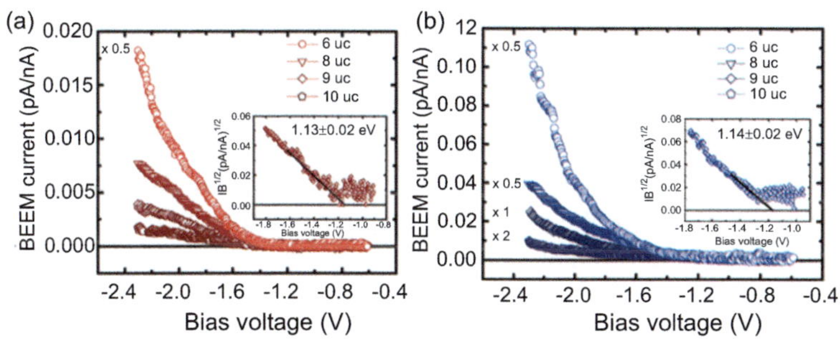

Figure 3.12 Thickness dependence of BEEM transmission at (a) 300 K and (b) at 120 K. The extracted SBH are presented in the inset respectively [54].

Electronic transport across the ferromagnetic phase transition in SRO thin films was studied across its interface with Nb:STO, in the same set of devices, by performing similar BEEM measurements at 120 K. The characteristic BEEM spectra are shown in Fig. 3.12b. Similar to our earlier observations, the BEEM transmission progressively decreases with increasing thickness of the films. We find that the hot electron transmission increases by an order of magnitude as compared to its corresponding values at 300 K. The extracted SBH at 120 K, at these interfaces, is found to be 1.14 ± 0.03 eV (inset, Fig. 3.12b), which is the same as that obtained in its paramagnetic phase. One can thus safely discard the origin of the enhanced BEEM transmission in SRO at 120 K, due to differences in the SBH across its magnetic phase transition.

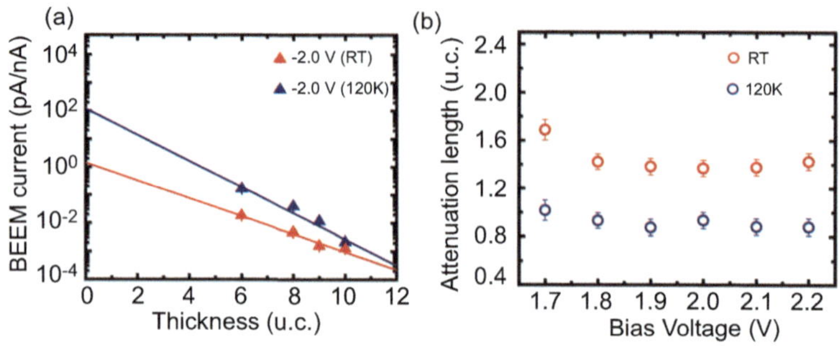

Figure 3.13 Plot of attenuation length with applied bias and energy dependence [54].

An exponential fit of the BEEM transmission for varying thicknesses of SRO films allows us to extract the (energy dependent) transport length scale known as *attenuation length* (Eq. 3.5). Above the Fermi energy (E_F), the transmission of hot electrons in SRO films is mostly governed by inelastic scattering, originating from the availability of increased phase space for hot electrons to decay into [45, 46]. The extracted attenuation length of the carriers reflects the combined effects of inelastic, elastic as well as quasi-elastic scatterings. We plot the transmissions for the paramagnetic and ferromagnetic phase of SRO at a certain bias voltage of −2 V, shown in Fig. 3.13a. An important observation is that the BEEM transmission starts to deviate appreciably at 120 K from that at 300 K, with decreasing thickness. This has not been observed earlier in other metallic systems using BEEM. Extrapolation of the data for zero SRO thickness shows two orders higher transmission, at the interface for ferromagnetic SRO. An exponential fit using Eq. 3.5, yields an attenuation length of 1.6 ± 0.2 uc in the paramagnetic phase and 0.88 ± 0.4 uc in the ferromagnetic phase at −2 V. In Fig. 3.13b, we plot the energy dependence of the attenuation length by fitting the data obtained in Fig. 3.12 with Eq. 3.5 for energies above the SBH. We find that the attenuation length decreases with increasing energy, consistent with the enhancement of the density of states (DOS) at higher energies [47]. What is surprising is that in spite of the enhanced transmission in the ferromagnetic phase; the attenuation length in SRO is shorter for all energies at 120 K than at 300 K. This apparent conundrum is explained by our ab initio calculations, taking into account the characteristics of geometries and electronic structures at the interface between SRO and $SrTiO_3$ along with the quasi-particle renormalization of attenuation length.

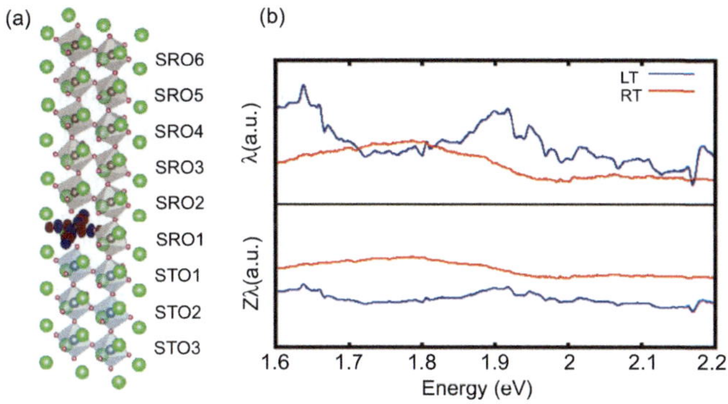

Figure 3.14 Side view of the slab and the calculated attenuation length [54].

The ab initio density functional calculations with a specific focus on correlating the structural and electronic properties at the interface to the BEEM transmission and to understand the origin of the flipping of the attenuation lengths at 120 K and 300 K. In order to simulate SRO thin films grown on $SrTiO_3$ (001) substrates, we fix the lattice parameter a to the experimental value of the $SrTiO_3$ substrate. Three, six and nine unit cells of SRO have been considered in the low temperature (LT) ferromagnetic and room temperature (RT) paramagnetic phases. As a paramagnetic phase with disordered local moments is difficult to simulate within our computational technique, we study the non-magnetic phase as the RT phase and ferromagnetic phase calculated at 0 K will represent the LT phase. Figure 3.14a shows the supercell with six layers of SRO on three layers of $SrTiO_3$. In the following discussions, we will follow the notation shown in Fig. 3.14a, where we denote STOL (or SROL) with $L = 1, 2, \ldots$, the L-th layer from the $SrTiO_3$/SRO interface.

To determine the energy-dependent attenuation length $\lambda_{LDA}(E)$, we calculate the mean value over the bands and k-points, at a given energy E, of the product of the modulus of the group velocity along the z-axis v_{nk}^{z} and the lifetime τ_{nk}, which is expressed as:

$$\lambda_{\mathrm{LDA}}(\mathrm{E}) = < v_{nk}^{z} \tau_{nk} > \propto \frac{< \dfrac{\partial \in_n (k)}{\hbar \partial k_z} >}{\int_0^E N(\in) d \in} \tag{3.6}$$

where the lifetime is proportional to the inverse of the cumulative density of states from the Fermi level to the energy E [48]. We have only considered the z-component, as the BEEM transmissions across the M–S interface are measured along this direction. As seen in Fig. 3.14b, the attenuation length λ_{LDA} (E) is almost constant in the range between 1.6 eV and 2.2 eV (above E_F) and is greater at 120 K than at RT. In recent works [49, 50], the importance of dynamical correlations in SRO has been discussed with the aid of dynamical mean field theory. Whereas the dynamical correlations and hence, the temperature dependent quasi-particle weight z is beyond the scope of LDA, we note that this weight is one in LDA and is a multiplicative factor of the spectrum and the velocity. Assuming the same quasi-particle weight for the e_g manifold and neglecting its influence on τ, we get a reduction of the attenuation length, which is more pronounced at lower temperatures. In Fig. 3.14b, we plot the attenuation length multiplying λ_{LDA} by z and observe that under these assumptions, the attenuation length at RT is greater than at 120 K, in agreement with the experimental findings.

To understand the experimentally observed enhancement in the interface transmissions at 120 K and 300 K, we calculate the electronic and geometric structures of SRO/$SrTiO_3$ interfaces. Considering the variation in the geometric and magnetic properties, we can divide the SRO thin films into three regions along the c-axis: the surface, the inner layer and the interface region. Our calculated densities of states of SRO along with that of $SrTiO_3$ are shown in Fig. 3.15a for 9 uc of SRO in the LT ferromagnetic phase and the RT non-magnetic phases. It is observed that between 1 eV and 2.2 eV (above E_F), the DOS in SRO is dominated by e_g electrons, whereas in $SrTiO_3$, the t_{2g} electrons dominate with a larger DOS between 1.2 eV and 1.4 eV (above E_F) in the non-magnetic phase of SRO. The exchange split e_g state produces a smaller DOS at LT between 1.2 and 1.4 eV (above E_F). While we observe the differences between LT and RT phases in all SRO layers, SRO1 layer at the interface needs an extra attention. A distinct feature of $d(x^2 - y^2)$ character (marked by the shaded area) is observed for RT phase in the SRO1 layer along with the renormalisation of e_g DOS between 1.2 eV and 1.3 eV (above E_F). This is not an energy shift but is due to an enlargement of the bandwidth. Our calculated hopping parameters, for example, for the 6 uc at RT are 580 meV between the x^2–y^2 orbitals (corresponding

Wannier function shown in Fig. 3.14a) in the SRO1 layer and 540–543 meV for other layers. Also, the hopping mismatch between the interface (575 meV) and the inner layers (554–561 meV) is reduced at LT. Thus, we can clearly see that the mismatch between the interface and the inner layers is larger at RT. We further note from our DOS calculations that the mismatch of the electronic states at the interface between SRO and $SrTiO_3$ is larger at RT than at LT. Such elastic scattering effects will lead to a reduction in the BEEM transmission at RT. All these differences in the electronic structures of the two phases strongly suggest the importance of the prefactor C in Eq. 3.5, which influences the BEEM transmission at LT and RT.

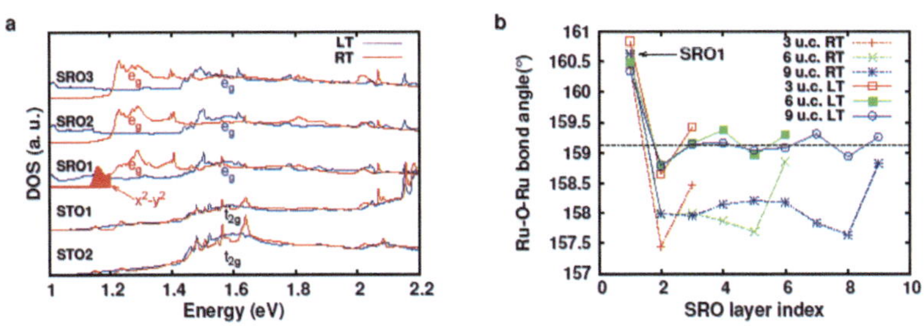

Figure 3.15 Layer-resolved electronic and structural properties [54].

The differences in the characteristics at the interface for LT and RT phases are quite evident also in the structural properties shown in Fig. 3.15b. The interface reconstruction makes this aspect more evident in the in-plane bond angles compared to the out-of-plane bond angles. For a slab of reasonable thickness (9 uc in our case), one can quantitatively identify the surface, the inner layers and the interface regions by the evolution of the Ru–O–Ru bond angle in the *ab* plane along the *c*-axis as shown in Fig. 3.15b. For 9 uc case, a large region of inner layers (properties similar to bulk) exists while this region is not distinctly observed in 3 uc case. Due to the interface reconstruction, the interface layer SRO1 is characterized by a Ru–O–Ru in-plane bond angle having a value between the SRO and $SrTiO_3$ bulk values. The properties of the inner layers are closer to the bulk. At the surface, we always obtain a bond angle larger than the bulk. Apart from the general differences between bond angles for different regions of the slab, an important observation can be made regarding the bond angles for the LT and RT phases.

It is clearly seen that the difference in bond angle between SRO1 and other SRO layers is larger for the RT phase compared to the LT one. For example, in the 6 uc case, the difference in the bond angle between SRO1 and SRO5 is 1.5° for the LT phase, while it is 3.0° for the RT phase (Fig. 3.15b). This influences the hopping of electrons (as evident from the calculated hopping parameters mentioned above) and hence the transport properties at the interfaces [40]. Therefore, we can conclude that the differences in the geometries and electronic structures between SRO1 and other SRO layers are less dramatic in the LT phase compared to the RT one, which may effectively lead to an enhanced transmission of hot electrons across the interface at LT. The larger mismatch at RT between the interface and the inner layers will yield a smaller transmission factor (*C* factor in Eq. 3.5) in the RT phase. Such a substantial influence of geometric and structural changes at the interface leading to strong temperature dependent transport characteristics across the magnetic phase transition in SRO is a remarkable finding, not observed before in any oxide heterostructures [51].

3.8 Engineering Interfacial Energy Profile across Non-polar Heterointerfaces

Complex oxide heterointerfaces are of immense interest in oxide electronics. Creating, engineering and characterizing such functional interfaces have been propelled by the evolution of sophisticated thin film techniques allowing the growth of atomically flat, lattice matched interfaces and probing them for their structures and electronic properties. As a result, many unique transport phenomena have been observed in devices with polar/non-polar interfaces, where the electronic properties are promoted by the presence of an electric field. In this context, although devices with non-polar oxide interfaces have been studied [52–57], little focus has been given to the study of emerging functional properties at their heterointerfaces. The functional interface of metallic SRO with an oxide semiconductor (Nb:STO) Schottky interface provides an excellent template to study this.

Growth-wise, it has been reported that the surface terminating plane of the underlying substrate can significantly influence the initial growth of the deposited SRO films and this has been proposed to be an interesting approach to fabricate ordered oxide nanostructures [51]. Trenches that are created in thin films of SRO are associated with a local surface termination of SrO on the Nb:STO substrate, where the initial growth of SRO is unfavorable (as explained in experimental section). Regions in between the trenches are areas of TiO_2 termination, where SRO film grows unhindered. Here we use the technique of BEEM to locally probe the electron transport across a thin SRO (8 uc) film where the local termination varies from SrO to TiO_2. Surprisingly, based on the transmission of hot electrons perpendicular to the SRO/Nb:STO interface we observe variations in transmission using BEEM, between the two regions; the transmission is higher on the trenches (SrO termination) and corresponds to a larger Schottky barrier height (SBH). Although this provided critical insights into energy band alignments of the buried interface, to establish its origin and to understand the underlying relationship between structure and electronic transport at such an interface, here we engineer the surface terminating plane in $SrTiO_3$ from TiO_2 to SrO and study its effect on tuning the interfacial band alignment. We do this by inserting a monolayer of SrO before depositing the SRO layer on Nb:STO.

Thin films of SRO were grown by pulsed laser deposition (PLD) while the surface quality was monitored by RHEED. A combined chemical and thermal treatment was applied to achieve a single TiO_2 termination of the Nb:STO substrates [58]. However, no chemical treatments have been reported to produce the opposite single-terminated SrO surfaces, while heat treatment of the as-received $SrTiO_3$ substrates usually results in mixed termination. The single-terminated SrO surfaces can be obtained by deposition of an SrO monolayer on a single-terminated TiO_2 surface. For epitaxial SrO monolayer growth, pulsed laser interval deposition was applied [59]. In this deposition technique, the total number of laser pulses for one monolayer has to be provided rapidly (50 Hz) to stabilize the correct SrO layer without multilevel islands. Concerning the deposition conditions, a single-crystal SrO target is ablated with an energy density of 1.3 J/cm^2. During growth, the substrate is held at 850 °C in an oxygen environment at 0.13 mbar. This results in crystalline SrO-terminated $SrTiO_3$ surfaces with perfectly straight step-edges. Subsequently, SRO thin films of 8 unit cells (uc) were deposited on both types of single-terminated Nb:STO substrates. The growth was carried out on a substrate kept at 600 °C in an oxygen environment at 0.13 mbar. Figures 3.16a and 3.16b show the RHEED specular spot intensity during the deposition of SRO on T–Nb:STO and for the growth of one monolayer of SrO followed by 8 uc of SRO, respectively. The oscillation of the RHEED intensity indicates a partially layer-by-layer growth mode in both cases. The corresponding RHEED diffraction images clearly show sharp two-dimensional spots, indicating diffraction from smooth crystalline surfaces. Figure 3.16c shows the atomic force microscopy (AFM) images of the 8 uc SRO grown on both types of single-terminated surfaces (T–Nb:STO/S–Nb:STO). The low level of surface roughness was confirmed for both cases as the micrographs and roughness analyses show smooth terraces with unit cell steps. However, for SRO films grown on T–Nb:STO, small regions can be observed with the formation of narrow trenches at the step-edges. This formation was previously reported to result from the coexistence of TiO_2 as well as SrO termination leading to disparate growth kinetics of the deposited SRO thin film [49, 60]. The presence of both terminations with well-defined lateral separation in a single T–Nb:STO sample enables us to study the local difference in transport behavior and compare it to the S–Nb:STO sample with full SrO termination.

All the measurements were performed at RT (300 K). The STM surface topography of the T–Nb:STO device, recorded at −1.4 V and 4 nA, is shown in Fig. 3.17a. It is observed that the surface topography replicates the AFM image and here, too, trenches and terraces are clearly visible. The average width of these trenches running along the sample is around 30–40 nm. As the technique of BEEM is spatially resolved, we can record a large number of BEEM spectra both inside and outside the trenches. BEEM transmission was recorded for many such trenches across several devices. The representative average BEEM spectra across SrO termination in T–Nb:STO is shown in Fig. 3.17. Figures 3.17b and 3.17c correspond to SrO and TiO_2 terminations, respectively. In both cases, we observe a sharp onset in I_B beyond a certain tip bias voltage which corresponds to the local Schottky barrier height at the metal–semiconductor interface.

It is observed that transmission in SrO terminated region is an order of magnitude higher than TiO_2 terminations. This can be understood by the fact that I_B exponentially decreases with increasing thickness of the metal layer [51, 60], which in trenches is thinner than in terraces. The local SBH at the different regions can be extracted from their respective I_B plots using an approximation of the B–K model (under standard assumptions of planar tunneling and free-electron dispersion relation in STM tip and base metal) by plotting the square root of the BEEM current (I_B) versus tip bias (V_T). The intersect of the straight line with the voltage axis gives the local SBH, which for SrO terminations is found to be 1.27 ± 0.03 eV and 1.11 ± 0.03 eV for TiO_2 terminations as shown in Figs. 3.17b and 3.17c, respectively. Thus, we find local differences in the energy band alignments at the M–S interface in trenches and terraces.

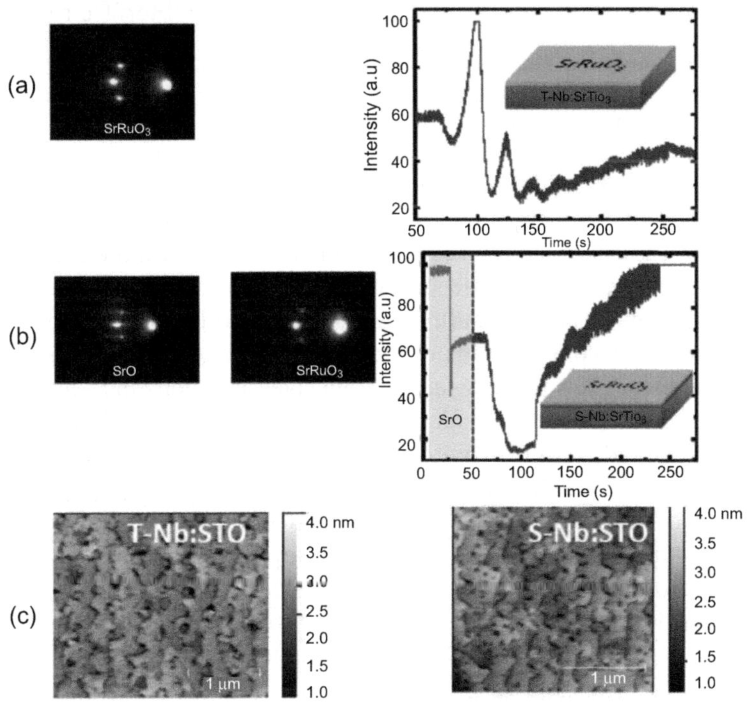

Figure 3.16 (a) RHEED data for SRO growth on non-engineered substrate T–Nb:STO. (b) RHEED data for the growth of a monolayer of SrO followed by 8 uc of SRO. (c) AFM scan image of T–Nb:STO and S–Nb:STO. Reprinted from Ref. [68], with permission from American Physical Society, Copyright 2016.

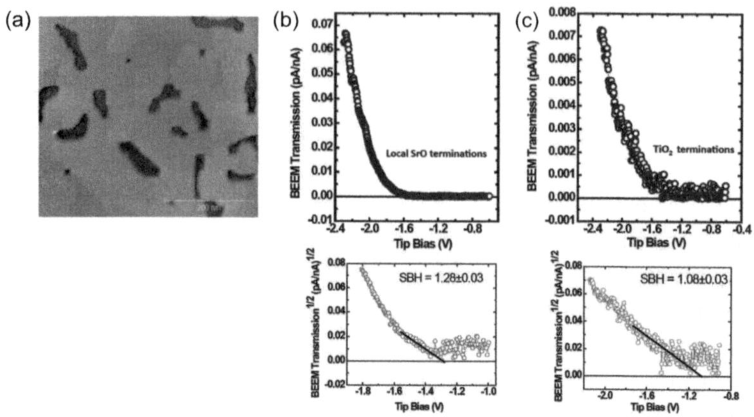

Figure 3.17 (a) STM topography of 8 uc of SRO grown on T–Nb:STO at V_T = −1.4 V and I_T = 4 nA. Occasional dark patches indicate thinner SRO grown on SrO terminations of the substrate. (b) Corresponding BEEM spectra for tip placed in dark patches (trenches), i.e., on SrO terminations of the substrate. The extracted corresponding local SBH is represented below. (c) BEEM spectra for tip placed on TiO_2 terminated regions, i.e., on thicker SRO growth. Corresponding local Schottky barrier height is represented below. Reprinted from Ref. [68], with permission from American Physical Society, Copyright 2016.

As the local SrO terminated regions in T–Nb:STO are mostly confined to the terrace edges and occur in small pockets, it is not apparently clear whether the growth of SRO in such confined regions could alter the electronic property at the interface by possible occurrences of defects and changes in local strain. In order to rule out any such contributions to the band alignment at the interface, we engineer the substrate terminating layer as is done in S–Nb:STO. We study electronic transport across it using BEEM and compare the extracted local SBH with that obtained for SRO grown inside the trenches of T–Nb:STO. Figure 3.18a shows the STM topography image of the top surface of SRO grown on S–Nb:STO. By locating the tip on the flat regions of the terraces, several BEEM spectra were obtained and an average of such spectra taken on several such regions is shown in Fig. 3.18b. By fitting the B–K model, we extract the local SBH (Fig. 3.18c). Our findings highlight two key features:

(a) A comparison of the BEEM transmission at SrO terminations in Figs. 3.17 and 3.18 shows that it is higher for T–Nb:STO than in S–Nb:STO. This deviation, of an order of magnitude, is a result of the variation of thickness in SRO on SrO terminations on either substrate. The growth rate in local SrO terminated pockets is lower than on the terraces of T–Nb:STO and thus the thickness of SRO grown inside the trenches is less than 8 uc However, for S–Nb:STO, a uniform 8 uc SRO is grown on the engineered interface, which is thicker than above and thus results in a lower BEEM transmission in this case.

(b) The local SBHs extracted from the different SrO terminations for both devices (T–Nb:STO and S–Nb:STO) yield the same value of 1.28 ± 0.03 eV.

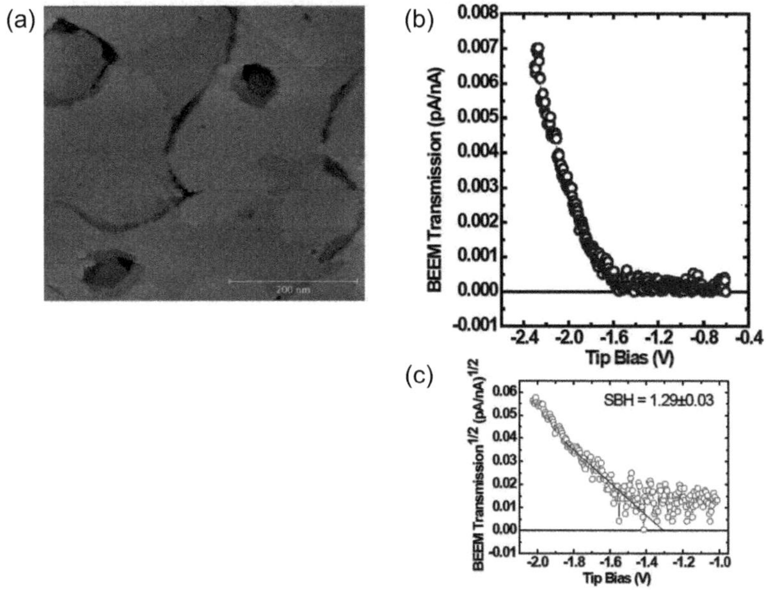

Figure 3.18 STM topography of 8 uc of SRO grown on S–Nb:STO at $V_T = -1.4$ V and $I_T = 4$ nA. (b) Corresponding BEEM spectra through 8 uc of SRO grown on S–Nb:STO. (c) The extracted local Schottky barrier height. Clearly, the local SBH matches with that extracted from SRO grown inside the trenches of T–Nb:STO. Reprinted from Ref. [68], with permission from American Physical Society, Copyright 2016.

It is fascinating to find that such a deliberate design of the oxide interface by merely changing the surface termination at the interface can influence the electronic transport across such non-polar interfaces. Further, this measurement confirms that the different energy band line-ups at the M–S interface is truly governed by the terminating layer of the substrate and is thus an intrinsic property of the interface itself.

To understand the intrinsic origin of the differences in the electronic structure of S–Nb:STO and T–Nb:STO, we propose the creation of an interfacial dipole that is different at both the interfaces, thus bringing in a change in the SBH. However, the occurrence of such an induced dipole is puzzling at the symmetric SRO/Nb:STO interface. One possible reason for this could be due to a small amount of intermixing of the adjoining Ru/Ti sites. This can be triggered by the different surface energies of the terminating planes of SrO and TiO_2 at the surface of Nb:STO [61]. This can result in an intermixing of Ru/Ti ions of S–Nb:STO due to the close matching of the ionic radii of Ru^{4+} and Ti^{4+}, hence lowering the surface energy of the S–Nb:STO interface. Considering such a scenario, first-principles DFT calculations were performed using the Quantum ESPRESSO package [62] using the Perdew–Burke–Ernzerhof generalized gradient approximation (GGA) functional [63]. The interface for T–Nb:STO is considered pristine and the S–Nb:STO interface is simulated by a 50% intermixing of Ru and Ti ions. The supercells are constructed by stacking the structural unit cells of SRO and $SrTiO_3$ along the (001) direction (z-direction). Both supercells are fully atomically relaxed until forces are less than 20 meV/Å with the constraint that the in-plane lattice constant of the supercell has the GGA calculated lattice constant of cubic $SrTiO_3$, a = 3.937 Å. This was done to simulate epitaxial growth of SRO on an $SrTiO_3$ substrate. Results of the relaxation for both cases are shown in Figs. 3.19 (a and b). For the interface at T–Nb:STO, only a small out-of-plane metal/oxygen displacement is present, indicating that the structure is close to the natural continuation of both perovskite structures at the interface. For the S–Nb:STO intermixed interface, however, large displacement of the Sr sites with respect to oxygen sites is readily apparent. The large difference in displacement profile between the two types of interface terminations is expected to lead to a significant difference in the electrostatic interface dipole between SRO and $SrTiO_3$ and should therefore be apparent in the electronic band line-up. Indeed, this is the case as can be seen in the local density of states (LDOS) deep inside the $SrTiO_3$ as shown in Fig. 3.19c. It is observed that the variation of the interface composition leads to a linear change of the $SrTiO_3$ conduction band offset with respect to the T–Nb:STO and hence to a linear change of the Schottky barrier height at this interface. Indeed, we see that the S–Nb:STO interface has an n-type SBH 0.19 eV larger than the T–Nb:STO interface. This is remarkably similar to the experimentally measured difference in n-SBH between the S-Nb:STO and T-Nb:STO [64].

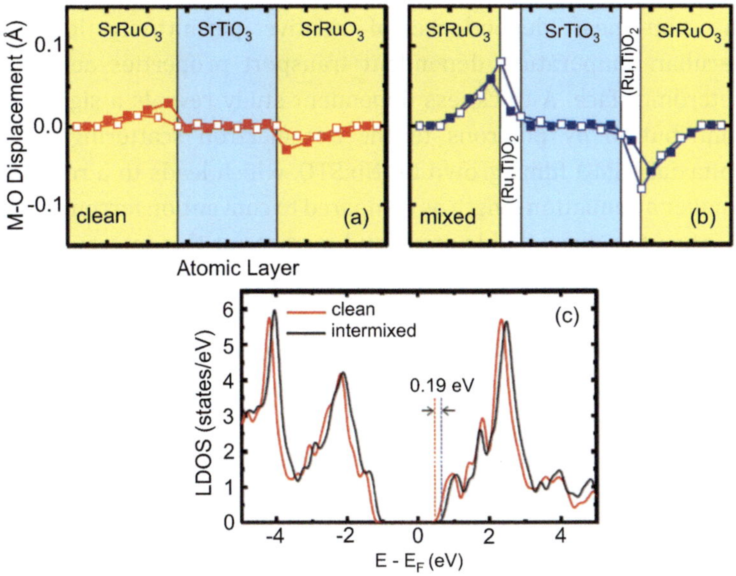

Figure 3.19 Layer-by-layer metal–oxygen displacements for the two SRO/$SrTiO_3$ supercells used to simulate the (a) T–Nb:STO and (b) S–Nb:STO interface. Solid symbols indicate the (Ru or Ti)O_2 displacements and the open symbols are for the SrO layers. (c) The local density of states in the center of the $SrTiO_3$ for both interface terminations. Vertical dashed lines indicate the position of the conduction band minimum (CBM) with respect to the Fermi level, E_F. Reprinted from Ref. [68], with permission from American Physical Society, Copyright 2016.

While intermixing at the interface successfully explains the difference in band alignments for the two devices, a different amount of oxygen vacancies at the S-Nb:STO and T-Nb:STO interfaces could also play an important role in the interfacial structural distortion [65]. The presence of oxygen vacancy in the SrO plane significantly reduces the electron repulsion in that direction (*c*-axis) and that results in an increased displacement between the atomic planes at the interface. This also impacts Ti-O-Ru bond angle, thereby altering the local electrostatic potential at the SRO/S-Nb:STO interface. Our study emphasizes the flexibility that substrate termination offers in the designing of electronic interfaces for diverse applications in oxide electronics.

3.9 Conclusions

Nb:STO is marked as an unconventional semiconductor, where the non-linear dependence of relative permittivity leads to peculiar temperature dependent transport properties across its heterointerface. A thickness dependent study reveals a significant contribution by polarons to the hot electron scattering across epitaxial LSMO films grown on Nb:STO, which leads to a relatively smaller attenuation length as compared to convention ferromagnets. Further, the unavoidable steps and terraces at this interface show a distinct electronic transport, with reduced electron transmission at the edges. For SRO/Nb:STO interface, the surface terminations selectively determine the arrangement of atoms at the interface and lead to a different electrostatic potential for either case. The hot electron attenuation lengths are found to reduce by at least a factor of two at the low temperature ferromagnetic phase as compared to room temperature in SRO. This is attributed to the renormalized density of states caused by low temperature structural distortions. The local inhomogeneity observed using such probe combined with its temperature dependent properties is shown to determine the observed macroscopic behavior of all oxide metal-semiconductor interfaces. These experiments demonstrate BEEM as a powerful tool to explore nanoscale transport in buried layers in an oxide heterostructure and across engineered interfaces as well as to study the band bending across such heterointerfaces, which are crucial to advance oxide electronics.

Acknowledgments

This work is supported by research programs of the Netherlands Organisation for Scientific Research: NWO-VIDI, NWO-FOM (nano), and the Rosalind Franklin Fellowship programme. We gratefully acknowledge all our collaborators S. Parui, T. T. T. M. Palstra, B. Noheda, T. Yajima, Y. Hikita, H. Y. Hwang, A. F. Kemper, T. P. Deveraux, J. D. Burton, E. Y. Tsymbal, C. Auitieri, B. Sanyal, A. Solmaz, M. Huijben, G. Rijnders, and Nanolab-NL, a Dutch national facility for nanotechnology research.

References

1. N. Hur, S. Park, P. A. Sharma, J. S. Ahn, S. Guha, and S. W. Cheong, *Nature*, **429**, 392 (2004).
2. J. Wang, J. B. Neaton, H. Zheng, V. Nagarajan, S. B. Ogale, B. Liu, et al., *Science*, **299**, 1719 (2003).
3. T. Kimura, T. Goto, H. Shintani, K. Ishikawa, T. Arima, and Y. Tokura, *Nature*, **426**, 55 (2003).
4. J. W. Reiner, F. J. Walker, and C. H. Ahn, *Science*, **323**, 1018 (2009).
5. A. Ohtomo, D. A. Muller, J. L. Grazul, and H. Y. Hwang, *Nature*, **419**, 378 (2002).
6. A. Ohtomo and H. Y. Hwang, *Nature*, **427**, 423 (2004).
7. A. P. Ramirez, *J. Phys.: Condens. Matter*, **9**, 8171 (1997).
8. J. H. Park, E. Vescovo, H.-J. Kim, C. Kwon, R. Ramesh, and T. Venkatesan, *Nature*, **392**, 794–796 (1998).
9. C. Zener, *Phys. Rev.*, **82**, 403 (1951).
10. B. Nadgorny, I. I. Mazin, M. Osofsky, J. R. J. Soulen, P. Broussard, R. M. Stroud, et al., *Phys. Rev.*, **63**, 184433 (2001)
11. M. Bowen, M. Bibes, A. Barthélémy, J.-P. Contour, A. Anane, Y. Lemaître, et al., *Appl. Phys. Lett.*, **82**, 233 (2003).
12. G. Koster, L. Klein, W. Siemons, G. Rijnders, J. Steven Dodge, C.-B. Eom, et al., *Rev. Mod. Phys.*, **84**, 253 (2012).
13. K. Maiti, *Phys. Rev. B* **73**, 235110 (2006).
14. R. Aso, D. Kan, Y. Shimakawa, and H. Kurata, *Adv. Funct. Mater.*, **24**, 5177 (2014).
15. F. Bern, M. Ziese, A. Setzer, E. Pippel, D. Hesse, and I. Vrejoiu, *J. Phys.: Condens., Matter* **25**, 496003 (2013).
16. J. Choi, C. B. Eom, G. Rijnders, H. Rogalla, and D. H. A. Blank, *Appl. Phys. Lett.*, **79**, 1447 (2001).
17. G. Rijnders, D. H. A. Blank, J. Choi, and C. B. Eom, *Appl. Phys. Lett.*, **84**, 505 (2004).
18. G. J. H. M. Rijnders, G. Koster, D. H. A. Blank, and H. Rogalla, *Appl. Phys. Lett.*, **70**, 1888 (1997).
19. K. G. Rana, PhD Thesis, *Electron Transport across Complex Oxide Heterointerfaces*, University of Groningen (2013).
20. W. J. Kaiser and L. D. Bell, *Phys. Rev. Lett.*, **60**, 1406 (1988).
21. M. Prietsch, *Phys. Rep.*, **253**, 163 (1995).

22. A. Spinelli, M. A. Torija, C. Liu, C. Jan, and C. Leighton, *Phys. Rev. B,* **81**, 155110 (2010).
23. Y. Hikita, M. Nishikawa, T. Yajima, H. Y. Hwang, *Phys. Rev. B,* **79**, 073101 (2009).
24. H. P. R. Frederikse, W. R. Thurber, and W. R. Hosler, *Phys. Rev.,* **134**, 2A (1963).
25. O. N. Tufte and P. W. Chapman, *Phys. Rev.,* **155**, 796 (1967).
26. K. G. Rana, V. Khikhlovskyi, and T. Banerjee, *Appl. Phys. Lett.,* **100**, 213502 (2012).
27. T. Shimizu and H. Okushi, *J. Appl. Phys.,* **85**, 7244 (1999).
28. F. M. Postma, R. Ramaneti, T. Banerjee, H. Gokcan, E. Haq, D. H. A. Blank, et al., *J. Appl. Phys.*, **95**, 7324 (2004).
29. T. Susaki, Y. Kozuka, Y. Tateyama, H. Y. Hwang, *Phys. Rev. B,* **76**, 155110 (2007).
30. Y. Hikita, M. Kawamura, C. Bell, and H. Y. Hwang, *Appl. Phys. Lett.,* **98**, 192103 (2011).
31. T. Yajima, Y. Hikita, M. Minohara, C. Bell, J. A. Mundy, L. F. Kourkoutis, et al., *Nat. Commun.,* **6,** 6759 (2015).
32. T. Shimizu, N. Gotoh, N. Shinozaki, and H. Okushi, *Appl. Surf. Sci.,* **117/118**, 400 (1997).
33. K. G. Rana, S. Parui, and T. Banerjee, *Phys. Rev. B,* **87**, 085116 (2013).
34. K. G. Rana, T. Yajima, S. Parui, A. F. Kemper, T. P. Devereaux, Y. Hikita, et al., *Sci. Rep.,* **3**, 1274 (2013).
35. L. F. Mattheiss, *Phys. Rev. B,* **6**, 4718–4740 (1972).
36. K. A. Muller and H. Burkhard, *Phys. Rev. B,* **19**, 3593 (1979).
37. S. Parui, K. G. Rana, L. Bignardi, P. Rudolf, B. J. van Wees, and T. Banerjee, *Phys. Rev. B*, **85**, 235416 (2012).
38. Y. Chen, B. G. Ueland, J. W. Lynn, G. L. Bychkov, and S. N. Barilo, *Phys. Rev. B*, **78**, 212301 (2008).
39. D. Louca and T. Egami, *Phys. Rev. B,* **59**, 6193–6204 (1999).
40. W. E. Pickett and D. J. Singh, *Phys. Rev. B,* **53**, 1146 (1996).
41. F. Sánchez, G. Herranz, J. Fontcuberta, M. V. García-Cuenca, C. Ferrater, and M. Varela, *Phys. Rev. B,* **73**, 073401 (2006).
42. G. Herranz, V. Laukhin, F. Sánchez, P. Levy, C. Ferrater, M. V. García-Cuenca, et al., *Phys. Rev. B,* **77**, 165114 (2008).

43. R. Aso, D. Kan, Y. Shimakawa, and H. Kurata, *Adv. Funct. Mater.,* **24**, 5177 (2014).

44. C. Autieri, M. Cuoco, and C. Noce, *Phys. Rev. B,* **89**, 075102 (2014).

45. F. Bern, M. Ziese, A. Setzer, E. Pippel, D. Hesse, and I. Vrejoiu, *J. Phys. Condens. Matter,* **25**, 496003 (2013).

46. A. Vailionis, H. Boschker, W. Siemons, E. P. Houwman, D. H. A. Blank, G. Rijnders, et al., *Appl. Phys. Lett.,* **83**, 064101 (2011).

47. A. Grutter, F. Wong, E. Arenholz, M. Liberati, A. Vailionis, and Y. Suzuki, *Appl. Phys. Lett.,* **96**, 082509 (2010).

48. A. Grutter, F. Wong, E. Arenholz, A. Vailionis, and Y. Suzuki, *Phys. Rev. B,* **85**, 134429 (2012).

49. T. Banerjee, J. C. Lodder, and R. Jansen, *Phys. Rev. B,* **76**, 140407(R) (2007).

50. E. B. Guedes, M. Abbate, K. Ishigami, A. Fujimori, K. Yoshimatsu, H. Kumigashira, et al., *Phys. Rev. B,* **86,** 235127 (2012).

51. V. P. Zhukov, E. V. Chulkov, and P. M. Echenique, *Phys. Rev. B,* **73**, 125105 (2006).

52. M. Kim and B. I. Min, *Phys. Rev. B,* **91**, 205116 (2015).

53. H. T. Dang, J. Mravlje, A. Georges, and A. J. Millis, *Phys. Rev. B,* **91**, 195149 (2015).

54. S. Roy, C. Autieri, B. Sanyal, and T. Banerjee, *Sci. Rep.,* **5**, 15747 (2015).

55. N. Nakagawa, H. Y. Hwang, and D. A. Muller, *Nat. Mater.,* **5**, 204 (2004).

56. N. Reyren, S. Thiel, A. D. Caviglia, L. Fitting Kourkoutis, G. Hammerl, C. Richter, et al., *Science,* **317**, 1196 (2007).

57. A. D. Caviglia, S. Gariglio, N. Reyren, D. Jaccard, T. Schneider, M. Gabay, et al., *Nature,* **456**, 624 (2008).

58. Guneeta Singh-Bhalla, C. Bell, J. Ravichandran, W. Siemons, Y. Hikita, S. Salahuddin, et al., *Nat. Phys.,* **7**, 80 (2011).

59. Y. Hotta, T. Susaki, and H. Y. Hwang, *Phys. Rev. Lett.,* **99**, 236805 (2007).

60. P. R. Willmott, S. A. Pauli, R. Herger, C. M. Schleptz, D. Martoccia, B. D. Patterson, et al., *Phys. Rev. Lett.,* **99**, 155502 (2007).

61. G. Koster, B. L. Kropman, G. J. H. M. Rijnders, D. H. A. Blank, and H. Rogalla, *Appl. Phys. Lett.,* **73**, 2920 (1998).

62. G. Koster, G. J. H. M. Rijnders, D. H. A. Blank, and H. Rogalla, *Appl. Phys. Lett.,* **74**, 3729 (1999).

63. M. Huijben, A. Brinkman, G. Koster, G. Rijnders, H. Hilgenkamp, and D. H. A. Blank, *Adv. Mater.*, **21**, 1665 (2009).

64. S. Roy, A. M. Kamerbeek, K. G. Rana, S. Parui, and T. Banerjee, *Appl. Phys. Lett.,* **102**, 192909 (2013).

65. R. I. Eglitis and D. Vanderbilt, *Phys. Rev. B* **77**, 195408 (2008).

66. P. Giannozzi, S. Baroni, N. Bonini, M. Calandra, R. Car, C. Cavazzoni, et al., *J. Phys. Condens. Matter,* **21**, 395502 (2009).

67. J. P. Perdew, K. Burke, and M. Ernzerhof, *Phys. Rev. Lett.,* **77**, 3865 (1996).

68. S. Roy, A. Solmaz, J. D. Burton, M. Huijben, G. Rijnders, E. Y. Tsymbal, et al., *Phys. Rev. B,* **93**, 115101 (2016)

Chapter 4

A Theoretical Overview of the Quantum Phenomena at Oxide Interfaces: The Role of Spin and Charge

Carmine Autieri and Biplab Sanyal
Department of Physics and Astronomy, Uppsala University, Box-516, 75120 Uppsala, Sweden
biplab.sanyal@physics.uu.se

4.1 Introduction

The structural reconstruction at the interfaces and the surfaces of the transition metal oxides [1] can modify their physical properties due to the breaking of the crystal symmetry. Moreover, the two-dimensionality of the surface reduces the electrons' kinetic energies thereby enhancing the effects of electron correlations, which can modify many important physical properties such as optical, transport, magnetic, and superconductivity, to name a few. The breaking of time reversal and/or inversion symmetry at the interfaces gives rise to unusual phenomena. It is remarkable to notice that one can utilize the novel properties occurring at the interfaces, which are absent

Oxide Spintronics
Edited by Tamalika Banerjee

ISBN 978-981-4774-99-4 (Hardcover), 978-0-429-46819-3 (eBook)
www.jennystanford.com

in bulk to design new characteristic features in a nanometer scale thereby opening up possibilities of new technological innovations. Specifically, the magnetic proximity effect and the reduction in the bandwidth can yield a substantial change in the magnetic properties of the materials as observed in many interfaces and surfaces of oxide perovskites [2–7].

Another interesting property occurs at the interfaces of polar and non-polar oxides, which gives rise to interface reconstruction. The polar discontinuity, known as *polar catastrophe* [8] arising from electrostatic boundary conditions gives rise to rough interfaces during growth. Specifically, at the $LaAlO_3/SrTiO_3$ interfaces, one may realize two situations, such as $AlO_2/LaO/TiO_2$ and $AlO_2/SrO/TiO_2$ depending on the interface termination. In both cases, the unreconstructed interfaces will generate finite electric fields and diverging electrostatic potentials. However, this problem is solved by interface reconstruction either by compensating the interface charges by mixed valence metal states providing an extra charge or by the introduction of oxygen vacancies at the interfaces.

Therefore, it is widely recognized that the interfaces of oxide heterostructures present very interesting phenomena, which require fundamental understanding from both experimental and theoretical sides. For the last few decades, the materials science community has experienced a tremendous advancement in the theoretical understanding of properties in the atomic and electronic scale by means of quantum mechanical calculations within density functional theory (DFT) [9, 10]. The oxides, in particular, have been studied in detail regarding the structural, electronic, and magnetic properties. A special focus has been in describing the correlated electrons in oxides by taking into account Coulomb interactions (U) in a simplified way (DFT + U) or in a more rigorous fashion (dynamical mean field theory). In summary, both experiments and microscopic theories have developed this subject hand in hand. In this chapter, we will present a few examples of unusual magnetic and transport properties occurring at the oxide heterointerfaces. The experimental results will be analyzed by DFT based ab initio electronic structure theory that gives us a fairly good microscopic understanding at the electronic scale.

In the next section, we will present the study of the interface control of $SrRuO_3/SrTiO_3$ [11] heterostructures and the magnetic properties of the ultrathin films [12]. More recently, a new case of long-range proximity effect was proposed for the $SrRuO_3/Sr_2RuO_4$ heterostructures [13]. The ferromagnet $SrRuO_3$ induces long-range ferromagnetism in the incipient ferromagnet Sr_2RuO_4 [14].

4.1.1 Interface Control across the Magnetic Phase Transition in $SrTiO_3/SrRuO_3$ Film

A nanoscale transport probe [ballistic electron emission microscopy (BEEM)] was employed to study complex correlations across the interface between $SrRuO_3$ and semiconducting Nb:$SrTiO_3$ (Nb:STO). Principally, this technique allows one to study any material system, whose electronic properties are susceptible to a temperature-driven phase transition, while simultaneously allowing to investigate its homogeneity across such buried interfaces with a semiconductor, at the nanoscale. This is an important approach for materials systems such as complex oxides where small changes in the geometrical properties with thickness or temperature can produce large changes in the electric and magnetic properties. Systematic experimental studies were performed to understand the effect of thickness and temperature on electronic transport across thin films of $SrRuO_3$ with Nb:STO. An order of magnitude increase in the transmission of the same device was observed across its ferromagnetic phase transition. Surprisingly though, it was found that this increase originates from an enhanced transmission at the interface between $SrRuO_3$ and Nb:$SrTiO_3$, rather than in the bulk of $SrRuO_3$ and manifests as two different transport lengths at room temperature and below T_C = 160 K [15]. These findings were combined with density functional theory calculations to quantitatively understand the role of interface reconstruction, octahedral distortions and exchange splitting to affect the electronic band structure of $SrRuO_3$ thin films and electronic transport.

4.1.1.1 Film growth and device scheme

High-quality films of $SrRuO_3$ were grown varying the thickness $d_{SrRuO_3} = 6$, 8, 9, and 10 unit cells (uc) on Nb:STO substrates using pulsed laser deposition (PLD). In order to probe electronic transport of the deposited $SrRuO_3$ films, they were patterned with standard

UV lithography and further ion beam etched to be defined as devices. BEEM uses the scanning tunneling microscope (STM) to inject a distribution of hot electrons with energies higher than the Fermi energy (E_F). An electrical contact between the STM tip and $SrRuO_3$ measures the tunnel current. The device design necessitates an additional contact at the rear-end of Nb:STO, for the collection of the transmitted electrons. These injected hot-electrons travel across the metallic base to reach the metal–semiconductor (M–S) interface; if their energy is sufficient to overcome the Schottky barrier, they can be collected into the semiconductor conduction bands [11]. As these hot electrons propagate through the $SrRuO_3$ thin films, they undergo scattering, which influences the collected current.

4.1.1.2 Thickness and temperature-dependent BEEM transmission

The hot electron distribution encounters elastic and inelastic scattering while transmitting through the $SrRuO_3$ metallic base. In this process, the momentum distribution at the M–S interface is broadened, as compared to the injection interface, adding to the reduction of the collected BEEM current (I_B) [16]. These effects get more pronounced as the thickness of the base layer is increased. The normalized BEEM transmission follows an exponential behavior as described by Ref. [17]:

$$\frac{I_B}{I_T} = C \times \exp\left[-\frac{d_{SrRuO_3}}{\lambda_{SrRuO_3}(E)}\right] \tag{4.1}$$

where $\lambda_{SrRuO_3}(E)$ is the energy-dependent hot electron attenuation length, I_B is the collected BEEM current, I_T is the injected tunnel current and C is a constant that is the kinematic transmission factor representative of electron transport across the M–S interface. The transmission for all film thicknesses is low up to a certain bias voltage beyond which, the transmission increases rapidly. This onset in bias voltage corresponds to the local Schottky barrier height (SBH) at the M–S interface. In accordance to the Bell–Kaiser (B–K) model, a plot of the square root of BEEM current versus the bias voltage yields the value of SBH, which in the paramagnetic case is 1.13 ± 0.03 eV [18]. As the thickness of $SrRuO_3$ films is increased, the local SBH extracted from measurements on films of different thicknesses are found to be the same. Electronic transport across the ferromagnetic phase transition in $SrRuO_3$ thin films was studied across its interface with Nb:STO, in the same set of devices, by performing similar BEEM measurements at 120 K. It is observed that the BEEM transmission progressively decreases with increasing thickness of the films. The hot electron transmission increases by an order of magnitude as compared to its corresponding values at 300 K. The extracted SBH at 120 K, at these interfaces, is found to be 1.14 ± 0.03 eV, which is the same as that obtained in its paramagnetic phase. One can thus safely discard the origin of the enhanced BEEM transmission in $SrRuO_3$ at 120 K, due to differences in the SBH across its magnetic phase transition.

4.1.1.3 Influence of interface and film thickness on electronic transport

An exponential fit of the energy dependence of BEEM transmission for varying thickness of $SrRuO_3$ films allows us to extract the (energy dependent) transport length scale known as *attenuation length* (Eq. 4.1). Above E_F, the transmission of hot electrons in $SrRuO_3$ films is mostly governed by inelastic scattering, originating from the availability of increased phase space for hot electrons to decay into [19, 20]. The extracted attenuation length of the carriers reflects the combined effects of inelastic, elastic as well as quasi-elastic scatterings. The transmissions for the paramagnetic and ferromagnetic phases of $SrRuO_3$ at a certain bias voltage of –2 V are shown in Fig. 4.1a. An important observation is that the BEEM transmission starts to deviate appreciably at 120 K from that at room temperature (RT) (300 K), with decreasing thickness. Extrapolation of the data for zero $SrRuO_3$ thickness shows two orders higher transmission, at the interface for ferromagnetic $SrRuO_3$.

Such an enhanced interface transmission has not been earlier observed in other metallic systems using BEEM. An exponential fit using Eq. 4.1, yields an attenuation length of 1.6 ± 0.2 uc in the paramagnetic phase and 0.88 ± 0.4 uc in the ferromagnetic phase. In Fig. 4.1b, the energy dependence of the attenuation length by fitting the data with Eq. 4.1 for energies above the SBH is shown. It is seen that the attenuation length decreases with increasing energy, consistent with the enhancement of the density of states (DOS) at higher energies. What is surprising is that in spite of the enhanced transmission in the ferromagnetic phase, the attenuation length in $SrRuO_3$ is shorter, for all energies at 120 K, than at RT. This apparent conundrum is explained by ab initio calculations, taking into account the characteristics of geometries and electronic structures at the interface between $SrRuO_3$ and $SrTiO_3$ along with the quasiparticle renormalization of attenuation length.

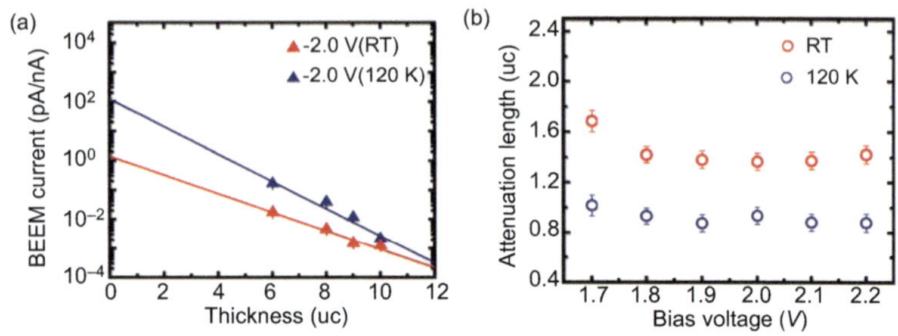

Figure 4.1 Plot of attenuation length with applied bias and energy dependence, (a) The BEEM transmission at −2 V is plotted against film thicknesses for both at 300 K and at 120 K. An exponential fit (solid lines) shows the dependence of BEEM current with film thickness. An extrapolation of the transmission indicates that the interface transmission has increased by two orders of magnitude at 120 K. (b) Energy dependence of attenuation length at 120 K and RT is plotted. Contrary to the higher BEEM transmission across $SrRuO_3$/Nb:STO devices at 120 K, the corresponding attenuation length is consistently shorter than the attenuation length measured at 300 K. The error bar in attenuation length comes as a fitting error from the deviation of the best fit. Reprinted from Ref. [11].

4.1.1.4 Ab initio study of $SrRuO_3$/$SrTiO_3$ (001) interfaces

Ab initio density functional calculations were performed with a specific focus on correlating the structural and electronic properties at the interface to the BEEM transmission and to understand the origin of the flipping of the attenuation lengths at 120 K and RT. In order to simulate $SrRuO_3$ thin films grown on $SrTiO_3$ (001) substrates, 3, 6, and 9 uc of $SrRuO_3$ have been considered in the low-temperature ferromagnetic and room temperature paramagnetic phases. As a paramagnetic phase, the non-magnetic phase was considered as the RT phase and ferromagnetic phase calculated at 0 K represents the low temperature (LT) phase. Figure 4.2 shows the supercell with six layers of $SrRuO_3$ on three layers of $SrTiO_3$. In Fig. 4.2, STOL (or SROL) with $L = 1, 2, ...$, denotes the L-th layer from the $SrTiO_3$/ $SrRuO_3$ interface.

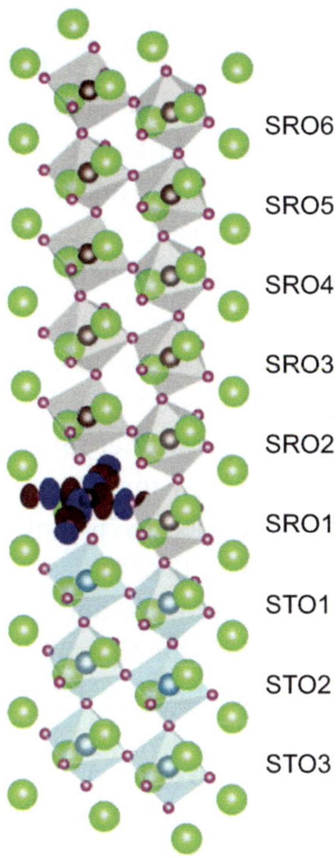

Figure 4.2 Side view of the slab with three layers of $SrTiO_3$ and six layers of $SrRuO_3$. Sr, O, Ru, and Ti atoms are shown as green, fuchsia, grey, and blue balls respectively. In the SRO1 layer, the Wannier function corresponding to the $x^2 - y^2$ orbital is shown. Red and blue contours are for isosurfaces of identical absolute values but opposite signs. Reprinted from Ref. [11].

4.1.1.5 Variation of the electronic and structural properties of the interface at LT and RT

To understand the experimentally observed enhancement in the interface transmissions at 120 K and RT, the electronic and geometric structures of $SrRuO_3$/$SrTiO_3$ interfaces have been determined. Looking at the geometric and magnetic properties, one can divide the $SrRuO_3$ thin films into three regions along the c-axis:

the surface, the inner layer and the interface region. The calculated densities of states of $SrRuO_3$ along with that of $SrTiO_3$ are shown in Fig. 4.3a for 9 uc of $SrRuO_3$ in the LT ferromagnetic phase and the RT non-magnetic phases. It is observed that between 1 eV and 2.2 eV, the DOS in $SrRuO_3$ is dominated by e_g electrons, whereas in $SrTiO_3$, the t_{2g} electrons dominate with a larger DOS between 1.2 eV and 1.4 eV in the non-magnetic phase of $SrRuO_3$. The exchange split e_g state produces a smaller DOS at LT between 1.2 eV and 1.4 eV. While the differences between LT and RT phases are observed in all SRO layers; SRO1 layer at the interface needs an extra attention. A distinct feature of $d_{x^2-y^2}$ character (marked by the shaded area) is observed for the RT phase in the SRO1 layer along with the renormalization of e_g DOS between 1.2 eV and 1.3 eV. The calculated hopping parameters, for example, for 6 uc at RT are 580 meV between the x^2–y^2 orbitals (corresponding Wannier function shown in Fig. 4.2 in the SRO1 layer and 540–543 meV for other layers. Also, the hopping mismatch between the interface (575 meV) and the inner layers (554–561 meV) is reduced at LT. Thus, one can clearly state that the mismatch between the interface and the inner layers is larger at RT. Furthermore, the DOS calculations indicate that the mismatch of the electronic states at the interface between $SrRuO_3$ and $SrTiO_3$ are larger at RT than at LT. Such elastic scattering effects will lead to a reduction in the BEEM transmission at RT. All these differences in the electronic structures of the two phases strongly suggest the importance of the prefactor C in Eq. 4.1, which influences the BEEM transmission at LT and RT phases.

The difference in the characteristics at the interface for LT and RT phases is quite evident also in the structural properties, which is shown in Fig. 4.3b. For a slab of reasonable thickness (9 uc in our case), one can quantitatively identify the surface, inner layers, and interface regions by the evolution of the Ru–O–Ru bond angle in the *ab* plane along the *c*-axis as shown in Fig. 4.3b. For 9 uc case, a large region of inner layers (properties similar to bulk) exists while this region is not distinctly observed in 3 uc case. Due to the interface reconstruction, the interface layer SRO1 is characterized by a Ru–O–Ru in plane bond angle having a value between the $SrRuO_3$ and $SrTiO_3$ bulk values. The properties of the inner layers are closer to the bulk. At the surface, one obtains a bond angle larger than the bulk due to open boundary conditions. Apart from the general differences between bond angles for different regions of the slab, an important observation can be made regarding the bond angles for the LT and RT phases.

It is clearly seen that the difference in bond angle between SRO1 and other SRO layers is larger for the RT phase compared to the LT one. For example, in the 6 uc case, the difference in the bond angle between SRO1 and SRO5 is 1.5° for the LT phase, while it is 3.0° for the RT phase (Fig. 4.3b). This influences the hopping of electrons (as evident from the calculated hopping parameters mentioned above) and hence the transport properties at the interfaces. Therefore, one can conclude that the differences in the geometries and electronic structures between SRO1 and other SRO layers are less dramatic in the LT phase compared to the RT one, which may effectively lead to an enhanced transmission of hot electrons across the interface at LT. The larger mismatch at RT between the interface and the inner layers will yield the transmission factor (*C* factor in Eq. 4.1) smaller for the RT phase. Such a substantial influence of geometric and structural changes at the interface leading to strong temperature dependent transport characteristics across the magnetic phase transition in $SrRuO_3$ is a remarkable finding, not observed before in any oxide heterostructure.

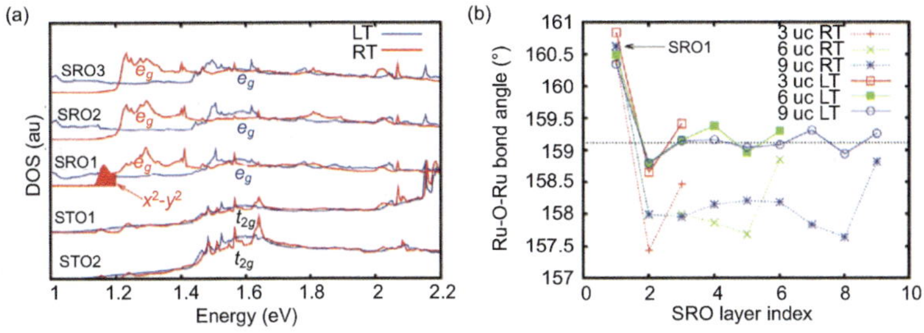

Figure 4.3 Electronic and structural mismatch at the interface. (a) Layer projected DOS (spin up + spin down) closer to the interface from STO2 to SRO3 for the 9 uc case. The LT (RT) phase is shown in blue (red). The shaded DOS below 1.2 eV in the RT SRO1 layer has x^2–y^2 character. The energy range is from 1.0–2.2 eV. (b) Ru–O–Ru bond angle in the *ab* plane as a function of the layer index for LT (solid lines) and RT (dashed lines) phases. The red, green, and blue lines represent respectively 3 , 6, and 9 uc cases. The black dashed line represents the calculated LT bulk value in the same volume setup. The SRO1 layer values are indicated by the arrow. Reprinted from Ref. [11].

The evolution of such unique features in electronic transport in complex oxide heterostructures, across a magnetic phase transition, opens new possibilities to manipulate the electronic and spin degrees of freedom by a selective choice and tailoring of their interfaces. This approach of tuning heterointerfaces by coupling structural, electronic, and magnetic properties can be extended to other material systems with promising prospects for future oxide electronic and spintronic devices.

4.1.2 Evolution of the Structural, Electronic, and Magnetic Properties in $SrRuO_3/SrTiO_3$ Ultrathin Films

When thin films of $SrRuO_3$ are epitaxially grown on $SrTiO_3$, films less than 2 uc thick were found to be insulating [21–23]. Moreover, a coexistence of ferromagnetic and antiferromagnetic phases was suggested [22]. This metal–insulator transition is not observed in any other strontium ruthenate compounds, but an insulating behavior was already observed in the isoelectronic Ca ruthenate series. Indeed, the $Ca_{n+1}Ru_nO_{3n+1}$ compounds tend to behave as antiferromagnetic (AFM) insulators, where the Ca atoms just modify the structural distortions turning the system into a Mott insulator. The Ca_2RuO_4 compound is a Mott insulator, which exhibits *G*-type AFM ordering and a metal–nonmetal transition occurring simultaneously with a first-order structural phase change [24–26].

The competition between the metallic ferromagnetic phase and the antiferromagnetic phase in $SrRuO_3$ thin films has been investigated. It has been argued that the reduced kinetic energy and the surface reconstruction might drive the system to a Hubbard AFM insulator. The orbital order properties at the surface were examined and compared with other orbital ordering found in ruthenate [27, 28]. The $SrRuO_3$ surface was analyzed by Mahadevan et al. [29] capturing the insulating phase. The experimental surface with SrO termination [30, 31] of Sr ruthenate systems has been studied using the best exchange correlation function for the $SrRuO_3$ compound [32]. Another study analyzed the quantum confinement of the electrons in the $SrRuO_3$ layers [23]. This explains the weakness of the ferromagnetic (FM) solution but cannot explain the insulating AFM state [33].

A thorough analysis of the structural, magnetic, and electronic properties by first-principles theory has been done up to 4 uc comparing the results with respect to other quasi-two-dimensional ruthenate. In this study, structural optimization was performed in local spin density approximation, LSDA + U separately for the ferromagnetic and antiferromagnetic cases. The energy difference ΔE ($\Delta E = E_{AFM} - E_{FM}$) per Ru atom between the FM and the G-type AFM phases was calculated.

4.1.2.1 Slab with one unit cell of $SrRuO_3$

A slab with three layers of $SrTiO_3$ and one layer of $SrRuO_3$ was simulated. If the geometry relaxation of the surface is not performed, the AFM solution converges to a metallic phase, which is not the ground state. Performing the full relaxation of the atomic positions, a strong reconstruction with a shrinking of the surface layer due to under-coordinated surface atoms was found, which was also reported in other perovskite oxide surfaces [34]. Moreover, it was observed that the oxygen atom of the SrO termination layer moved outwards as was found in the surface $SrTiO_3$ [35, 36]. These effects produce buckling in the SrO layer, which is expected as the SrO layer is very sensitive to the surface or interface [37]. The distance of oxygen plane from the strontium plane at the surface is 0.206 Å for the AFM surface and 0.216 Å for the FM surface. This geometrical structure of the SrO surface layer is very similar to the quasi-two-dimensional Sr ruthenate and their heterostructures [37]. As the Sr–O bond changes at the surface, one may also expect modification in the covalency of the oxygen. The covalency plays a crucial role as is shown in a similar compound $CaRuO_3$ [38, 39] and influences the value of magnetization.

Now, we investigate the DOS near the Fermi level projecting on the d_{xy} and $d_{\gamma z} = d_{xz} + d_{yz}$ in Fig. 4.4. The two-dimensionality of this surface with 1 uc produces a smaller bandwidth in comparison to the bulk phase. For the ferromagnetic phase, a half-metallic behavior is found with a total magnetic moment of 2 μB typical of LSDA + U for $SrRuO_3$ [40] at U = 2.0 eV, but the system becomes insulating for U = 3.4 eV. Instead, the AFM phase is always insulating and it is the ground state for every value of U. At U = 2.7 eV, the gap in the FM phase is very small around 0.02 eV, while it is 0.6 eV in the AFM phase. In both cases, the low energy region is dominated by the t_{2g} electrons. The electrons are strongly confined to their layer in this system [23, 41] and the influence of the Ti electrons on the $SrRuO_3$ is negligible. In the FM phase, the entire t_{2g} sector is occupied in the majority spin channel and just one electron is present in the down spin channel. In the AFM phase, a smaller bandwidth with respect to the FM case is found and hence an increased band gap is observed. The d_{xy} band on the surface is very tight, so the d_{xy} electrons are more localized than other electrons. Surprisingly, an inversion of the crystal field with respect to the bulk $SrRuO_3$ [33] is observed. Basically, the d_{xy} does not contribute to the magnetization while the two d_{yz} bands are occupied in the up spin channel and are unoccupied in the down channel. This is also called *xy* ferro-orbital order and was already observed in Ca_2RuO_4 (Fig. 4.2 of reference [27]). The structural reconstruction and the consequent inversion of crystal field are fundamental to stabilize the AFM phase. Indeed, the superlattice $(SrRuO_3)_1/(SrTiO_3)_1$ is ferromagnetic [41] because there is no strong structural rearrangement, but for the large strain the superlattice $(SrRuO_3)_1/(SrTiO_3)_1$ is found to be antiferromagnetic *G*-type [42]. In this case, the effect of the tensile strain is replaced by the shrink of the *c*-axis, because the tensile strain in superlattice and the shrinking at the surface have the same consequence on a fundamental ratio *c/a*.

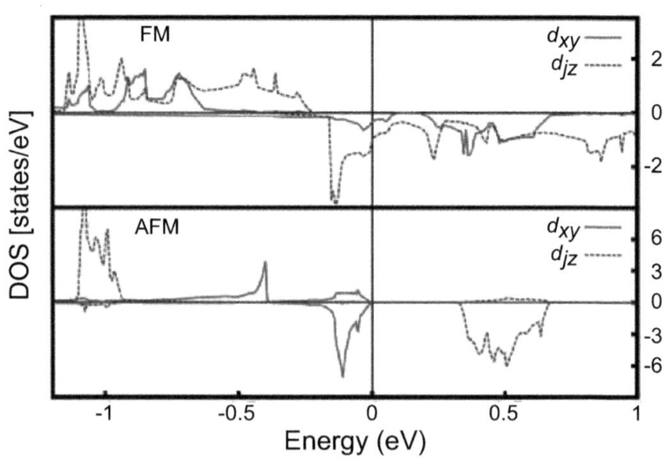

Figure 4.4 d_{xy} (solid line) and d_{jz} (dashed line) layers projected DOS for the FM and AFM $SrRuO_3$ with 1 uc at *U* =2 eV. At the surface, the occupied orbital is d_{xy} in the minority spin channel of the AFM phase. Majority (minority) spin is represented in the upper (lower) half of the panel. The DOS of the AFM phase is similar to the DOS of Ca_2RuO_4 at low temperature. Reprinted from Ref. [27], with permission the American Physical Society, Copyright 2003.

4.1.2.2 Comparison between the single layer $SrRuO_3$ and the bulk Ca_2RuO_4

The strong similarity between the DOS of the single layer $SrRuO_3$ and bulk Ca_2RuO_4 suggests to investigate and also compare other properties of the two systems. In Table 4.1, the geometrical properties are reported and the magnetization of the $SrRuO_3$ and Ca_2RuO_4 bulk compared with the single layer $SrRuO_3$. Due to the broken symmetry, the surface presents more values of the Ru–O bonds. First, we can observe the compression of the octahedra in the single layer that favors the *xy* ferro-orbital order. In both Ca_2RuO_4 and a single layer of $SrRuO_3$, the shortest bond is the Ru–O(2) bond, the bond along the *c*-axis. Both the bulk Ca_2RuO_4 and the single layer of $SrRuO_3$ show the Ru–O(l) bonds equal to 2.00 A and the Ru–O(2) equal to 1.97 A, while the Ru–O bonds in the bulk $SrRuO_3$ are all equal to 1.98 Å. The bond between the Ru and the oxygen atom of the surface is 1.95 Å due to the surface reconstruction. We define the rotation angle as half of the complementary angle of the Ru–O–Ru in plane bond angle and similarly we define the tilt angle as half of the complementary of the Ru–O–Ru out of plane bond angle. The rotation angle of the single layer $SrRuO_3$ is 12.6° and closer to the Ca_2RuO_4 respect to the bulk $SrRuO_3$. Instead, the small value of the tilt angle, 1.2°, can be attributed to the presence of $SrTiO_3$ that in the bulk has zero tilt angle. Also in other ruthenate surfaces, it is reported a strong surface reconstruction that gives us a large rotation angle of 12° and a small tilt of 2.5° [45]. Due to the many similarities with Ca_2RuO_4, it will be interesting to also investigate whether the metal–insulator transition as a function of the temperature happens in the single-layer $SrRuO_3$ above room temperature.

4.1.2.3 Multilayer slabs

Also in the case of multilayers of $SrRuO_3$, we find a structural rearrangement at the surface which shrinks along the *c*-axis and the outer oxygen. We analyze in details the case of three $SrRuO_3$ layers. The $SrRuO_3$ region is composed of a surface layer, an inner layer and the inner layer interfaced with the $SrTiO_3$. We report in Table 4.2 the comparison of the geometrical properties among the experimental volume, the theoretical volume, the surface with 1 uc and the surface with 3 uc. The Jahn–Teller distortion creates a long (l) and a short (s) Ru–O bond in the ab plane. We calculate the quantity $\left|\frac{l-s}{l+s}\right| |v|$ that is an indication of the strength of the Jahn–Teller effect. We can observe that in all the case, the tilt angle closer to the interface is always smaller than 7° because the $SrTiO_3$ compound prefers to have the tilt angle equal to zero. In the FM phase, the Jahn–Teller effect increases at the surface, while the rotation and tilt angles decrease at the surface. The situation is more interesting for the AFM phase. The Jahn–Teller effect at the surfaces is close to the FM phase, but in the bulk and inner layers, the Jahn–Teller effect increases by one order of magnitude. This is due to a structural change in the AFM bulk. We always have two equal bonds and a very short bond, but the shortest Ru–O bond is not along the c-axis as in the single layer, but it is in the ab plane. This effect also creates the inversion of the tilt and rotation angle with the tilt angle that becomes larger than the rotation angle in bulk and inner layers. In the bulk, the system is not able to reproduce the compression of the octahedra along the c-axis. This kills the xy ferro-orbital order and inhibits the antiferromagnetic phase in the bulk. The rotation angles at the surface are larger than 12° closer to the Ca_2RuO_4.

Table 4.1 Geometrical properties of the octahedra for the single layer in LSDA + U compared with the experimental results of $SrRuO_3$ [43] and Ca_2RuO_4 [44]*

	$SrRuO_3$	Single layer	Ca_2RuO_4
long Ru–O(1)	1.9876	2.0013; 2.0007	2.0182
short Ru–O(1)	1.9809	2.0006; 1.9999	2.0042
Ru–O(2)	1.9835	1.9765; 1.9451	1.9731
Rotation	8.6	12.6	11.9
Tilt	8.6	1.2	11.2
Magnetization	1.6	2.0	2.0

*The surface reconstruction has a significant impact on the monolayer slab. The unit of the bond angle is degree, the unit of the bond length is angstrom, and the magnetization is in Bohr magneton.

Table 4.2 Geometrical properties of the system calculated in LSDA + *U* for the experimental volume, the theoretical volume, 1 uc surface and 3 uc surface

	Experimental volume	Theoretical volume	1 uc	3 uc*
$\frac{l-s}{l+s}$ **FM**	0.0004	0.0004	0.0006	0.0009/0.0006/ 0.0005
$\frac{l-s}{l+s}$ **AFM**	0.0119	0.0097	0.0002	0.0011/0.0075/ 0.0092
Rotation FM	10.2	10.5	9.5	9.8/10.5/9.7
Tilt FM	9.9	10.0	4.5	8.7/9.3/6.2
Rotation AFM	9.7	9.9	12.6	12.6/10.2/9.6
Tilt AFM	12.7	12.7	1.2	8.4/11.7/6.9

*The first value in this column is for the surface layer, the middle is for the inner layer and the last is for the inner layer interfaced with $SrTiO_3$.

Table 4.3 Energy differences ΔE in meV per formula unit between ferromagnetic and antiferromagnetic phases as a function of the number of unit cells for U = 2.0, 2.7, and 3.4 eV

	$U = 2.0$	$U = 2.7$	$U = 3.4$
1	-8	-66	-145
2	77	37	-16
3	89	50	-9
4	107	90	55

The stability of the magnetic phases is shown in Table 4.3 as a function of the Coulomb repulsion parameter and number of layers. A larger number of layers produce a larger bandwidth in this system. The AFM phase is very stable for one uc being the ground state for every value of *U*. One can see that the surface layer gives a strong contribution to the stabilization of the AFM phase with $\Delta E \approx -150$ meV for U = 3.4 eV. For 2 uc, ΔE jumps to a higher value and this jump can be attributed to the loss of the orbital order. With 1 uc, the *xy* ferro-orbital order exists. When more $SrRuO_3$ layers are added, the orbital order disappears in the inner layers and the system tends to become ferromagnetic. For 2 uc and 3 uc, the ground state depends on the value of U because the two magnetic phases are really close in energy. For 4 uc, the ferromagnetic phase becomes definitively the ground state for every value of U used in this calculation. In the case of 4 uc, two inner layers are not interfaced with $SrTiO_3$ and these layers strongly prefer to be ferromagnetically coupled and hence the system with 4 uc becomes ferromagnetic. The increase in U enlarges the band gap and stabilizes the AFM insulating phase. The critical thickness depends on the Coulomb repulsion. For $U = 2.0$ and 2.7 eV, the system is insulating just for 1 uc, while the system is insulating until 3 uc for $U = 3.4$ eV. The eventual first-order metal–insulator transition, analogue to the transition in Ca_2RuO_4, should have a critical temperature strongly dependent on the number of unit cells with the critical temperature that should be higher for the thinner samples.

In conclusion, the surface formation generally leads to the shrinking of the c-axis due to the under-coordinated oxygen atoms at the surface, the outward movement of the oxygen atom and a consequent inversion of the crystal field of the surface layer. The bulk $SrRuO_3$ is an itinerant ferromagnet, but the reduction of the dimensionality and kinetic energy change the magnetic phase in the thin films. Also, the inversion of the crystal field largely determines their electronic and magnetic properties. A metal to insulator transition takes place by reducing the slab thickness, together with the ferromagnet to antiferromagnet transition in an intermediate spin scenario. The metal–insulator transition is simply due to the electron localization and consequently to the reduction of the bandwidth. The large Coulomb repulsion and the low thickness favor the insulating G-type antiferromagnetic phase, which also present an increase in the Jahn–Teller effect in the inner layers of $SrRuO_3$. For the single unit cell, the antiferromagnetic phase is extremely stable, while increasing the number of layers, the phase stability depends on the Coulomb repulsion. At 4 uc, the ferromagnetic phase becomes definitely stable for every value of the Coulomb repulsion.

Because the phase with FM layers and the AFM surface layer is not energetically stable, the coexistence experimentally found can be due to spatial inhomogeneity of the system. The $SrRuO_3$ surface presents many similarities (rotation angles, projected DOS and d_{xy}

ferro-orbital ordering) with the Ca_2RuO_4 compound. However, the inner layers of the $SrRuO_3$ and the bulk are not able to stabilize the d_{xy} ferro-orbital order favoring the ferromagnetic phase. Due to the many similarities between the Ca_2RuO_4 and the $SrRuO_3$ thin film, the $SrRuO_3$ thin film could also present a first-order metal–insulator transition above room temperature.

4.2 Vertical Shift in Exchange Bias

In this section, we will discuss an unusual phenomenon of the large vertical shift in exchange bias occurring in a specific oxide heterostructure. Exchange bias is one such outcome of interfacial coupling across two different magnetic states [46], usually antiferromagnetic and ferromagnetic phases. In perovskite-based heterostructures, magnetic interactions are particularly fascinating as they can show interface ferromagnetism between two antiferromagnets or between an antiferromagnet and a paramagnet [47]. Competing magnetic interactions, which give rise to proximity coupling such as exchange bias have found technological applications in magnetoresistive sensors, however, its microscopic origin often raises debates particularly regarding the coupling configurations at the interface [46].

Orthorhombic $YMnO_3$ is a ferroelectric antiferromagnet with $T_{FE} \approx 30$ K and $T_N \approx 42$ K. The reported ferroelectric properties of $YMnO_3$ thin films indicate the occurrence of the coexistence of cycloidal and *E*-type antiferromagnetic orderings [48] or of commensurate *E*-type antiferromagnetic ordering at low temperatures [49]. Orthorhombic $YMnO_3$ is also energetically close to antiferomagnetic *A*-type ordering [50]. Since the magnetic properties of $YMnO_3$ lie at the verge between different magnetic orderings, it is tempting to speculate that interfaces could modify this ordering as well the electronic and structural properties.

Magnetic heterostructures showing a horizontal shift of the hysteresis loop along the field axis have been commonly observed for bilayer combinations [5, 51]. However, an uncommonly observed effect of exchange coupling across an AFM/FM interface is the shift along the magnetization axis or the vertical shift [52]. Element specific magnetic studies with X-rays of FeF_2/Co and CoO/Fe [53, 54,

55] layered structures confirmed the existence of this vertical shift and revealed its relation to specific uncompensated moments in the antiferromagnet. The vertical shift is the irreversible magnetization of the non-ferromagnetic layer [56] and it was conjectured to be either correlated or not correlated to the bias field, but no convincing experiment could be made possible to identify due to the extremely small magnitude of the vertical shift.

However, a recent study on heteroepitaxial interfaces involving $YMnO_3$ and the weak anisotropic ferromagnet $La_{0.7}Sr_{0.3}MnO_3$ (LSMO) showed an unusually large vertical shift. It was also shown that this vertical shift was correlated with the horizontal shift [57, 58]. Samples with a larger horizontal shift showed a smaller vertical shift and vice versa as in other systems [59]. In core/shell structured Fe nanoparticles, large exchange bias and vertical shifts were attributed to the frozen moments in a spin glass–like phase in the shell of the nanoparticles [60]. However, the horizontal shift in $YMnO_3$/$La_{0.7}Sr_{0.3}MnO_3$ is negligible compared with the vertical shift. The vertical shift was found together with the horizontal shift [61, 62], but this system basically shows a pure vertical shift in the hysteresis loop. $YMnO_3$/$La_{2/3}Sr_{1/3}MnO_3$ heterostructure is the missing element in the phenomenology of the magnetic interface and can be a key point in the understanding of the magnetic interface phenomena. In this section, we focus on the coupling between AFM $YMnO_3$ and FM $La_{2/3}Sr_{1/3}MnO_3$ [63] to understand what is the specificity of this system to create a pure vertical shift.

4.2.1 First-Principles Calculations of Heterostructures

First, we discuss the case of a heterostructure with two $YMnO_3$ layers, two LSMO layers, and two interface layers from the structural and magnetic point of view. Further, using the results of this heterostructure, we analyze in more detail the properties of a larger supercell with five $YMnO_3$ layers.

4.2.1.1 Superlattice: structural properties

In these calculations, the experimental values [49] of $YMnO_3$ for the in-plane lattice parameters of the supercell have been used. For the *c*-axis, the bulk value has been chosen and at the interface, the average out-of-plane separation between $YMnO_3$ and LSMO has been used to get the total *c* value of the supercell. The following notations will be used for the heterostructure, we define L0 as the interface layer and LI as the first neighbor layer. In the supercell with five $YMnO_3$ layers, the second and third neighbor layers of the interface are denoted as L2 and L3 respectively, as shown in Fig. 4.5.

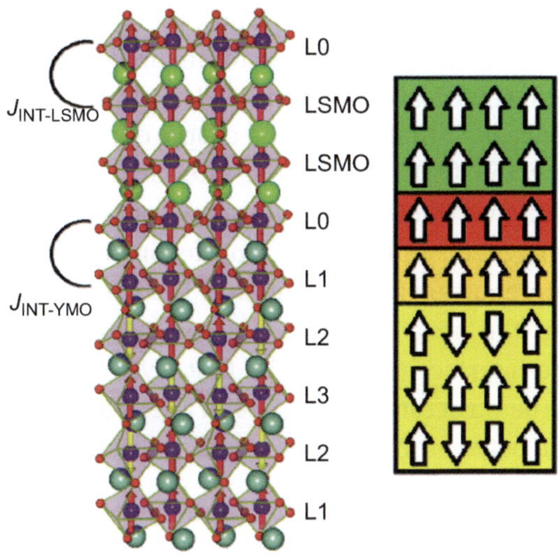

Figure 4.5 Ground state geometry and magnetic structure of LSMO/YMO heterostructure with five $YMnO_3$ layers (left) and the corresponding schematic picture showing different regions and their magnetic configurations (right). MnO_6 octahedra are shown with red (yellow) arrows indicating up (down) moments on Mn atoms. For simplicity, we do not show the direction of the moment in the *ab* plane. Y, La, and Sr atoms are shown as dark blue, dark green, and light green balls. $J_{INT-YMO}$ ($J_{INT-LSMO}$) is the effective magnetic superexchange along the *c*-axis between interface Mn atoms and Mn atoms of LI (LSMO) layer. Reprinted from Ref. [63].

To start with, a small supercell with two $YMnO_3$ layers is considered. In this interface model, a layer of YO, a layer of MnO_2 and another layer of $La_{1-x}Sr_xO$ that close the cage around the MnO_6 octahedra are taken. However, the exact stoichiometry of the $La_{1-x}Sr_xO$ layer at the interface is unknown. From first-principles calculations, the energies for different magnetic phases and different Coulomb repulsion parameters U_{INT} (Coulomb repulsion for the interface) have been calculated. The most stable configuration is with LaO layers and SrO layers in agreement with similar results for other kinds of supercells [64, 65] and the interface with only La atoms has always the lowest energy. Therefore, at the interface, one has MnO_6 octahedra enclosed in a cage with Y on one side and La on the other side. Probably, the La atoms are energetically favored because of their atomic radii being very similar to the Y radii. As Y and La are electronically equivalent, there is no charge doping or double exchange at the interface. Thus, the interface layer can be metallic or an insulator depending on the magnetic phase. On the other hand, the inner layers of LSMO will be always metallic because of the presence of Sr that produces a charge doping.

Assuming the LaO layer at the interface, the results for the supercell with five $YMnO_3$ layers are shown at $U_{INT} = 2$ eV and $U_{INN} = 3$ eV in the magnetic ground state reported in Fig. 4.5, where U_{INN} is the Coulomb repulsion for the interface. The magnetic ground state, discussed in the next subsection, is composed of ferromagnetic layers at the interface (L0) and the first layer of $YMnO_3$ (LI). It is found that the structural properties of the interface layer L0 are intermediate between $YMnO_3$ and LSMO.

4.2.1.2 Superlattice: magnetic properties

Besides the most stable magnetic phases, *G*-type, *A*-type, and *E*-type magnetic phases were also studied at the interface. The spin configuration is $3d^4\uparrow$ according to Hund's rule both for the interface and $YMnO_3$ inner layers independently from the magnetic phase, while it is $3d^{3.5}\uparrow$ in the LSMO due to the charge doping. We report the energy as a function of U_{INT} in Fig. 4.6 for different magnetic phases. The four following magnetic phases are close to the ground state:

(a) A fully ferromagnetic phase (FM-FM)
(b) An antiferromagnetic *E*-type in $YMnO_3$ and ferromagnetic phase in LSMO and interface (AFM-FM)
(c) A ferromagnetic phase with spin up in $YMnO_3$ and a ferromagnetic with spin down in LSMO and interface (FMu-FMd)
(d) An antiferromagnetic *E*-type in $YMnO_3$ and an antiferromagnetic *A*-type in LSMO and interface (AFM-AFM)

It is found that in the range of typical U_{INT} for this system (between 1.8 eV and 4.1 eV), the heterostructure is completely ferromagnetic also in $YMnO_3$. Above 4.1 eV, an E-type solution in $YMnO_3$ and an A-type solution in LSMO are observed as the magnetic ground states. It should be noted that the heterostructure is orthorhombic. Instead, the LSMO bulk system is pseudocubic and does not present any instability towards the A-type magnetic order. Finally, at very low values of U_{INT}, one finds the ferromagnetic $YMnO_3$ phase with opposite moments with respect to the ferromagnetic LSMO and interface.

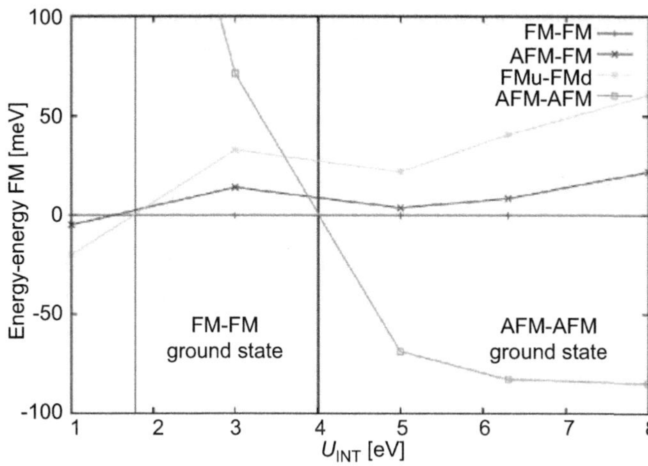

Figure 4.6 Total energies of the magnetic phases of $YMnO_3$/LSMO with six layers as a function of Coulomb repulsion at the interface. The difference between each magnetic phase and the fully ferromagnetic solution is shown. The fully ferromagnetic solution is the ground state between 1.8 eV and 4.0 eV. Reprinted from Ref. [63].

Remarkably, AFM E-type in $YMnO_3$ and FM LSMO and interface is the sum of the two bulk ground states; however, it is never the ground state. Although U_{INT} is a local property of the interface layer, we can see that it influences the properties of the whole heterostructure determining a change of the magnetic properties. A small variation of U_{INT} can also have long-range effects on the magnetic properties.

When the bigger supercell with five $YMnO_3$ layers is analyzed, qualitatively similar results are found. To investigate how the ferromagnetism goes deep in the $YMnO_3$ region, the magnetic properties as a function of U_{INN} are analyzed, as shown in Fig. 4.7. Two ferromagnetic layers at U_{INN} = 2–3 eV are present in the ground state. These two ferromagnetic layers correspond to the two L1 layers in Fig. 4.5. When U_{INN} = 4 eV, all the five $YMnO_3$ layers become ferromagnetic. In summary, it is found that in two different supercells and in the typical range of Coulomb repulsion for this system, the first neighbor layer of the interface is ferromagnetic. This is in agreement with recently observed few nanometers of ferromagnetic layer at the interface of $YMnO_3$/LSMO [58]. The $YMnO_3$ develops an interface-induced ferromagnetism as in other perovskite systems at the borderline between different magnetic orders [5, 66]. The superexchange parameters for the supercell with five $YMnO_3$ layer have been calculated. One obtains $J_{INT-LSMO}$=−39 meV, which indicates a strong magnetic coupling between the interface and LSMO. Instead, the coupling $J_{INT-YMO}$ between L0 and L1 is just −1.2 meV.

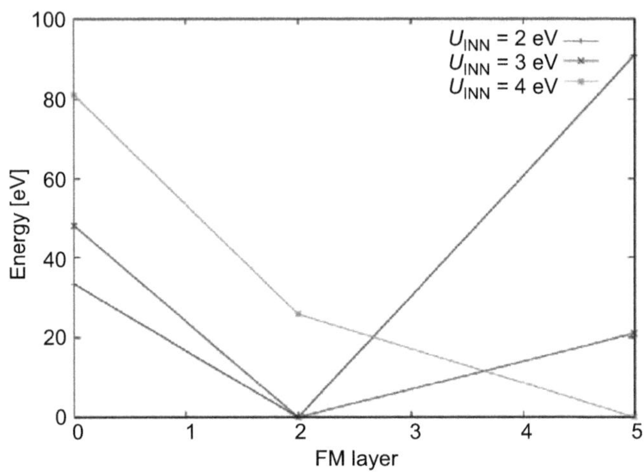

Figure 4.7 Total energy as a function of the number of ferromagnetic YMnOa layers for the heterostructure with five $YMnO_3$ layers. The energy of the ground state is set to zero. In the configuration with no FM layers, five *E*-type AFM layers are present. In the configuration with two FM layers, there are two ferromagnetic L1 layers. In the configuration with five FM layers, the heterostructure is fully ferromagnetic. Reprinted from Ref. [63].

4.2.1.3 Superlattice: electronic properties

The calculated DOS for the heterostructure is shown in Fig. 4.8. As expected, the LSMO region shows a metallic behavior (Fig. 4.8a). In the L0 layer, there are three electrons in the t_{2g} manifold and one occupied electron in the e_g. There are two e_g metallic bands at the Fermi level in the L0 layer (Fig. 4.8b). Remarkably, the most occupied e_g orbital is $x^2 - y^2$ as shown in Fig. 4.9 although the layer is metallic and also the $3z^2 - r^2$ orbital gives us a strong contribution to the DOS at the Fermi level. However, this is completely different from $YMnO_3$ where the most-occupied orbital has $3l^2 - r^2$ character in the bulk and heterostructure in both FM and AFM phases. The interface layer is half-metallic while both the FM and AFM $YMnO_3$ layers are insulators as observed in the local DOS of these atoms respectively in Figs. 4.8c and 4.8d. Electrons on atoms with spin majority down are more localized and the bandwidth is narrower in the DOS.

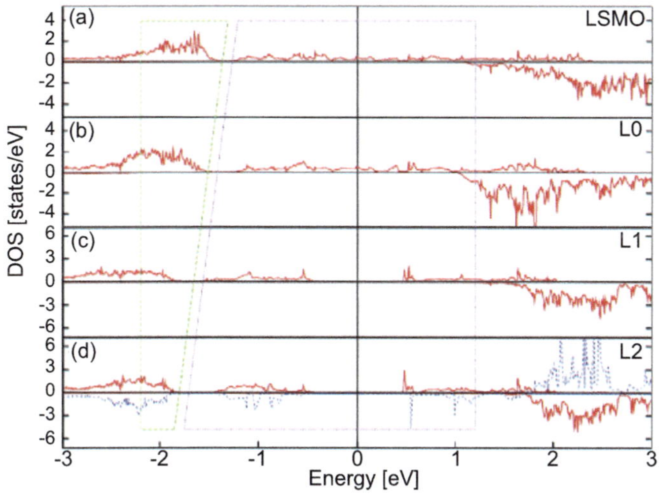

Figure 4.8 Layer-resolved DOS of the superlattice with five $YMnO_3$ layers. The Fermi energy is set to zero. Spin up (down) contributions are shown in the positive (negative) *y*-axis. Panels (a–d) respectively show the local DOS of Mn atoms in the LSMO, L0, L1, and L2 layers. The DOS of the L3 layer is very similar to L2. The entire system shows a half metallic behavior. Solid red (dashed blue) line represents the DOS of the Mn atoms where the spin majority is up (down). In the figure, the region is indicated where the main orbital contribution to the DOS for the majority spin is t_{2g} (green dashed line) and e_g (pink dotted line). Reprinted from Ref. [63].

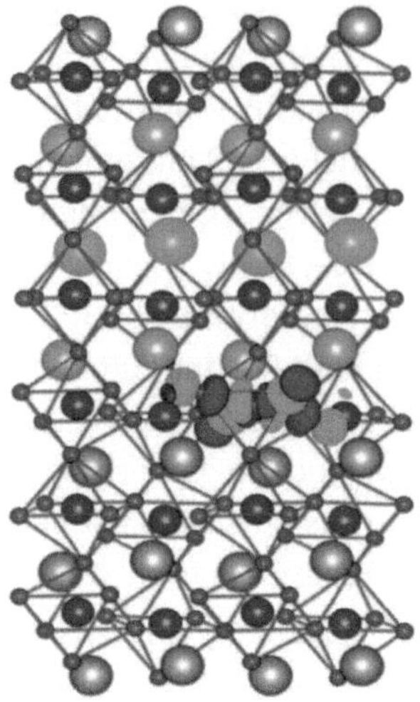

Figure 4.9 Occupied e_g-like Wannier state on Mn atoms at the interface for the heterostructure with two $YMnO_3$ layers. The orbital is of $x^2–y^2$ nature. Reprinted from Ref. [63].

4.2.2 Bulk $YMnO_3$

An unusual FM phase was reported in $YMnO_3$ in the previous section. To understand more about this new phase, the FM and AFM phases are studied and compared to the experimental volume. Three pseudocubic axes can be considered along the Mn–Mn directions for $YMnO_3$.

There are three inequivalent directions for $YMnO_3$, which are called 1, 2, and c. Along the direction 1, the coupling is ferromagnetic while along the directions 2 and c, it is antiferromagnetic.

One can map the $YMnO_3$ system in an Ising model where σ_i = {−1, +1} represent spin down and spin up, respectively:

$$H = \Sigma_{<i,j>} J_{i,j}\, \sigma_i \sigma_j \quad (4.2)$$

where the sum is over pairs of adjacent spins. In Fig. 4.10, the values of magnetic superexchange $J_{i,j}$ as a function of U and with a fixed ratio between the Hund coupling and Coulomb repulsion J_H/U = 0.15 are shown. Depending on the U value, there are three magnetic phases: E-type at low U, A-type at intermediate U and FM at high U. The results are in agreement with reference [67]. The experimental magnetic phase is reproduced for $U < 4$ eV. The ferromagnetic phase of $YMnO_3$ has the same group symmetry of the dielectric phase *(Pbnm).* In this new phase, the inversion symmetry is recovered.

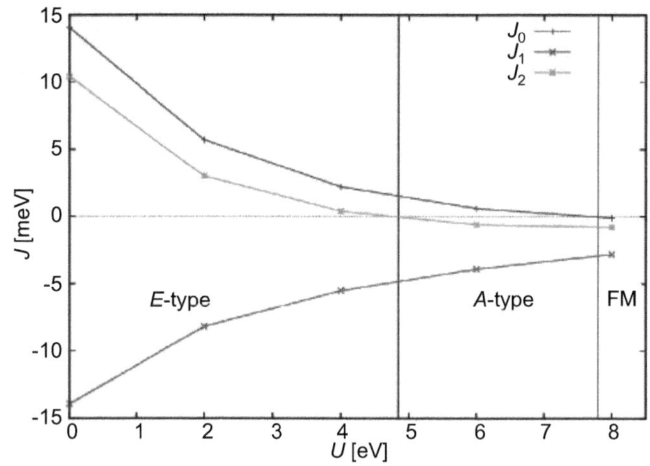

Figure 4.10 Effective superexchange for $YMnO_3$ at experimental positions as a function of U. There are three independent values of J. If J is negative, one has FM coupling, if J is positive, it is an AFM coupling. When J_1 is negative and the others are positive, the E-type magnetic order occurs. When J_1 and J_2 are negative and J_c is positive, one realizes the A-type magnetism. Finally, when they are all negative, the FM phase is obtained. Reprinted from Ref. [63].

Electrons are more localized in the AFM E-type phase than in the FM phase. Jahn–Teller effect (JT) is reduced in the FM phase and $(1 - s)/(1 + s)$ goes from 0.077 in the E-type to 0.067 in the FM phase. The orbital order is $3l^2 - r^2$ in both magnetic phases and also hopping parameters are not modified.

4.2.2.1 Hopping parameters

The low-energy physics of $YMnO_3$ can be captured by e_g states. The hopping parameters are calculated for the low energy band structure in the basis of e_g-like Wannier functions. The hopping parameters of the non-magnetic phase for the e_g manifold are reported in Table 4.4. The hopping in the ab plane is almost similar, small differences arise from different bond angles and Mn–Mn distances. In a perfect tetragonal symmetry, one should find $\left|t_{p,q}^{1,2}\right|$ along the direction 1 equal to $\left|t_{q,p}^{1,2}\right|$ along the direction 2. A large hopping $t_{p,p}^{1,2}$ = −243 meV between $x^2 - y^2$ like Wannier function in the *ab* plane is found. The hopping parameter between $3z^2 - r^2$ like Wannier function along the *c* direction is found to be −307 meV. These values of hoppings are in good agreement with values for other Mn-perovskites. It is observed that the second-neighbor hoppings are strongly asymmetric along the two directions *a* and *b*. It was shown that the orbital order is $3l^2 - r^2$ and considering the larger value of *b*, the charge of the Mn atoms points towards the direction *b*. This is the main reason why the hopping parameters are bigger along the *b*-axis creating a large anisotropy in the plane.

Table 4.4 Hopping parameters in $YMnO_3$ in the e_g-like basis at experimental positions*

	ε_p	ε_q	Δ^1	Δ^2
	1591	2019	438	438
	First neighbor			
Direction	$t_{p,p}^{1,2}$	$t_{p,q}^{1,2}$	$t_{q,p}^{1,2}$	$t_{q,q}^{1,2}$
1	-243	203	137	-109
2	-244	−138	-203	-112
c	-53	-79	-79	-307
	Second neighbor			
Direction	$t_{p,p}^{1,2}$	$t_{p,q}^{1,2}$	$t_{q,p}^{1,2}$	$t_{q,q}^{1,2}$
a	16	6	5	-6
b	59	3	4	-28

*The basis is composed by $|x^2 - y^2, 1\rangle$, $|x^2 - y^2, 2\rangle$, $|3z^2 - r^2, 1\rangle$, and $|3z^2 - r^2, 2\rangle$. Hopping integrals, $t_{j,j'}^{i,i'}$, are from a site *i* with orbital *j* to neighboring site *i'* with orbital *j'*. The connecting vector is $\mathbf{T} = l\mathbf{x} + m\mathbf{y} + n\mathbf{z}$ where the vectors **x**, **y**, and **z** connect two first-neighbor Mn atoms first neighbor; the hopping integrals are tabulated up to the first neighbor. ε_j^i is the energy on site for the orbital *j* on the site *i* and Δ^i is the on-site inter-orbital hopping of the atom *i*. *p* and *q* are used to denote respectively the orbitals $x^2 - y^2$ and $3z^2 - r^2$ in the hopping notation. The second-neighbor hopping parameters are also reported along the two strongly different *a* and *b* axes. All energies are in meV.

4.2.2.2 Electronic properties

The electronic properties of the AFM and FM phases of bulk $YMnO_3$ have been studied. The influence of the Jahn–Teller effect and the change of symmetry on the DOS in the FM phase are shown. Both magnetic phases show insulating behavior (see Fig. 4.11). In the AFM case, one has the t_{2g} at 1 eV below the Fermi level, while in the FM case is as 2 eV below the Fermi level. Instead of the classical e_g orbital, there is a linear combination of them. In both magnetic phases, the $3l^2 - r^2$ orbitals just below the Fermi level and the $s^2 - z^2$ orbitals just above the Fermi level are found [68] for the AFM phase. In the FM case, the bandwidth is larger because electrons are more delocalized.

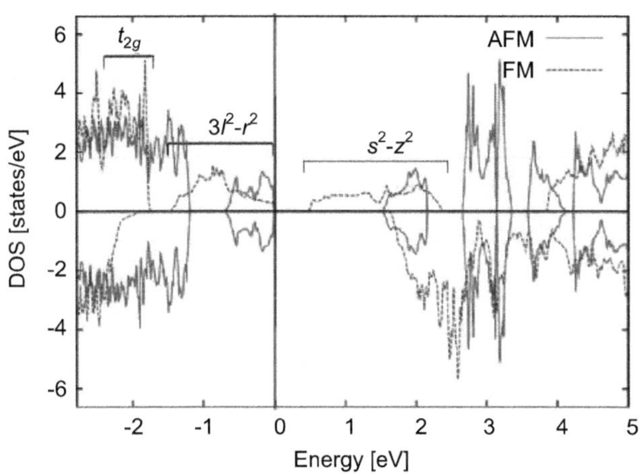

Figure 4.11 Total DOS per formula unit for $YMnO_3$ in the *E*-type (solid line) and FM (dashed line) phase. The Fermi level is set to zero. A majority (minority) of spin contributions are shown in the positive (negative) *y*-axis. In the figure, the main orbital contribution to the DOS for the majority spin of the FM phase is indicated. Reprinted from Ref. [63].

4.2.2.3 Magnetocrystalline anisotropy

As the magnetic field in the experimental case is applied in the *ab* plane, one may be interested to study the anisotropy for the same situation. For the ferromagnetic *Pbnm* phase, the formula for the anisotropy energy E_A in the *ab* plane for this system is

$$E_A(\phi) = K_2 \cos^2 \phi + K_4 \cos^4 \phi \tag{4.3}$$

where ϕ is the angle between the a-axis and the magnetization in the ab plane. As the hard axis is along the c-axis, one can neglect it in Eq. (4.3) for anisotropy energy. The calculated ab initio value of the magnetocrystalline anisotropy in the ferromagnetic phase is $K_2 = 0.561$ meV [58] per formula unit for the leading term. The higher order anisotropy term K_4 is also reported in Table 4.5 and is found to be negligible. As the value of K_2 is positive, the easy axis is along the b-axis when $\phi = 90°$. The calculated value of K_2 is quite big in comparison to the values found in the literature [69–71] and is also two orders of magnitude greater than usual magnetocrystalline constants. This large value originates from a significant anisotropy $b/a = 1.111$ between the lattice parameters combined with the $3l^2 - r^2$ orbital order.

Table 4.5 Magnetocrystalline anisotropy constants for $YMnO_3$ in a fully ferromagnetic bulk phase

	K_2	K_4
meV/(formula unit)	0.561	0.008
10^4 J/m^{-3} = 10^5 erg/cm^{-3}	160	2

To simulate the hysteresis, one needs to take into account the interaction of the magnetic system with the external magnetic field. The effect of the magnetic field on the energy is represented by the Zeeman energy E_H, which is expressed as:

$$E_H(\phi) = -MH\cos(\phi - \phi_F) \tag{4.4}$$

where ϕ_F is the direction of the external magnetic field H and $\phi - \phi_F$ is the angle between the magnetic field and the magnetic moment M. Here, the interface and shape anisotropies are not considered. The calculated total magnetic moment $M = M_{spin} + M_{orbital}$ is $M = 3.62$ µB per Mn atom. Adding Eqs. (4.3) and (4.4), one gets the expression for the total energy as:

$$E(\phi) = K_2\cos^2\phi - MH\cos(\phi - \phi_F) \tag{4.5}$$

For $\phi_F = 90°$, there is only one stable solution for $H > H_c = 2K_2/M$ that is the coercive field in this limit case. In this simple model, this critical magnetic field coincides with the magnetic anisotropy field, that is, the magnetic field needed to saturate the magnetization along the hard axis. Using the total energy (Eq. 4.5), the hysteresis curve for $YMnO_3$ is calculated. It is found that the coercive field H_c is of the order of $2 - 5 \times 10^4$ Oe, while in the experiment about $YMnO_3$/$La_{0.7}Sr_{0.3}MnO_3$ heterostructure [57], the magnetic field was up to 0.5×10^4 Oe. In this study, the magnetic anisotropy field is 5.4×10^4 Oe.

In LSMO, the expression for the magnetocrystalline anisotropy is

$$E(\theta) = K_2^{\mathrm{LSMO}} \sin^2\theta \tag{4.6}$$

where θ is the angle between the direction of the magnetization and the [001] direction. A very small anisotropy $K_2^{\mathrm{LSMO}} = 0.18 \times 10^4$ J/m^{-3} was measured in LSMO [72]. Following Eq. (4.6), the easy axis is along the *c*-axis and the hard axes are in the *ab* plane. However, at the interface, an orthorhombic structure of LSMO is found and hence one may expect a small magnetocrystalline anisotropy with a different easy axis. However, this effect is neglected here and the easy axes plane as the *ab* plane is considered as suggested by the experiments.

4.2.3 Hysteresis Loops for Two Independent Ferromagnetic Phases

In Ref. [57], a vertical shift m_{shift} in $YMnO_3$/$La_{0.7}Sr_{0.3}MnO_3$ heterostructures at 10 K was found experimentally and this vertical shift is attributed to the properties of the diluted antiferromagnetic state in a magnetic field. However, the properties of the diluted antiferromagnetic state disappear at low temperature in bulk $YMnO_3$ and an *E*-type AFM phase take place [49]. Instead, the nature of the large vertical shift might be found in the interdiffused ferromagnetic layer at the interface composed also by an unusual $YMnO_3$ ferromagnetic phase. It is shown that the $YMnO_3$ L1 layer near the interface becomes ferromagnetic. Therefore, the m_{shift} effect can be explained by the large coercivity of ferromagnetic $YMnO_3$. The magnetic field used to reverse the spin in ferromagnetic LSMO is too small to reverse the spin in ferromagnetic $YMnO_3$.

More clearly, the FM layer of $YMnO_3$ is a hard ferromagnet and is weakly ferromagnetically coupled to the interface and LSMO. The region composed by the interface and LSMO is a soft ferromagnet. One may assume that the ferromagnetic $YMnO_3$ is magnetically decoupled from the interface. The total energy of the system can be written in the following way:

$$E(\phi_S,\phi_H)=n_S\left(K_2^S\cos^2\phi_S-M^S H\cos(\phi_S-\phi_F)\right)$$
$$+n_H\left(K_2^H\cos^2\phi_H-M^H H\cos(\phi_H-\phi_F)\right) \qquad (4.7)$$

as the sum of the energy of a soft ferromagnet (*S*) and a hard ferromagnet (*H*). n_S and n_H represent respectively the numbers of atoms in the soft and hard ferromagnetic regions. When a relatively small magnetic field is applied, the hard ferromagnet does not reverse when the magnetic field is applied in the opposite direction, determining a shift of the magnetization. The magnetic configuration coming from Eq. (4.7) is shown in Fig. 4.12. The m_{shift} is the magnetization of the pinned magnetic moment in the AFM region. If these results are compared with the experimental values of the vertical shift, one observes that the FM phase of heterostructure should be composed of few interface layers of $YMnO_3$ within the approximation of a sharp interface. What seems to be important in the determination of m_{shift} is the roughness of the interface, which is not considered here. A large roughness gives us a large ferromagnetic interdiffused layer and therefore a large m_{shift} as experimentally observed [57]. Another important condition is the weak coupling between the L0 and L1 layers, that permits the L0 layer to rotate avoiding to influence the L1 layer.

The peculiarity of this system is the weak exchange bias. The reduced horizontal shift arising from weak exchange bias can be attributed to the weak coupling between the rotating FM phase (L0) and the AFM phase (L2). Indeed, the exchange bias is proportional to the magnetic coupling between these phases [73]. However, this is a second neighbor exchange interaction and hence, leads to a weak exchange bias. A large vertical shift comes from a large value of pinned magnetic moment in the AFM region and hence, the magnetic coupling between AFM and FM is weak. The large vertical shift is associated with a small exchange bias when the creation of the pinned magnetic moments destroys the antiferromagnetic order. To have the simultaneous presence of large magnetic moment and large exchange bias, one needs a net magnetic moment in the AFM phase and an AFM phase strongly coupled to the FM phase. This is observed in other systems where the net magnetic moment is provided by non-collinear magnetic structures in the AFM phase or in a ferromagnetic one [61, 62].

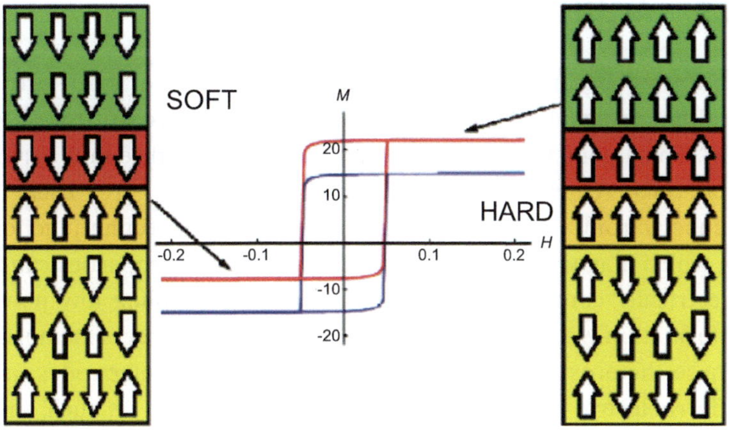

Figure 4.12 Microscopic mechanism of a pure vertical shift. The positive direction of the magnetic field is the direction of the field cooling. Magnetic configurations are shown with a magnetic field in the positive (negative) direction on the right (left) side. The hard ferromagnetic layer (shown in orange) does not reverse its spin when a small magnetic field is applied in the negative direction. The results are obtained using K_2^S = 0.002 meV and show the hysteresis loop without hard ferromagnet (blue line) and with hard ferromagnet (red line). Schematic magnetic configurations are consistent with Fig. 4.5. The unit of the external magnetic field is Tesla along the horizontal axis and is an arbitrary unit along the vertical axis to represent magnetization. Reprinted from Ref. [63].

This phenomenon of pure vertical shift is realized mainly due to three conditions. First, one needs an FM layer in the AFM side of the interface. Second, this FM layer should have to be magnetically weakly coupled to the FM side of the interface. Finally, one needs a large magnetocrystalline anisotropy of the FM layer relative to the FM side of the interface. The first two conditions are possible just when the *A*-type and the FM phases are energetically close to the ground state as in the case of $YMnO_3$ (shown in Fig. 4.10). The third condition is the requirement of a strongly anisotropic system described above. The conclusions support the idea that horizontal and vertical shifts observed in the hysteresis loops are correlated and tend to exclude each other if the creation of the pinned magnetic moments destroys the antiferromagnetic order. Thus this study, in general, would instigate revisits on various other systems showing such pure vertical shifts.

4.3 Artificial Multiferroics

Recently, it became possible to fabricate composites of ferroelectrics and ferromagnets in the form of an artificial heterostructure [74–77]. This opens the route to design room temperature multiferroic materials. Examples of such heterostructures are used in nanoelectronic devices include artificial multiferroic tunnel junctions in which magnetic and spin dependent transport properties can be electrically controlled in a non-volatile way, enabling low power information write operations in magnetic random access memories. These heterostructures combine ferromagnetic electrodes and a ferroelectric insulator (FE) as the tunnel barrier. As such, their tunnel resistance is expected to depend on the magnetic and ferroelectric orders, i.e., they display tunnel magnetoresistance and tunnel electroresistance effects, respectively, giving rise to four resistance states. In addition, a novel magnetoelectric phenomenon at FE/FM interfaces has recently emerged as the pivotal point in such nanojunctions, with potentially emerging interfacial phases revealing unusual electronic properties [78]. These interfacial magnetoelectric effects may show up as a modulation of the spin polarization by ferroelectricity at FE/FM interfaces and the corollary induction of a finite magnetic moment in the FE. Interface mechanisms at the origin of such magnetoelectric coupling were first theoretically predicted by Duan et al. for the Fe/$BaTiO_3$ (BTO) system [79].

A crucial aspect for the Fe/BTO heterostructures is the oxygen concentration at the interface. The first theoretical structure [79] was without oxygen at the interface. Without oxygen, the charge transfer gives rise to the formation of Ti^{3+} at the interface for large Coulombic repulsion [80] that was not observed experimentally [81]. A more realistic model with the formation of 1 FeO layer at the interface (Fe/FeO/BTO) was investigated theoretically and experimentally [81]. Finally, a recent work shows the formation of an antiferromagnetic FeO phase with two or more FeO layers (Fe/$(FeO)_2$/BTO) probably due to a different annealing process [6]. In this section, we will present the last two cases.

4.3.1 The Ferroelectric–Ferromagnetic Interface in Fe/BTO

In the case of Fe/FeO/BTO, the ferroelectric is interfaced directly with the ferromagnetic Fe because the single FeO layer is ferromagnetically coupled with the other Fe atoms [82].

4.3.1.1 The Fe/FeO/BTO case

Experimental data suggest that the interfacial region between the BTO and the Fe layers consists of a perovskite-type unit cell, however, slightly chemically different from the BTO structure. Atomically resolved electron energy loss spectroscopy based elemental profiles indicate the presence of multiple features typical of Fe^{3+} and Fe^{3+} contributions at the interface. This finding indicates the presence of oxidized iron at the interface, most probably, over solely one atomic column [81]. In the frame of the experimental and simulation results, the oxidized interface is most likely to describe the nanofabricated multiferroic tunnel junctions. Density functional theory calculation confirms this picture. This kind of interface would also explain the observation of solely Ti^{4+} cations at the interface by energy loss near edge structures and provide theoretical and experimental evidence that induced magnetism on interfacial Ti atoms originate from Ti^{4+} cations.

In summary, atomically resolved investigations were performed to determine a realistic interface model for the BTO/Fe interface and used it as an input in first-principles calculations to understand in detail the magnetic and electronic properties of the interface. The analysis evidences the subtle, but fundamental role of an intermediate FeO monolayer between BTO and Fe. Spin-resolved DOS and atomically resolved magnetic moment calculations show a fair agreement with experimental results.

4.3.2 Possible Electrical Manipulation of the Exchange Bias

In the case of large annealing temperature, the BTO pumps oxygen in the interface layers of the metallic phase. The result is the creation of transition metal oxides at the interface as was shown in the case of Fe/BTO [6] and Cr/BTO [83]. In the following, we will analyze in more detail the case of the Fe/$(FeO)_2$/BTO that is an exchange bias bilayer system interfaced with a ferroelectric material.

4.3.2.1 The Fe/$(FeO)_2$/BTO case

To understand the origin of the exchange bias experimentally observed at the Fe/BTO interface [6], first-principles calculations of the interfacial magnetic moments for Fe/BTO and Fe/$(FeO)_2$/BTO heterostructures were performed. The supercells were built up with five layers of BTO and nine monolayers of Fe along the *c*-axis and an in-plane plane $\sqrt{2}a \times \sqrt{2}a$ supercell with two Ti atoms as shown in Fig. 4.13b. The supercell was constructed so that at each interface, O atoms of the TiO_2 termination occupy atop sites of Fe. Such termination was proved to be the most stable [79, 81]. The structural relaxations of the system were performed in LDA + U method optimizing the internal degrees of freedom by requiring that the forces are less than 10 meV $Å^{-1}$.

It is already known that the Fe/BTO heterostructure has a ferromagnetic order [79, 80, 84]. When one layer of oxygen is added at the interface to create one FeO layer, the ground state is still ferromagnetic. However, oxygen atoms at the interface receive two electrons from the iron interface, which creates Fe^{2+} and hence, the magnetic moment is increased similar to bulk FeO. The magnetic moment in the FeO layer increases from 3.2–3.7 μ_B and the magnetic moment on Ti is reduced to 0.1 μ_B. The average magnetic moment of the iron atoms in the inner layers increases from 2.73–2.82 μ_B due to the large moment of the FeO layers that induce a larger magnetic moment in the Fe inner layers. When two layers of oxygen atoms are added at the interface, the ground state is composed by one AFM interface layer while the other FeO layer is ferromagnetic. The energy difference $E_{AF} - E_{FM}$ is 209 meV per interface Fe atom for the Fe/BTO heterostructures, while it becomes –51 meV after the addition of two FeO layers. For the latter case, the magnetic order is composed of ferromagnetic chains, antiferromagnetically coupled to each other as shown in Fig. 4.13a. Another AFM phase was found at the surface of these heterostructures [85]. However, the electrons in the AFM surface are localized by the reduction of the dimensionality and hence by the reduction of the bandwidth. In the present case, the localization of the Fe 3d electrons is due to the Fe^{2+} valence.

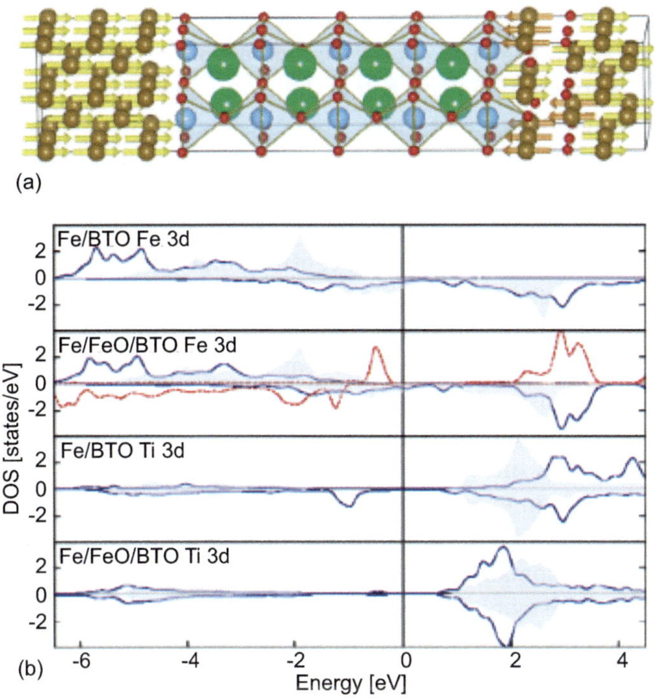

Figure 4.13 (a) Density of states resolved for Ti 3d and Fe 3d orbitals in Fe/BTO and Fe/$(FeO)_2$/BTO heterostructures. In each panel, majority and minority spin channels are plotted in the upper and lower panels respectively. Solid lines refer to the interface atoms while the dashed lines refer to the AFM coupled Fe atoms in Fe/$(FeO)_2$/BTO. The shaded plots are referred to atoms in the inner layer. The Fermi level is set to zero. (b) Ground state magnetic structure for the Fe/$(FeO)_2$/BTO system with two FeO layers. The yellow arrows indicate the spins upon Fe atoms, while the orange arrows indicate the spin down. The Ba, Ti, O, and Fe atoms are shown as green, blue, red, and brown balls respectively. Reprinted from Ref. [6], with the permission from AIP Publishing, Copyright 2014.

The exchange bias generates an effective magnetic field that increases the magnetization. This effect is observed at the interface between Fe and FeO in the Fe/$(FeO)_2$/BTO heterostructure, while it is not observed in Fe/FeO bilayers that show superparamagnetic behavior at the interfaces [86]. It is proposed that the FeO/BTO interface enhances the magnetocrystalline anisotropy, dominating over the thermal energy responsible for the superparamagnetic phase as already observed in Co/CoO bilayers [87]. As shown in Fig. 4.13b, the Fe interface is half-metallic in Fe/BTO while in Fe/$(FeO)_2$/

BTO, the AFM coupled Fe atoms show an insulating character as in bulk FeO [88]. The Ti 3d orbitals are mainly placed at 2 eV above the Fermi energy with a peak below the Fermi level in the spin down channel that produces a finite magnetic moment in Fe/BTO, while the magnetic moment on Ti is suppressed in $Fe/(FeO)_2/BTO$.

In conclusion, enhanced magnetic moment and exchange bias effects in ferroelectric–ferromagnetic $Fe/BaTiO_3$ bilayers have been observed. This reflects the presence of uncompensated spins in the Fe–BTO interface. Polarized neutron reflectometry measurements reveal that this effect originates from the presence of antiferromagnetic FeO layers at the interface between ferromagnetic Fe and $BaTiO_3$. First-principles density functional calculations support the formation of antiferromagnetic FeO layers at the interface, while the absence of FeO renders the system ferromagnetic. The enhanced magnetic moment of Fe in experiments is also observed in theory due to the moment induced by FeO at the interface. One can envisage that as ferroelectric domain structure in BTO is manipulated by an electric field, electrical manipulation of exchange bias would also be possible in such systems.

From the above examples, it is evident that the oxide heterostructures provide a unique playground for designing new functionalities. The manipulation of properties at the interface utilizing spin, charge, orbital, and lattice degrees of freedom gives us enormous possibilities to control in a precise way many different physical properties crucial for technological innovations such as oxide based electronics. It is expected that more and more novel functionalities and a better understanding of the phenomena will emerge with time due to the availability of advanced experimental techniques and microscopic theories.

References

1. H. Y. Hwang, Y. Iwasa, M. Kawasaki, B. Keimer, N. Nagaosa, and Y. Tokura, *Nat. Mater.*, **11**, p. 103 (2012).
2. U. Luders, et al., *Phys. Rev. B*, **80**, 241102 R (2009).
3. B. R. K. Nanda, S. Satpathy, and M. S. Springborg, *Phys. Rev. Lett.*, **98**, 216804 (2007).
4. H. Boschker, et al., *Phys. Rev. Lett.*, **109**, 157207 (2012).

5. M. Gibert, P. Zubko, R. Scherwitzl, J. Iriiguez, and J. M. Triscone, *Nat. Mater.,* **11**, p. 195, (2012).
6. A. Paul, C. Reitinger, C. Autieri, B. Sanyal, W. Kreuzpaintner, J. Jutinioosikni, et al., *Appl. Phys. Lett.,* **105**, 022409 (2014).
7. F. Bern, M. Ziese, I. Vrejoiu, X. Li, and P. A. van Aken, *New J. Phys.,* **18**, 053201 (2016).
8. N. Nagakawa, H. Y. Hwang, and D. A. Muller, *Nat. Mater.,* **5**, p. 204 (2006).
9. P. Hohenberg and W. Kohn, *Phys. Rev.,* **136**, B864 (1964).
10. W. Kohn and L. J. Sham, *Phys. Rev.,* **140**, A1133 (1965).
11. S. Roy, C. Autieri, B. Sanyal, and T. Banerjee, *Sci. Rep.,* **5**, 15747 (2015).
12. C. Autieri, *J. Phys. Condens. Matter.,* **28**, p. 42 (2016).
13. S. R. Lee, M. S. Anwar, Y. J. Shin, M. C. Lee, Y. Sugimoto, M. Kunieda, et al., Preprint available at https://arxiv.org/abs/1609.03010.
14. C. Autieri, P. Gentile, and M. Cuoco. Preprint
15. G. Koster, et al., *Rev. Mod. Phys.,* **84**, p. 253 (2012).
16. S. Parui, P. S. Klandermans, S. Venkatesan, C. Scheu, and T. Banerjee, *J. Phys. Condens. Matter,* **25**, 445005 (2013).
17. W. J. Kaiser and L. D. Bell, *Phys. Rev. Lett.,* **60**, p. 1406 (1988).
18. K. G. Rana, T. Yajima, S. Parui, A. F. Kemper, T. P. Devereaux, Y. Hikita, et al., *Sci. Rep.,* **3**, p. 1274 (2013).
19. T. Banerjee, J. C. Lodder, and R. Jansen, *Phys. Rev. B,* **76**, 140407(R) (2007).
20. M. Prietsch, *Phys. Rep.,* **253**, p. 163 (1995).
21. D. Toyota, et al., *Appl. Phys. Lett.,* **87**, 162508 (2005).
22. J. Xia, et al., *Phys. Rev. B,* **79**, 140407(R) (2009).
23. Y. J. Chang, et al., *Phys. Rev. Lett.,* **103**, 057201 (2009).
24. C. S. Alexander, et al., *Phys. Rev. B,* **60**, R8422 (1999).
25. E. Gorelov, M. Karolak, T. O. Wehling, F. Lechermann, A. I. Lichtenstein, and E. Pavarini, *Phys. Rev. Lett.,* **104**, 226401 (2010).
26. X. Ke, et al., *Phys. Rev. B,* **84**, 201102 (2011).
27. J. H. Jung, et al., *Phys. Rev. Lett.,* **91**, 056403 (2003).
28. H. T. Jeng, et al., *Phys. Rev. Lett.,* **97**, 067002 (2006).
29. K. Gupta, B. Mandal, and P. Mahadevan, *Phys. Rev. B,* **90**, 125109 (2014).
30. G. Rijnders, et al., *Appl. Phys. Lett.,* **84**, p. 505 (2004).

31. R. Matzdorf, Z. Fang, Ismail, J. Zhang, T. Kimura, Y. Tokura, et al., *Science,* **289**, p. 746 (2000).
32. C. Etz, et al., *Phys. Rev. B,* **86**, 064441 (2012).
33. A. T. Zayak, et al., *Phys. Rev. B,* **74**, 094104 (2006).
34. M. Shishkin and T. Ziegler, *Surf. Sci.,* **606**, pp. 1078–1087 (2012).
35. Z. Q. Li, et al., *Phys. Rev. B,* **58**, p. 8075 (1998).
36. M. Q. Cai, et al., *J. Chem. Phys.,* **124**, 174701 (2006).
37. C. Autieri, M. Cuoco, and C. Noce, *Phys. Rev. B,* **89**, 075102 (2014).
38. K. Maiti, *Phys. Rev. B,* **73**, 235110 (2006).
39. A. T. Zayak, et al., *Phys. Rev. B,* **77**, 214410 (2008).
40. J. M. Rondinelli, et al., *Phys. Rev. B,* **78**, 155107 (2008).
41. M. Verissimo-Alves, et al., *Phys. Rev. Lett.,* **108**, 107003 (2012).
42. M. Gu, Q. Xie, X. Shen, R. Xie, J. Wang, G. Tang, et al., *Phys. Rev. Lett.,* **109**, 157003 (2012).
43. C. W. Jones, et al., *Acta Crystallogr. Sect. C: Cryst. Struct. Commun.,* **45**, p. 365 (1989).
44. O. Friedt, et al., *Phys. Rev. B,* **63**, 174432 (2000).
45. B. Hu, G. T. McCandless, M. Menard, V. B. Nascimento, J. Y. Chan, E. W. Plummer, et al., *Phys. Rev. B.* **81**, 184104 (2010).
46. J. Nogues and I. K. Schuller, *J. Magn. Magn. Mater.,* **192**, p. 203 (1999).
47. K. Ueda, H. Tabata, and T. Kawai, *Science,* **280**, p. 1064 (1998).
48. H. Wadati, J. Okamoto, M. Garganourakis, V. Scagnoli, U. Staub, Y. Yamasaki, et al., *Phys. Rev. Lett.,* **108**, 047203 (2012).
49. D. Okuyama, S. Ishiwata, Y. Takahashi, K. Yamauchi, S. Picozzi, K. Sugimoto, et al., *Phys. Rev. B,* **84**, 054440 (2011).
50. F. Jimenez-Villacorta, J. A. Gallastegui, I. Fina, X. Marti, and J. Fontcuberta, *Phys. Rev. B,* **86**, 024420 (2012).
51. C. He, A. J. Grutter, M. Gu, M. D. Browning, Y. Yakamura, B. J. Kirby, et al., *Phys. Rev. Lett.,* **109**, 197202 (2012).
52. J. I. Hong, T. Leo, D. J. Smith, and A. E. Berkowitz, *Phys. Rev. Lett.,* **96**, 117204 (2006).
53. H. Ohldag, A. Scholl, F. Nolting, E. Arenholz, S. Maat, A. T. Young, et al., *Phys. Rev. Lett.,* **91**, 017203 (2003).
54. H. Ohldag, H. Shi, E. Arenholz, J. Stohr, and D. Lederman, *Phys. Rev. Lett.,* **96**, 027203 (2006).
55. M. Gruyters, and D. Schmitz, *Phys. Rev. Lett.,* **100**, 077205 (2008).

56. U. Nowak, K. D. Usadel, J. Keller, P. Miltenyi, B. Beschoten, and G. Güntherodt, *Phys. Rev. B,* **66**, 014430 (2002).

57. C. Zandalazini, P. Esquinazi, G. Bridoux, J. Barzola-Quiquia, H. Ohldag, and E. Arenholz, *J. Magn. Magn. Mater.,* **323**, 2892 (2011).

58. A. Paul, C. Zandalazini, P. Esquinazi, C. Autieri, B. Sanyal, P. Korelis, et al., *J. Appl. Crystallogr.,* **47**, p. 1054 (2014).

59. E. C. Passamania, C. Laricaa, C. Marquesa, J. R. Provetia, A. Y. Takeuchib, and F. H. Sanchez, *J. Magn. Magn. Mater.,* **299**, p. 11 (2006).

60. R. K. Zheng, G. H. Wen, K. K. Fung, and X. X. Zhang, *J. Appl. Phys.,* **95**, p. 5244 (2004).

61. J. de la Venta, M. Erekhinsky, S. Wang, K. G. West, R. Morales, and I. K. Schuller, *Phys. Rev. B,* **85**, 134447 (2012).

62. S. J. Yuan, et al., *Phys. Rev. B,* **88**, 024413 (2013).

63. C. Autiei and B. Sanyal, *New J. Phys.,* **16**, 113031 (2014)

64. C. Ma, Z. Yang, and S. Picozzi, *J. Phys. Condens. Matter,* **18**, p. 7717 (2006).

65. G. Colizzi, A. Filippetti, and V. Fiorentini, *Phys. Rev. B,* **76**, 064428 (2006).

66. A. Hoffmann, et al., *Phys. Rev. B,* **66**, 220406(R) (2002).

67. S. Picozzi, K. Yamauchi, G. Bihlmayer, and S. Blügel, *Phys. Rev. B,* **74**, 094402 (2006).

68. K. Yamauchi, F. Freimuth, S. Blügel, and S. Picozzi, *Phys. Rev. B,* **78**, 014403 (2008).

69. Landolt-Bornstein, Vol. III/19a, ed. by H.P.J. Wijn (Springer Berlin 1986).

70. B. D. Cullity and C. D. Graham, *Introduction to Magnetic Materials,* John Wiley & Sons, Hoboken (2005).

71. G. H. O. Daalderop, P. J. Kelly, and M. F. H. Schuurmans, *Phys. Rev. B,* **41**, 11919 (1990).

72. Y. Suzuki, et al., *J. Appl. Phys.,* **83**, p. 7064 (1998).

73. W. H. Meiklejohn and C. P. Bean, *Phys. Rev.,* **102**, p. 1413 (1956)

74. H. Zheng, J. Wang, S. E. Lofland, Z. Ma, L. Mohaddes-Ardabili, T. Zhao, et al., *Science,* **303**, p. 661 (2004).

75. N. Reyren, S. Thiel, A. D. Caviglia, L. Fitting Kourkoutis, G. Hammerl, C. Richter, et al., *Science,* **317**, p. 1196 (2007).

76. J. Chakhalian, J. W. Freeland, H. U. Habermeier, G. Cristiani, G. Khaliullin, M. Veenendaalvan, et al., *Science,* **318**, p. 1114 (2007).

77. S. M. Wu, S. A. Cybart, P. Yu, M. D. Rossell, J. X. Zhang, R. Ramesh, et al, *Nat. Mater.,* **9**, p. 756 (2010).
78. M. Bibes, J. E. Villegas, and A. Barthélémy, *Adv. Phys.,* **60**, p. 584 (2011).
79. C. G. Duan, et al., *Phys. Rev. Lett.,* **97**, 047201 (2006).
80. D. Di Sante, et al., *J. Phys. Condens. Matter,* **25**, 066001(2013).
81. L. Bocher, et al., *Nano Lett.,* **12**, pp. 376–382 (2012).
82. G. R.adaelli, et al., *Nat. Commun.,* **5**, p. 3404 (2014).
83. M. Cantoni, C. Autieri, S. Picozzi, M. Asa, C. R.inaldi, R. Bertacco. Preprint.
84. S. Valencia, et al., *Nat. Mater.,* **10**, p. 753 (2011).
85. M. Fechnr, et al., *Phys. Rev. B,* **78**, 212406 (2008).
86. X. Lin, et al., *J. Appl. Phys.,* **76**, p. 6543, (1994).
87. V. Skumryev, et al., *Nature,* **423**, p. 850 (2003).
88. M. Cococcioni, et al., *Phys. Rev. B,* **71**, 035105 (2005).

Chapter 5

Domain Walls in Multiferroic Materials and Their Functional Properties

Saeedeh Farokhipoor
Department of Materials Science & Metallurgy, University of Cambridge, 27 Charles Babbage Road, Cambridge CB3 0FS, UK
s.farokhipoor@rug.nl

5.1 Introduction to Ferroic and Multiferroic Materials

5.1.1 Ferroics

Ferroics are materials that display spontaneous ordering of a physical quantity below a certain temperature, the so-called *Curie temperature*, T_C, or, in general, *ordering temperature*, T_0. Three types of ferroic materials have been known for a long time: "ferroelectrics" present spontaneous polarization, "ferromagnets" show spontaneous magnetization and "ferroelastics" display spontaneous strain. Thermodynamically, these are the complementary variables of the ordering electric, magnetic, and

Oxide Spintronics
Edited by Tamalika Banerjee

ISBN 978-981-4774-99-4 (Hardcover), 978-0-429-46819-3 (eBook)
www.jennystanford.com

strain fields, respectively. The most interesting behavior takes place when the appearance of spontaneous polarization, magnetization or strain is associated with the loss of one or more symmetry elements while keeping some others unaffected, that is, when the transition takes place in between two phases that are related by symmetry by a group-subgroup relationship. Then, within the Landau approach, the emergent physical properties are called the *order parameters* of the ferroelectric, ferromagnetic, or ferroelastic phase transition, respectively. Ferroics can be also classified by their response under time and spatial reversal, as presented in Fig. 5.1. The essential and distinct characteristic of a ferroic material is the formation of domains, or regions with different orientations of the order parameter, which will be discussed in more details in Section 5.3.

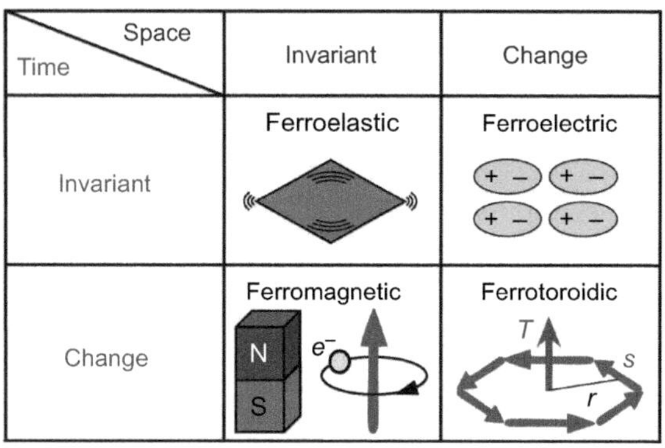

Figure 5.1 All forms of ferroic materials under time and spatial reversal investigation. Reprinted from Ref. [1], with permission from Springer Nature, Copyright 2007.

In recent years, the so-called *multiferroic* materials, those which display ferroelectricity and magnetism in the same phase, have obtained a lot of attention because of their potential to present large magnetoelectric effect. This is of fundamental interest, but also of technological importance because it would allow to control the magnetization (polarization) of devices with an electric (magnetic) field. According to symmetry arguments and using a Landau approach, the magnetoelectric terms in the Gibbs free energy would be

$$\Phi = \Phi_0 - \alpha_i k E_i H_k - 1/2\beta_{ijk} E_i H_k - 1/2\gamma_{ijk} H_i E_j E_k$$

and thus, for the magnetoelectric effect to be present, that is the magnetoelectric tensor being different from zero, the material has to be invariant under both space and time reversal. A multiferroic that is ferroelectric and ferromagnetic in the same phase fulfils this requirement because the polarization and the electric field are invariant under time reversal, but not under space reversal and the opposite is true for the magnetization and the magnetic field and thus the magnetoelectric terms survive in the Gibbs free energy. Moreover, since both the dielectric and magnetic susceptibilities of ferroelectrics and ferromagnets, respectively, are the largest of all materials, the largest intrinsic magnetoelectric constants are expected among these families of materials [1].

The current general interest in multiferroics was triggered by a paper published in 2000 by Hill/Spaldin [2]. In this review, mainly multiferroics were discussed. The cause of their scarcity was reported to be that ferroelectrics basically require empty d-shells, whereas magnetism requires partially filled d (or f) shells. The most direct way to circumvent these contradicting requirements can be to consider a "perovskite" structure providing ferroelectricity from the *A*-site cation and magnetism from the *B*-site cation, e.g., $Bi(Cr,Fe)O_3$ [3] or $Pb(Fe_{1/2}Nb_{1/2})O_3$ [4], owing their ferroelectricity due to lone-pair X^{3+} ion in the *A*-site. But even if this difficulty can be overcome, in most of these cases, the magnetic and electrical properties are not coupled due to large differences in associated energies (reflected in very different ferromagnetic and ferroelectric ordering temperatures). In other words, even though multiferroics could in principle offer the largest magnetoelectric couplings, few multiferroics are magnetoelectrics. In fact, an ideal material that is a robust ferroelectric, a strong ferromagnet, and displays a large magnetoelectric coupling, all in the same phase and at room temperature, has not been found yet. A way to ensure strong magnetoelectric coupling has been seeking for compounds in which the breaking of space inversion symmetry is induced by a particular magnetic structure. This is the case in conical or spiral spin structures (e.g., $TbMnO_3$ or $CoCr_2O_4$). The issue in this case is the low multiferroic ordering temperatures.

Although large piezoelectric and dielectric response, interesting magnetic properties and extensive first principle calculation in oxides, have generated the tendency to focus on the multiferroic oxides family. But Scott pointed out that apriori there is no reason to anticipate the most exciting physics for device application in oxides and that high-temperature transition multiferroic temperatures are expected by turning to non-oxide materials [5], (also see Refs. [6, 7] for multiferroic fluorides). More studies on fluorides and oxyfluorides families have been reviewed by Palstra et al. [8]. However, experimental convenience during oxide growth, specifically, the feasibility of obtaining high-quality specimens in thin film form and the possibility of strain engineering have kept the field, mainly focused on oxides. We now discuss ferroelectrics in some more detail and present the general features of ferromagnets, ferroelastics, and ferrotoroids.

5.1.1.1 Ferroelectricity

Ferroelectrics are a particular type of dielectrics that show polarization even without an applied field, so-called *remanent polarization*. In ferroelectrics, the polarization as a function of the applied field shows non-linear and hysteresis behavior (very much like in ferromagnets) and thus the existence of a binary state (double potential well) at zero field and below their ordering temperature, shown in Fig. 5.2, presenting the hysteresis loop in bulk single crystal of BiFeO3 [9]. For comparison purpose, the same type of characteristic can be found in thin film form in Ref. [10]. The macroscopic polarization arises from the parallel alignment of ionic or molecular dipoles in existing in all unit cells of the material. The existence of a dipole in the unit cell is only allowed in certain crystals that lack inversion symmetry. The definition of a ferroelectric, which is based solely on symmetry arguments has recently been discussed by Setter et al. [11]. With increasing temperature, the dipoles can fluctuate and partially or totally counterbalance the dipolar ordering. At sufficiently large temperatures, the dipoles become randomly oriented and the macroscopic polarization disappears. The long-range ordering temperature in ferroelectrics is called *ferroelectric Curie temperature*. Above this, the material becomes paraelectric. This temperature dependent behavior of the macroscopic polarization in a polar dielectric material is called *the pyroelectric effect*, being the pyroelectric response the largest at the phase transition. Besides, ferroelectrics can be easily polarized, giving rise to the largest possible dielectric susceptibilities, which makes them distinguishable from other dielectrics in nature. The dielectric permittivity also peaks at the temperature that signals the ferroelectric–paraelectric transition. A dielectric anomaly can also occur at a ferroelectric–ferroelectric transition, both phases involving polar modes with different symmetries, which are also common. An example of the latter is the orthorhombic–rhombohedral transition of the archetypal perovskite ferroelectric $BaTiO_3$ [12, 13]. In the case of the nowadays very popular $BiFeO_3$, the ferroelectric ordering temperature occurs at 825 °C and it is a first-order (discontinuous, involving latent heat and thermal hysteresis) paraelectric–ferroelectric phase transition, accompanied by a cubic-rhombohedral structural phase transition. Showing pyroelectric behavior is a necessary requirement for ferroelectricity, but on the other hand, not all of the pyroelectrics are ferroelectric except for those that can switch their polarization under an external electric field. Moreover, ferroelectrics are of interest for storage (statics), but also for transducers (dynamics). They are used as non-volatile RAM memories in low memory applications like ID cards. Moreover, all ferroelectrics are piezoelectrics. This makes them a good candidate for transforming mechanical energy into electrical energy and vice versa. They can be used in memory function (ferroelectric RAM for computers and RFID cards), infrared camera, ultrasound generators (in sonars or medical apparatuses), gyroscope, thermistors, actuators, fuel injection on diesel engines, vibration dampers (in helicopter blades or cars), micropositioners (in microscopes), voltage generators (in gas lighters), shock detectors (in airbags), and many others.

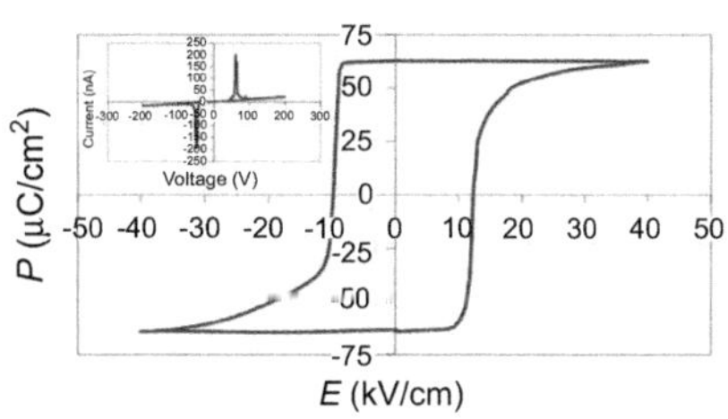

Figure 5.2 Hysteresis loop of the bulk single crystal of $BiFeO_3$ at room temperature. Reprinted from Ref. [9], with permission from AIP Publishing, Copyright 2007.

Classically, ferroelectrics have been classified into two large classes: displacive and order–disorder ferroelectrics. In *displacive ferroelectrics,* the ions that shift below the ordering temperature to form the dipole, remain in the high symmetry position above the ordering temperature. In *order–disorder ferroelectrics,* the displacement of the cations in the paraelectric phase also averages to zero, such that there is no net macroscopic polarization, however locally and/or instantly they are displaced from the centrosymmetric position. It is now understood that all ferroelectrics share both displacive and order–disorder character in more or less degree depending on the compound and the nature of the interactions [14]. The para–ferro phase transition can be described from the dynamical point of view using the soft-mode picture: an ion vibrating in the paraelectric phase around the high symmetry position will decrease its vibration frequency as the transition temperature is approached. Eventually, the frequency of the lowest energy transversal optical phonon reaches zero and there is no restoring force for the ion to return to the equilibrium position of the paraelectric phase. At that temperature, the ion takes a new (and non-centrosymmetric) equilibrium position and the phase transition to a polar phase has taken place [15–17].

The design, prediction, and understanding of new materials by theoretical calculation have also played a crucial role in focusing the interest in chemically and structurally simple materials. $PbTiO_3$, $BaTiO_3$, $TbMnO_3$, $BiFeO_3$, and related materials are very popular examples of this family in which the origin of ferroelectricity is best understood [18]. The origin of the appearance of ferroelectricity in oxide family can be explained by different mechanisms. The family of metal oxides (perovskites) typically shows displacive ferroelectricity and soft-mode behavior. This happens due to their covalent bonding interactions, e.g., in $BaTiO_3$ (the electron density of the filled oxygen 2p orbitals partially transfers to an empty Ti^{4+} $3d^0$ orbital on the octahedral *B*-site, which gives rise to a polar lattice distortion). Another type of displacive distortion is induced by the Pb and Bi lone pairs, as in $PbTiO_3$ or $BiFeO_3$ [18–21]. In this case, the active ns^2 lone pair in *A*-site plays the dominant role resulting in off-centered distortions that give rise to a permanent polarization. In general, there seems to be always an order–disorder component even in these mainly displacive ferroelectrics and this seems to be stronger in the Bi and Pb oxides in which multiple-valley potentials are still present above T_C [22].

Besides, "proper" ferroelectric with polarization as a primary order parameter (discussed up to now), there is another category of ferroelectrics, so-called *improper ferroelectrics*, for which polarization is a secondary order parameter, occurring as a result of a primary symmetry breaking of a different character. An example of this is the inversion symmetry breaking induced by the magnetic spiral spin structure is some antiferro- or ferri-magnets, such as $TbMnO_3$, $TbMn_2O_5$, or $CoCr_2O_4$. In another category of metal oxides, the hexagonal manganites, a different and less usual type of ferroelectricity is observed, giving rise to the so-called *geometric ferroelectrics*, which are driven by electrostatic interaction and ionic size differences causing ionic site shifts associated with the polyhedral tilt. As we will discuss later in Section 5.3, this family has very distinguishable properties regarding their domains and domain walls.

5.1.1.2 Ferromagnetism

Ferromagnetic materials are materials exhibiting a long-range magnetic ordering at the atomic level. This causes domains in which unpaired electron spins line up parallel with each other. The magnetic field in the domain is large, but the bulk material will be unmagnetized because the domains are randomly oriented through the material. The domains can be lined up with each other through an externally applied magnetic field, causing the material to be magnetized. A part of the magnetization can remain to some extent even after the removal of the external magnetic field, which is defined as *remanent magnetization*. During increasing and decreasing the applied magnetic field, the material undergoes the so-called *magnetic hysteresis loop*. Because of that, as in ferroelectrics, ferromagnets can display two different states of antiparallel magnetization at zero applied magnetic field and thus can be used as binary bits of information. Ferromagnets also have a temperature where the long-range ordering appears with decreasing temperature, also called the *Curie temperature*. In some materials, the spins line up either antiparallel with each other or follow a certain symmetric pattern (sinusoidal, spiral, etc.) such that the average magnetization in a domain is zero. Such materials are called *antiferromagnets*. In general, the response of ferromagnets to the magnetic field is more complex than that of ferroelectrics to the electric field. Ferromagnets with a very strong anisotropy are rare and they have typically much broader domain walls, stronger pinning effects and more complex domain formation than general ferroelectrics [23–25].

5.1.1.3 Ferroelasticity

The spontaneous strain is the determining order parameter in ferroelastic materials [26]. Unlike ferroelectrics or ferromagnets, their order parameters are not vectors, but second rank tensors. Twins (or 90° domains) can form in pure ferroelastic material, such as non-polar $CaTiO_3$ with 90° domains with ferrielectric twin walls [27, 28]. The twins also form in ferroelastic–ferroelectric materials, such as tetragonal $PbZr_{1-x}Ti_xO_3$ (PZT) or $PbTiO_3$ with two perpendicular strain components [26]. In the case of rhombohedral $BiFeO_3$, due to the coupling between polarization and strain, all the non-180° domain walls (109° and 71° domain walls) are both ferroelectric as well as ferroelastic. Since domain walls are regions of symmetry breaking, they offer additional possibilities for multiferroic behavior. A complete study on the possible coupling between ferroelectric and ferroelastic order parameters (those that satisfy the rules of mechanical and electrical compatibility among adjacent domains) has been provided by Fousek and Janovec [29, 30]. This issue will be further discussed in Section 5.5.2 (domain walls).

5.1.1.4 Ferrotoroidicity

As mentioned earlier, ferroelectrics, ferromagnets, and ferroelastics have been extensively studied because of their potential applications. However, there is one more type of ferroic ordering, which has been named explicitly by Fiebig et al. in 2007 for the first time, named "ferrotoroidicity" [1]. Ferrotoroidic is a vortex of magnetic moments, which can generate a toroidal moment [31]. It is worth noting that due to the presence of the spin vortex-like state, the toroidic moment by itself is asymmetric under both time and space reversal [32], hence, the linear magnetoelectric effect should be symmetry allowed. Although they provide non-zero components of linear magnetoelectric tensor [1], they can still be almost zero due to interaction strengths. Nevertheless, recent interesting theoretical and experimental reports have focused on revealing this type of ordering in existing or novel materials [33–35]. In the next section, the effect of epitaxial strain on the ferroic family will be discussed.

5.2 Ferroelectrics and Multiferroics in Thin Film Form

In modern technology, such as in solar cells, LCD displays, memories, and other applications that require device integration, thin films of the active layers are preferred to the bulk form. Thin films of ferroelectric and multiferroic materials are of great interest in storage devices (including ferroelectric FETs, so-called FeFETs), micro-actuators, infrared sensors, electro-optical devices, spin valves, and several others. In some of these applications, the thickness of the thin layer is in the micrometer range, thicknesses at which the thin film properties are very similar to those of the bulk material. In these cases, the main issues related to thin film deposition are associated to the control of the chemistry at the interfaces with the substrate or the electrodes in order to avoid dead-layers, second magnetic phases or spurious effects of any kind. However, for other applications that require larger precision during integration (as memory cells, spin valves or FeFETs), the interfaces need to be extremely flat, which becomes easier as the layers are thinner and are able to follow the substrate structure throughout the whole thicknesses, while being grown one atomic layer at a time. Moreover, for spin valves or FeFETs, the device requires that the layers are just a few nanometers thick. In this case, the structure of the thin films can adapt to the substrate, if this is chosen with a lattice parameter differing only slightly from that of the film. The stress that originates from the mismatch between the lattice parameters of the film and the substrate, causes the lattice deformation (strain) in the films. Therefore, for the growth of very flat films, it is most often necessary to deal with the effects of epitaxial strain on the thin film properties. Most interestingly, one can deliberately use the effect of epitaxial strain to impose structural changes in a material and, hence, to modify their properties. This is particularly effective in the case of ferroelectric and magnetic materials, whose defining interactions are largely dependent on the atomic structure and unit cell sizes. In order to take advantage of epitaxial strain, the thin film thickness should stay below a certain value, known as *critical thickness* for strain relaxation. The larger the lattice mismatch, the smaller the film thickness needed to keep the imposed strain before strain relaxation occurs.

Applied stress and the resulting induced strain in the material cost energy, which is proportional to the film thickness. As the film thickness grows, the strain energy competes with the energy needed to relax the lattice by means of dislocations, disclinations, atomic vacancies, or crystallographic domains and domain boundaries [36, 37]. Depending on the type of materials and the kinetics involved during the growth, either one or a combination of the mentioned mechanisms take place. While in Si- or Ge-based semiconductors, dislocations are most common, in oxides oxygen vacancies are often formed to relax the lattice. Moreover, in ferroelastic materials and in particular in ferroelastic perovskites, twin formation is energetically less costly than the formation of dislocations [26 and references therein]. The density of chemical and structural defects present after total or partial relaxation depends, thus, on the lattice mismatch and thickness. It is important to mention that the temperature dependence of the lattice mismatch plays an important role and needs to be taken into account when investigating the relaxation process. Since in ferroelectric materials, the thermal expansion is unusual (often negative thermal expansion is observed close to the ferroelectric–paraelectric phase transition), strain relaxation can be complex and highly dependent on the differences between the growth temperature and the Curie temperature. This gives rise to a more intricate relation between misfit strain and critical thickness for strain relaxation than in the case of non-polar semiconductors [38].

In a "fully relaxed" film, the grown material is analogous to the bulk structure. In other words, fully relaxed films are those in which the misfit stress or lattice mismatch, is too large to be accommodated by lattice strain. Relaxation can happen, while keeping partial coherence with the substrate, via dislocations or domain walls in the thin films. Dislocations and domain walls can display certain periodicity, giving rise to diffuse scattering in X-ray and neutron diffraction (see Fig. 5.3). The appearance of these relaxation features introduces local disorder, which locally changes the strain field and hence, hinders the homogeneous intrinsic strain effect.

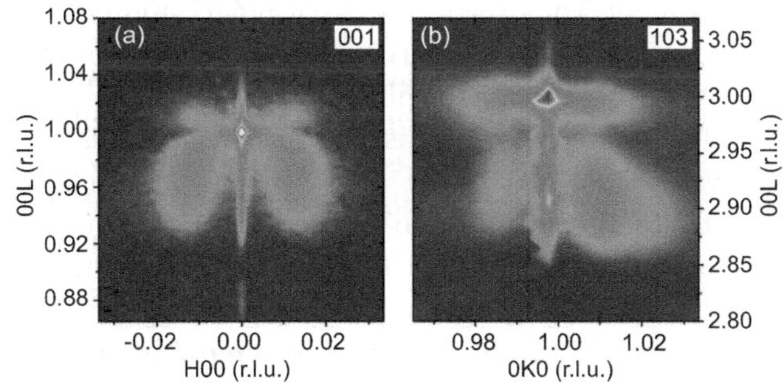

Figure 5.3 Logarithmic reciprocal maps in (a) 001 scattering plane and (b) 103 reflections for a 5 nm $PbTiO_3$ film directly grown on a $DyScO_3$ substrate under tensile strain, showing a modulation in reciprocal space indicating stripe domains with the polarization rotated away from the tetragonal axis. Reprinted from Ref. [39], with permission from American Physical Society, Copyright 2006.

5.2.1 Ferroelectrics and Ferromagnet Sunder Epitaxial Strain

The application of ferroelectric thin films for memory devices raised fundamental questions related to how and why the ferroelectric properties depend on the thickness of the thin films [40–46]. Epitaxial strain allows applying stress values an order of magnitude higher than what can be applied mechanically (using a press or diamond anvil) and chemically (by cation substitution) in bulk form. This can produce significant changes in the crystal structure and thus, epitaxial strain can be used to pronouncedly improve the ferroic oxides properties. Moreover, epitaxial strain modifies (typically lowers) the symmetry of materials by adding new symmetry elements according to the directions of the stress tensor. In the case of ferroelectrics, this symmetry lowering has important consequences for the dielectrics properties: rhombohedral $BiFeO_3$, with space group R3c and tetragonal $PbTiO_3$, with space group P4mm, both become monoclinic (space group Cm) upon being epitaxially strained [47–49]. This monoclinic phase is reminiscent of that found for the high strain large piezoelectric compositions of $PbZr_{1-x}Ti_xO_3$ (PZT), the most technologically important piezoelectric [50, 51]. Monoclinic ferroelectrics with space groups Cm and Pm [52] lose their polar axes and retain a mirror plane as the only symmetry element. Thus, the ferroelectric polarization is not forced by symmetry to align along a particular direction, but instead, it is free to rotate within the mirror plane [50, 53, 54]. This freedom to rotate has been shown to increase the piezoelectric responses of oxide perovskites [55, 56].

According to Rossetti and co-workers [57], a monoclinic Pm or Cm ferroelectric phase is not thermodynamically stable, but it can become stabilized by strain or electric field. The monoclinic compositions in PZT ceramic samples are found at the boundary between the rhombohedral R3c and tetragonal P4mm phases. In dense ceramics, the coexistence of crystallites of these two phases imposes the elastic energy and induces the monoclinic distortion as a way to accommodate the internal stresses. A similar mechanism is in place in single crystal thin films when the R3c (e.g., $BiFeO_3$) or P4mm (e.g., $PbTiO_3$) ferroelectrics are subjected to epitaxial strain, as described above, opening the door to design large strain piezoelectrics.

Another way of improving the ferroelectric properties via strain is offered by the possibility of tuning the ordering temperature up to surprisingly large values.

In thin films of the lead-free ferroelectric $BaTiO_3$ under epitaxial strain, an enhancement of the ferroelectric Curie temperature by 500 °C is reported, while the remanent polarization is increased by 250% [58]. Even more extraordinarily, epitaxial strain can also be used to induce ferroic order in a non-ferroic material: epitaxial strain on thin films of $SrTiO_3$, a quantum paraelectric, gives rise to a ferroelectric–paraelectric transition located very close to room temperature [59–62].

Achieving high values of strain (6.7% fo$BiFeO_3$ on $YAlO_3$ substrates and 4.5% for $BiFeO_3$ on $LaAlO_3$ substrates) [62, 63] can induce crystal deformations never observed before and monoclinic phases with extremely large c/a ratios, so-called *the supertetragonal phase*. In $BiFeO_3$ on $LaAlO_3$ and $YAlO_3$, the supertetragonal phase and the pseudo-rhombohedral phase (a monoclinic phase with a unit cell very close to the rhombohedral unit cell of the bulk material but with a slightly elongated *c*-axis and an orthogonal in-plane lattice) both have very different c/a ratios and compete in stability [64, 65]. This has made it possible to achieve a mixed phase with periodic stripes alternating the two phases at the nanoscale. The stripes, in turn, exist confined within narrow and long bands. These stripes are best visible during atomic force imaging due to the large height contrast between the two coexisting phases. It was found that it is possible to write and erase those stripes with the electric field [66, 67]. This produces large differences in the average film height when a bias voltage is applied to a metallic AFM tip in physical contact with the sample surface (a technique known as *piezoforce microscopy*), giving rise to a huge average film deformation and a very large effective piezoelectric response [63]. Interestingly, this is evocative (albeit different) of the martensitic-like theory of Khatchaturyan et al. [68] to explain the monoclinic phases observed for PZT compositions showing the larger piezoelectric responses. In this model, very dense tetragonal domains form close to the phase boundary with the rhombohedral phase. Due to the decrease in anisotropy, domain wall formation energy is low and domains are of nanometer sizes. The twinning that arises in order for the 90° domains to be able to share the equivalent (101) planes make a combination of twined tetragonal domains similar to a monoclinic homogeneous phase and it is not possible to resolve the difference with the standard diffraction techniques [69].

This is not only specified for ferroelectrics, in superlattices of Dy with non-magnetic Lu, the ferromagnetic Curie temperature could be increased by a factor of two [70]. In the case of $SrRuO_3$ on a $SrTiO_3$ substrate, by applying in-plane strain of about 0.67%, the ferromagnetic Curie temperature and saturation magnetic moment were enhanced by 7% and 20%, respectively [71]. In the case of LaCoO3 grown on a different substrate with a various range of lattice mismatch (–1.3%–2.6%), a non-magnetic material in bulk was modified to a ferromagnetic material with a relatively elevated Curie temperature [72].

5.2.2 Multiferroics under Epitaxial Strain

As mentioned in the previous sections, the strain introduced to the system by the mismatch between the film and the substrate lattice parameters can boost the properties with respect to the bulk. The alternation in the properties can vary from change in the ferroic's Curie temperature or determination of crystal structure, which gathers two or more ferroic orders in a single phase. For instance, a huge amount of strain dictates tetragonality to $TbMnO_3$ thin films instead of its parent orthorhombic structure [73] having a direct influence on its properties. Moreover, the possibility of gathering two ferroic phases in a single phase with a very strong coupling is very interesting for application such as magnetoelectric device. Also, in other type of manganites as well as $TbMnO_3$, it has been proposed that ferromagnetic response of the thin film can be correlated with the unit cell deformation originating from applied strain [74, 75]. Owing to the inherent connection between magnetic order and ferroelectricity in orthorhombic manganites, these can pave the way to tune the magnitude of magnetoelectric coupling and their ferroelectric response. Not only, the effect of strain introduced by the choice of substrate, the way that strain is applied to the system via growth has a crucial role as well. For instance multiferroics $BiFeO_3$ and $TbMnO_3$ grown with very low laser repetition rate form different type of domains due to different type of twining [73, 76]. This type of in-plane twining determines specific stoichiometry at the domain walls, raising very unique properties at the local scale [77, 78].

In addition, materials under strain can alter from a non-ferroic state to a multiferroic one. A popular example of a non-ferroic material that turned multiferroic is $EuTiO_3$, which is an antiferromagnetic quantum paraelectric. Taking advantage of the biaxial strain, this material turned to a ferromagnet and ferroelectric at the same time [79]. Another example of pronounced influence on the properties of a material by introducing epitaxial strain has been observed in room temperature multiferroic, $BiFeO_3$. This material is a ferroelectric, ferroelastic, and antiferromagnetic simultaneously, which can show a net magnetic moment in thin film form [80]. In spite of the previous discussion, considerations have to be taken to account as strain does not always modify and enhance the properties in a positive way, as it can accommodate via change the film stoichiometry (producing cation or anion vacancies in single or cluster form), octahedral rotation or domain/domain wall formation. For example, thin films of $BiFeO_3$ on cubic $SrTiO_3$ substrates are strained such that it is not the unit cell but a combination of the different domain variants or twins that matches the substrate lattice. This is extensively discussed in Section 5.3.

5.3 Domain Formation in Ferroic and Multiferroic Materials

The presence of polarization in ferroelectrics dictates a charge density resulting in a depolarization field, which has an opposite sign as the inherent polarization. This depolarization field should be screened by external charge suppliers, to prevent canceling out initial ferroelectricity of the material [46]. Absorbents from the atmosphere on the surface of the sample [81], good electrodes or presence of dead-layer [82–84] can supply the compensation of the polar charges, therefore giving rise to ferroelectricity in very thin epitaxial films.

One intrinsic way to cancel out the depolarization field is dividing the ground polar state to extremely small regions with different polar orientations in such a way that arises to zero average polarization [85]. We call these small regions with inverse polarity, but still the same energy state "domains." However, if the capability of a ferroelectric film to break up into multiple domains comes with the ability to switch the event, the material is locally polar and can be counted as ferroelectric, as discussed in Section 5.2.1. The same applies to ferromagnetic and ferroelastic materials. Moreover, if one type of ferroic couples with one or the other two ferroics, the resultant domain is a multiferroic domain, e.g., in the case of hexagonal manganites, structural domains couple with either magnetically and/or electrically ordered domains, these domains are determined as multiferroic domains [86].

For instance, in the case of $BiFeO_3$ with its rhombohedral symmetry at room temperature, there are eight energetically equal polarization direction regions. This results in different regions in the material with eight different domains, as shown in Fig. 5. 4. In bulk, there is no way to control the domains, but in thin film form, both electrical and physical boundary conditions, such as the electrical properties of the substrate and the lattice mismatch between film and substrate or the thickness, can be tuned in a way to control the number of domains as well as preferred polarization direction inside the domains. Possible energy cost of lattice deformation, which scales linearly with the thickness explains the tunability of domain formations in thin film forms. If the elastic energy becomes too large for large enough thicknesses, the film can relax, hence, forming domains. However, creating domains and domain walls (the area between two domain) also cost energy. Therefore, the system has to balance the cost of creating a domain wall versus the energy cost of increasing the thickness of a strained layer. That balance determines the domain wall density or, in other words, the domain size (w) as a function of thickness (d). Roitburd found out that for relatively large thicknesses, the ferroelastic domain width was proportional to the square root of the thickness [87], the same dependence that Kittel found for ferromagnetic domains [88]. Although, the fast growth rate determines deviation from linear thickness dependence [89]. It has to be mentioned that in a similar way, the uncompensated surface state develop the formation of twins in ferroelastics [90]. This allows great opportunity to modify the domain formation in epitaxial thin films.

Another popular ferroelectric is $PbTiO_3$, the films accommodate themselves by alternating their shorter (compared to the substrate) and longer (compared to the substrate) lengths along the in-plane axis, resulting in the formation of a/c-domains that arrange themselves in an orthogonal fashion, and are therefore known as the 90° domains. In this case, the lattice parameter of the substrate is in between the two in-plane lattice parameters of the film. The film tries to relax the imposed strain by forming the domains, which is energetically more favorable than relaxation through lattice strain.

When polarization couples with other order parameters, more type of domain morphologies can be obtained under highly confined geometries, e.g., in perovskite ferroelectrics, which are both ferroelectric as well as ferroelastic, a ferroelectric closure like structure can be obtained, confirmed by experimental evidence as shown in Fig. 5.5 [92].

Another type is vortices which imply a flux closure structure. The ferroelectric vortices are predicted theoretically [93]. Due to their switchability, they can be a good candidate for memory devices with high-density bits [94]. Although, it is experimentally difficult to observe static ferroelectric vortices in conventional tetragonal ferroelectrics [85]. Small spontaneous strain and high domain wall energy are the two requirements for obtaining stable closure structures. One of the good candidates is $BiFeO_3$ (a room temperature ferroelectric, ferroelastic, antiferromagnetic, and at the same time, ferrodistortive) because its domain walls provide a large energy (due to coupling between polarization to antiferrodistortive and magnetic order parameters) [95] and its piezoelectric deformation is small. The coupling and controlling of several order parameters in nanoscale closure structures in thin films of $BiFeO_3$ suggests another way to high-density bits memories [96]. Figure 5.6 presents two type of closure structure in thin films of $BiFeO_3$.

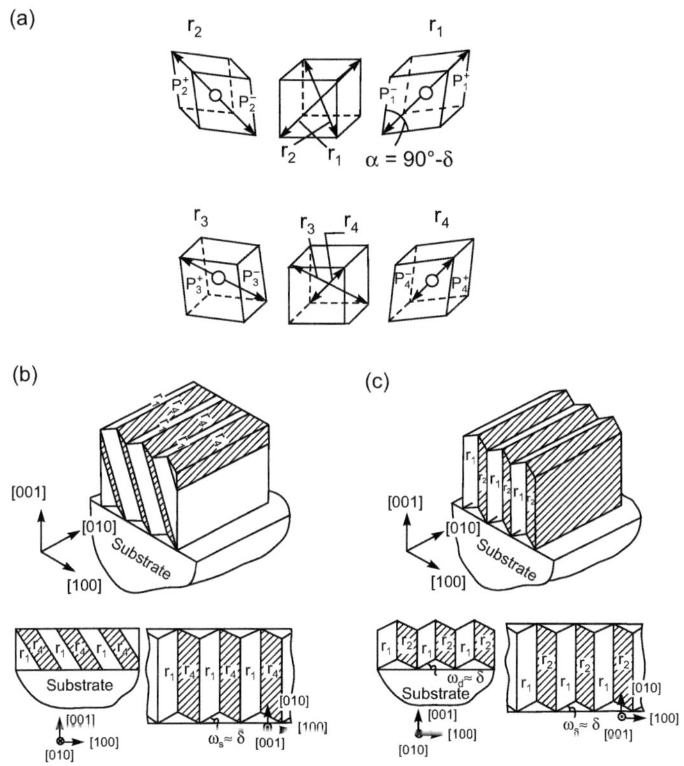

Figure 5.4 (a) Four structural (r_1, r_2, r_3, r_4) and eight polarization variants in a prototypical rhombohedral perovskite phase, (b) perspective, side, and plan view of 71° and (c) 109° DWs. Reprinted from Ref. [91], with permission from AIP Publishing, Copyright 1998.

The importance of considering the domain configuration (on top of mechanical and electrical boundary conditions) has been proven to be very crucial in calculation; e.g., if a single-domain state would be considered instead of a multiple-domain state, the calculated T_C can be an order of magnitude far from experimental data [97, 98].

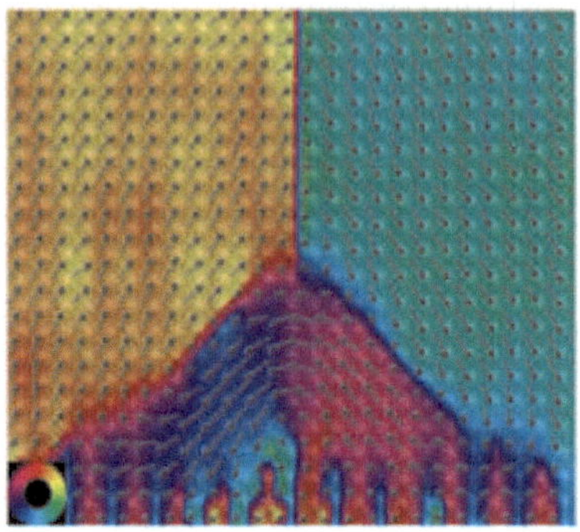

Figure 5.5 Experimental evidence in thin films of $BiFeO_3$ where the intersection between domain walls and interface shows closure-like polar arrangements. Reprinted from Ref. [92], with permission from American Chemical Society, Copyright 2011.

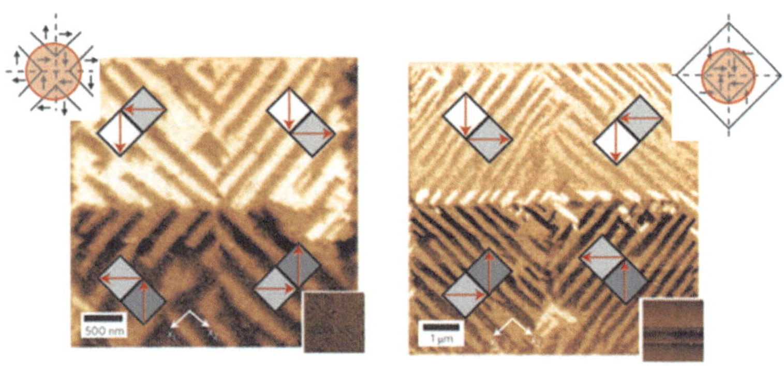

Figure 5.6 Two types of in-plane PFM image of a single closure domain in a thin film of BFO can be explained by a theoretical model in insets. Reprinted from Ref. [96], with permission from Springer Nature, Copyright 2009.

One of the considerable characteristics of these vortices in $BiFeO_3$ thin films is that they have been found to show higher conduction level compare to those of at the domain walls and in the domains [99]. The conduction property of domain walls will be discussed in more detail in Section 5.5.1. By defining the concept of domains, we can move on to the domain wall section.

5.4 Domain Walls in Ferroic and Multiferroic Materials

5.4.1 Substrates and Surface Interfaces

For the growth of high-quality metal oxide films, not only a proper substrate is important but also the way that the substrate gets treated is just as important. In principle, initiating a well-founded growth requires a specific termination and a flat surface, e.g., even different $SrTiO_3$ orientations need to be treated differently [100–102]. Notably, selective choice of one type of surface termination can give rise to conductive nanorods on an insulating surface [103] or modified substrate treatment results in different topography as well as selectivity in the type of ferroelectric domains in $BiFeO_3$ thin films for instance [104]. Additionally, one specific termination can dictate the presence or absence of 2DEG in $LaAlO_3/SrTiO_3$ heterostructure [105]. In the other approach, one can tune the physical properties by introducing artificial changes on the surface charge state, such as alternating the surface-bound charges by switching phenomenon in ferroelectrics, through different screening mechanisms [106].

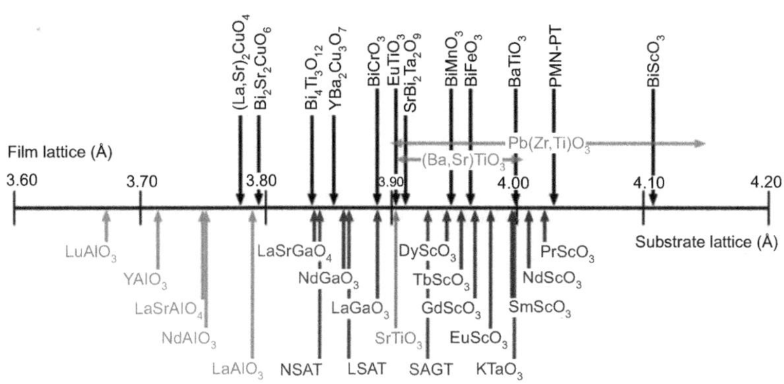

Figure 5.7 A horizontal axis presenting the lattice parameters of some perovskites and the commercially available substrates with the same range of lattice parameters, shown above and below the line, respectively.

As mentioned in the first section, except the perovskite in the big family of oxides, the other structural type of oxides (non-perovskite families; such as hexagonal manganites), fluorides, and others are potentially strong candidates for strain engineering to tailor the physical properties in the right way. But in the case of non-perovskites, lack of a proper substrate with the proper misfit is a limit. Beyond the most commonly used substrate ($SrTiO_3$), $DyScO_3$ and few more, shown in Fig. 5.7 [107], there are not that many more substrates commercially available for the wider variety of strain engineering necessary for non-perovskite family growth.

5.4.2 Domain Walls

In between two domains, there is what we call a domain wall, which in ferroelectrics is much smaller (a few unit cells) than in ferromagnetics (tens/hundreds nanometers) due to the existence of high anisotropy originating from ferroelastic order parameters.

The domain walls are very interesting for two reasons: (a) they display large strain gradients concentrated in a nanoscale region and (b) they break the crystal symmetry, which is the origin of distinguishable properties at the walls in comparison with the domains. For instance, in thin films of $BiFeO_3$, the rhombohedral distortions, dictates three possible angles between the domain walls: 71, 109, and 180 degrees, giving rise to three types of domain walls. The detailed analysis of the possible domain walls in this material is reported by Streiffer et al. [91]. Moreover, due to the strong coupling between the polarization and strain, the unit cells are also deformed along the [111] direction. Therefore, these are ferroelectric as well as ferroelastic domain walls. With this short forepart, we start with a general explanation of domain walls and we continue this section by looking at transport properties (especially conduction) at domain walls and will end it with domain walls, which possess more than one order parameter (multiferroic domain walls).

Generally, there are two types of domain walls: (i) Ising-like (non-chiral ones), Fig. 5.8a and (ii) Heisenberg-like (chiral ones), Figs. 5.8 (b, c). The magnetic domain walls, which are not able to change their magnitude, comprise the first category, while the ferroelectric domain walls that support the converse situation are non-chiral. However, Salje et al. [108] showed in multiferroics coupling between order parameters can result in partially chiral domain walls called mixed Ising–Neel walls by Lee et al. [109], Fig. 5.8d. In the case of

$BiFeO_3$, there are more than two coupled ordered parameters, resulting in chiral domain walls.

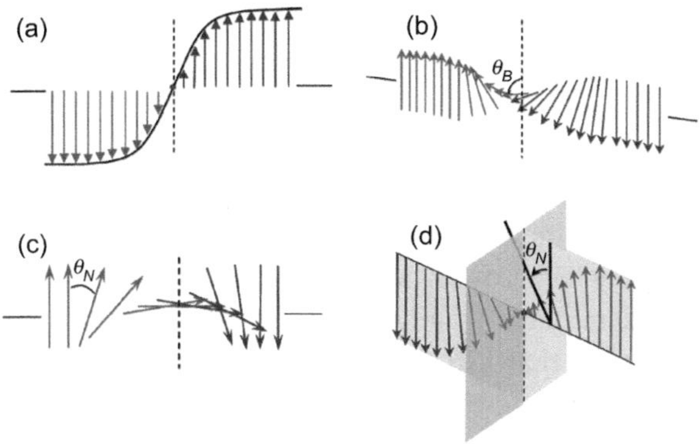

Figure 5.8 Different domain wall structures. (a) Ising wall, (b) and (c) Heisenberg wall, (d) mixed Ising–Néel wall. Reprinted from Ref. [109], with permission from American Physical Society, Copyright 2009.

Within the category of ferroelectric domain walls, their charged state can vary from neutral to charged (owing head-to-head polarization vectors in the adjacent domains or tail-to-tail configuration, known as head–head and tail–tail DWs, respectively), originating from the various orientations of the polar vectors existing inside the ferroelectric domains. This affects the width of the domain walls by an order of the magnitude, being bigger for charged DWs in comparison with neutral ones [110]. In 2008, Jia et al. evidenced the existence of charged domain walls by means of high-resolution transmission electron microscopy for the first time [111]. On the other hand, to screen the depolarization field produced by neighboring polar domains, a reasonable amount of carriers has to be present in the walls, which matches the domain walls with respect to their thicknesses as follows: neutral, head–head, and tail–tail, ordering from the narrowest to broadest, respectively [110].

Notably, one can estimate the concentration of the domain walls, which is closely related to the size of the domain that is dependent on ferroics boundary conditions as well as domain walls energy. For epitaxial ferroelectric and ferroelastic domains, the Kittel formula for the total energy of domains and DWs has been extended by Pertsev [112]. Although Kittel neglected the thickness of a domain wall in his calculations, one can estimate the width of the domain walls by refinement Kittel's formula. In 1959, Zhirnov et al. illustrated that ferromagnetic domain walls are thicker compared to very sharp ferroelectric and ferroelastic ones [113]. Afterwards, Schilling et al. showed that the size of magnetically ordered domains was wider than ferroelectrically ordered ones [114], as mentioned in Section 5.3.

Apart from the thickness dependence issue and its influence on tuning the properties, this narrowest known feature in materials domain wall has its own phase transition, which was realized for the first time by Lajzerowicz and co-workers [115, 116]. In addition, space group symmetry arguments in non-ferroelectric materials can be used to predict that 128 classes of investigated materials can have simultaneously non-zero net magnetization and polarization along the domain walls [29]. One of the studied space groups was R3c (the same space group as $BiFeO_3$), which allows magnetization at the walls [29]. Symmetry analysis is necessary but is not sufficient, because the next step is the investigation of the magnitude of the studied symmetry-allowed property. For instance, a coupling between magnetic and ferroelectric ordering temperatures is allowed by symmetry in $ErMnO_3$ but the difference between two ordering temperatures is too large, which suppresses the magnitude of coupling to non-observable value. Therefore, the underlying phase transition in a domain wall causes changes in electronic structure and concomitant changes in transport properties, which will be discussed in the next section.

5.5 The Role of Vertical Interfaces in Multiferroics

5.5.1 Conduction through Domain Walls in Multiferroics

Salje and co-workers were the first to observe significant different property in conduction level that evidenced superconductivity in twin walls in WO_3, which was explained by the ability of twin walls to attract defects from preferential doping along the walls, causing the appearance of selective conduction paths along the twin walls [117]. The investigation of transport properties was continued by looking at non-doped multiferroics. The first observation was on thin films of $BiFeO_3$ [118], which mentioned that artificially written domain walls show higher conduction level compared to domains and the conduction level is tunable by changing the chemistry state at the walls in La-doped thin films of $BiFeO_3$ [119]. After one year, Choi et al. exactly showed the opposite effect in improper ferroelectric $YMnO_3$ [86]. Catalan et al. interpreted this contrast by considering para-phase-like behavior besides polarization consideration [85]. Later, a difference in the conduction level between domains and domain walls has been observed in an as-grown state of thin films of $BiFeO_3$ [78, 120]. Figure 5.9 shows the piezoforce microscopy. Figures 5.9 (a, b) show conductive-AFM images and Figs. 5.9 (c, a) show line profile of the two set of images at the domain walls. Adjacent domains are shown in Fig. 5.9d.

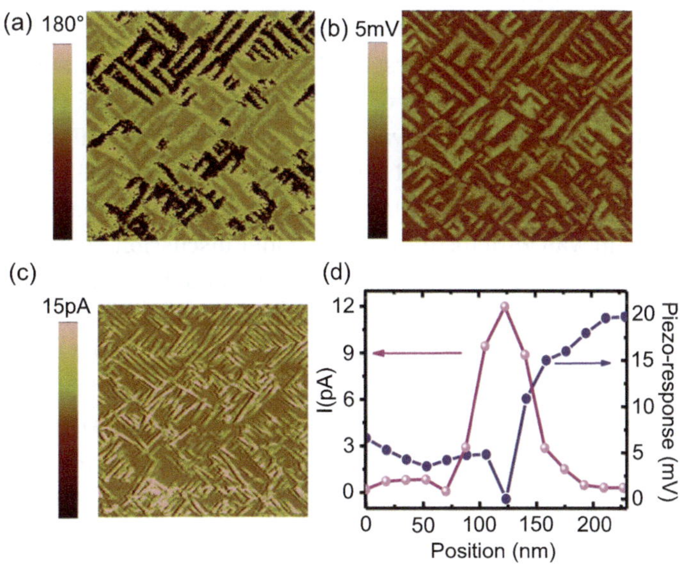

Figure 5.9 In-plane amplitude (a) and phase (b) piezoforce microscopy images of a 4 μm × 4 μm area of $BiFeO_3$ films grown on a $SrTiO_3$ substrate. (c) Conducting AFM image taken at RT in the same area. (d) Line scan across one domain wall in both the piezoresponse amplitude and the current images. Reprinted from Ref. [78], with permission from American Physical Society, Copyright 2011.

However, conduction phenomena are not specified to multiferroic $BiFeO_3$ and conduction at domain wall has been also reported in thin films of $Pb(Zr, Ti)O_3$ [121] and also in improper ferroelectric single crystals of $ErMnO_3$ [122] and $HoMnO_3$ [123]. For more details on conduction through domain walls, also see Ref. [124]. Apart from, conduction phenomena, another transport property has also been investigated. One of the earliest experiments has been performed when (antiferromagnetic) $BiFeO_3$ is put in contact with a soft ferromagnetic (permalloy) layer and the exchange bias observed. Interestingly, the bias of the magnetic hysteresis loop is directly related to the amount of 109° DWs in the $BiFeO_3$ layer [125]. In the same type of domain walls, the photovoltaic effect has also been observed [126], more investigation on the same sort of samples illustrated that domain walls function as nanoscale generators of the photovoltaic current [127]. Also, the magnetoresistance effect as large as 60% in an array of 109° DW was measured under temperature and field variations [128].

The observed conductivity of $BiFeO_3$ DWs, which inspired the field towards nanoelectronics, has been accounted for in two ways: on one hand, the reduction of the band gap at the wall (the local electronic changes at the wall) is responsible, and on the other hand, electrostatic potential changes due to structural variations affect conduction [90]. Yet, the definition of the local band gap at the domain wall with its finite size is under discussion. Nonetheless, the reduction of the band gap can be attributed to the possibility of complete hybridization between Fe 3d and O 2p and thus, a straightening of the Fe–O–Fe bond angle, resulting in the band gap reduction at the walls [85]. Lubk and co-workers have calculated the potential steps at the twin walls of $BiFeO_3$ due to rotation of the polar vector normal to the walls [90], this potential drop creates a large electric field where electrons and oxygen vacancies concentrate in two opposite sides of the wall, which has clear implication on charge aggregation around the wall and hence, the higher level of conduction at the walls. Moreover, the smaller band gap in $BiFeO_3$ and magnetoelectric coupling between polarization and spin lattice can also explain a higher conduction level in thin films of $BiFeO_3$ [85]. Catalan et al. summarized some of the existing reasons, which are responsible for conduction at the walls in thin films of $BiFeO_3$ and other oxides such as octahedral rotations, electrostatic steps because of rigid rotation of the polar vector and increased carrier density at the wall [129]. One more consideration, which has to be taken into account is the mutual interplay between vacancy content (oxygen anion and/or cation) and conductivity at the walls. The role of oxygen vacancies has been presented for the first time by electronic and elemental analysis by means of electron-energy-loss spectroscopy and TEM analysis for $BaTiO_3$ twin boundaries [111], this has been also emphasized by Salje et al. [130]. Notably, this fact has been extensively investigated in the other type of physics field and a case study of the $LaAlO_3/SrTiO_3$ interface also confirms the role of oxygen vacancies on conduction level [105].

5.5.2 Multiferroics Domain Walls

As already discussed, multiferroics are a class of materials that unite multiple ferroic properties such as ferroelasticity, ferroelectricity, ferromagnetism, or ferrotoroidicity, leading to a richer and more complex behavior. Schmid et al. made a survey on the symmetry conditions for the coupling of different order parameters in the same phase in bulk materials [131]. Since domain walls are regions of symmetry breaking, they offer additional possibilities for multiferroic behavior. Janovec et al. discussed the allowed couplings inside the domain walls [29]. One example is the case of multiferroics with two order parameters: ferroelectricity and ferroelasticity. A complete study on the two order parameters that satisfy the rules of mechanical and electrical compatibility among adjacent domains has been provided by Fousek and Janovec [30]. Interestingly, a domain wall can be multiferroic while the two adjacent domains carry a single-order parameter or even none. In the case of hexagonal manganites, the domain walls couple three order parameters (ferroelectric DWs, which carry a net magnetic moment, are topologically protected) [132, 133]. However, the strength of the coupling between order parameters determines the width of these DWs as mentioned earlier. For instance, if the coupling is weak, the correlation length for two different parameters is completely different and the corresponding ferroic wall thickness completely differs from one to another. $YMnO_3$ can be fitted to this category by owing ferroelectricity and ferromagnetism as its order parameters [134]. However, in the converse situation, the strong order parameters dictate its corresponding width to the shared domain wall of the weak order parameters. Equally thick ferroelectric domain walls as the ferromagnetic domain walls in $TbMnO_3$ are evidence for this prediction [135]. In the case of $BiFeO_3$, there are two order parameters with two different correlation lengths, but yet each will be affected by the moderate coupling between the ordering parameters, which results in a broader ferroelectric domain wall in $BiFeO_3$ compared to normal ferroelectrics [89].

5.6 Future Challenges

As mentioned earlier, the interdependence between domain size and domain walls density matters from an application point of view. This means, smaller domains results in higher domain wall concentrations, which can be interesting for device application. Therefore, an introduction of "domain/domain wall engineering" should matter for both fundamental science and device implementation. Wada et al. was first to propose this concept in 2006 [136]. For instance, the conducting domain walls are very good candidates for nanoelectronic device applications. Moreover, conduction through negatively charged domain walls (head–head) are potential candidates to be replaced for well-known 2DEG heterostructures ($LaAlO_3/SrTiO_3$) [137]. Also, the domain walls in orthorhombic manganites are ferromagnets with considerably a large magnetic moment while the domains adopting antiferromagnetic order as expected in bulk [138]. The density of these domain walls can be changed by strain engineering via thickness variations where they can be used as nanoreactors [77]. Moreover, domain wall configurations can dynamically as well as statically be designed in a way to influence the ferroelectric properties [139]. Also, the combination of strain-engineered oxide thin films with organic materials is growing towards the spintronic applications. Surprisingly, DWs are playing a very deterministic role in this respect [140]. Additionally, the effect of nano-walls in tailoring macroscopic properties is a to be explored-must for future applicable devices.

All these can be enormously extended by involving all possible multiferroic systems (including non-perovskite and fluorides and others) as soon as the suitable substrates become commercially available.

References

1. Van Aken, B. B., Rivera, J. P., Schmid, H., and Fiebig, M. B. B. (2007), Observation of ferrotoroidic domains, *Nature*, **449**, pp. 702–705.
2. Hill, N. A. (2000), Why are there so few magnetic ferroelectrics? *J. Phys. Chem. B*, **104**, pp. 6694–6709.
3. Baettig, P. and Spaldin, N. A. (2005), Ab initio prediction of a multiferroic with large polarization and magnetization, *Appl. Phys. Lett.*, **86**, p. 012505.
4. Yokosuka, M. (1993), Ferroelectric and antiferroelectric properties of solid-solution ceramics $Ba(Ca_{1/3}Nb_{2/3})O_3$–$PbZrO_3$, *J. Appl. Phys.*, **32**, p. 1142.
5. Catalan, G. and Scott, J. F. (2009), Physics and applications of bismuth ferrite, *Adv. Matter*, **21**, p. 2463.
6. Abrahams, S. C. and Ravez, J. (1992), Dielectric and related properties of fluorine-octahedra ferroelectrics, *Ferroelectrics*, **21**, p. 135.
7. Ravez, J. (1997), The inorganic fluoride and oxyfluoride ferroelectrics, *J. Physique III*, **7**, p. 1129.
8. Nenert, G. and Palstra, T. T. M. (2007), Prediction for new magnetoelectric fluorides, *J. Phys.: Condens. Matter*, **19**, p. 406213.
9. Lebeugle, D., Colson, D., Forget, A., and Viret, A. M. (2007), Very large spontaneous electric polarization in $BiFeO_3$ single crystals at room temperature and its evolution under cycling fields, *Appl. Phys. Lett.*, **91**, p. 022907.
10. Wang, J., Neaton, J. B., Zheng, H., Nagarajan, V., Ogale, S. B., Liu, B., et al. (2003), Epitaxial $BiFeO_3$ multiferroic thin film heterostructures, *Science*, **299**, p. 1719.
11. Setter, N. (2016), What is a ferroelectric—a materials designer perspective? *Ferroelectrics*, **500**, p. 164.
12. Wul, W. and Goldman, I. M. C. R. (1946), *Acad. Sci. Russ.*, **46**, p. 139.
13. Wul, W. and Goldman, I. M. C. R. (1947), *Acad. Sci. Russ.*, **5l**, p. 21.
14. Jona, F. and Shirane, G. (1993), *Ferroelectric Crystals*, Dover Publications Inc.
15. Cochran, W. (1959), Crystal stability and the theory of ferroelectricity, *Phys. Rev. Lett.*, **3**, p. 412.
16. Anderson, P. W. (1960), Physics of dielectrics, *Proc. of Conf. on Physics of Dielectrics*, Skanavi. G. I. (ed.), Moscow: Academy of Sciences, p. 290.
17. Cowley, R. A. (1962), Temperature dependence of a transverse optic mode in strontium titanate, *Phys. Rev. Lett.*, **9**, p. 159.

18. Cohen. R. E. (1992), Origin of ferroelectricity in perovskite oxides, *Nature,* **358**, p. 136.
19. Wu, Z. and Cohen, R. E. (2005), Pressure-induced anomalous phase transitions and colossal enhancement of piezoelectricity in $PbTiO_3$, *Phys. Rev. Lett.,* **95**, p. 90376.
20. Hill, N. A. (2002), Density functional studies of multiferroics, *Annu. Rev. Mater.,* **32**, p. 1.
21. Ravez, J. (2000), Ferroelectricity in solid state chemistry, *Acad. Sci. Paris, IIc Chem.,* **3**, p. 267.
22. Kiat, J. M., Gianguido, A., Baldinozzi, G., Dunlop, M., Malibert, Ch., Dkhil, B., et al. (2000), An harmonicity and disorder in simple and complex perovskites: a high energy synchrotron and hot neutron diffraction study, *J. Phys. Condens. Matter,* **12**, p. 8411.
23. Van den Brink, J., Nussinov, N., and Oles, A. (2011), Frustration in systems with orbital degrees of freedom, *Introduction to Frustrated Magnetism*, Chapter 23, Lacroix, C., Mendels, S., and Mila, F., (eds.), Springer Series in Solid-State Sciences.
24. Blundell, S. (2001), *Magnetism in Condensed Matter*, Oxford University Press.
25. Spaldin, N. A. (2010), *Magnetic Materials; Fundamentals and Applications*, 2nd edition, Cambridge University Press.
26. Salje, E. K. (2012) *Phase Transitions in Ferroelastic and Co-elastic Crystals*, Cambridge University Press.
27. Goncalves-Ferreira, L., Redfern, S. A. T., Artacho, E., and Salje. E. K. H. (2008), Ferrielectric domain walls in $CaTiO_3$, *Phys. Rev. Lett.*, **101**, p. 097602.
28. Salje, E. K. H., Alison Aird, A., Locherer, K., Hayward, S., Novak, J., and Chrosch, J. (1999), Ferroelastic twin walls for nanotechnological applications, *Ferroelectrics,* **223**, p. 1.
29. Janovec, V., Richterová, L., and Prívratská, J. (1999), Polar properties of compatible ferroelastic domain walls, *Ferroelectrics,* **222**, p. 73.
30. Fousek, J. and Janovec, V. (1969), The orientation of domain walls in twinned ferroelectric crystals, *J. Appl. Phys.*, **40**, p. 135.
31. Ginzburg, V. L., Gorbatsevich, A. A., Kopaev, Y. V., and Volkov, B. A. (1984), On the problem of super diamagnetism, *Solid State Commun.,* **50**, p. 339.
32. Ascher, E., Reider. H., Schmid, H., and Stossel, H. (1966), Some properties of ferromagnetoelectric nickel-iodine boracite, $Ni_3B_7O_{13}I$, *J. Appl. Phys.,* **37**, p. 1404.

33. Toledano, P., Khalyavin, D. D., and Chapon, L. C. (2011), Spontaneous toroidal moment and field-induced magnetotoroidic effects in $Ba_2CoGe_2O_7$, *Phys. Rev. B,* **84**, p. 094421.

34. Toledano, P., Ackermann, M., Bohaty, M., Becker, P., Lorenz, T., Leo, N., et al. (2015), Primary ferrotoroidicity in antiferromagnets, *Phys. Rev. B,* **92**, p. 094431.

35. Mettout, B., Toledano, P., and Fiebig, M. (2010), Symmetry replication and toroidic effects in the multiferroic pyroxene $NaFeSi_2O_7$, *Phys. Rev. B,* **81**, p. 214417.

36. Matthees, J. W. and–Blakeslee, A. E. (1974), Defects in epitaxial multilayers: I. Misfit dislocations, *J. Cryst. Growth*, **27**, p. 118.

37. Pompe, W., Gong, X., Suo, Z., and Speck, J. S., (1993), Elastic energy release due to domain formation in the strained epitaxy of ferroelectric and ferroelastic films, *J. Appl. Phys.,* **74**, p. 6012.

38. Speck, J. S. and Pompe, W., (1994), Domain configurations due to multiple misfit relaxation mechanisms in epitaxial ferroelectric thin films. I. Theory, *J. Appl. Phys.*, **76**, p. 446.

39. Catalan, G., Janssens, A., Rispens, G., Csiszar, S., Seeck, O., Rijnders, G., et al. (2006), Polar domains in lead titanate films under tensile strain, *Phys. Rev. Lett.,* **96**, p. 127602.

40. Scott, J. F. and Paz de Araujo, C. A. (1989), Ferroelectric memories, *Science,* **246**, p. 1400.

41. Scott, J. F. (1998), The physics of ferroelectric ceramic thin films for memory applications, *Ferroelectr. Rev.,* **1**, p. 1.

42. Scott, J. F. (2000), *Ferroelectric Memories*, Springer, Berlin.

43. Dawber, M., Rabe, K. M., and Scott, J. F. (2005), Physics of thin-film ferroelectric oxides, *Rev. Mod. Phys.*, **77**, p. 4.

44. Schlom, D. G., Chen, L. Q., Eom, C. B., Rabe, K. M., Streiffer, S. K., and Triscone, J. M. (2007), Strain tuning of ferroelectric thin films, *Annu. Rev. Mater. Res.,* **37**, p. 589.

45. Rabe, K. M., Ahn, C. H., and Triscone, J. M. (2007), *Physics of Ferroelectrics: A Modern Perspective*, Springer.

46. Lichtensteiger, C., Zubko, P., Stengel, M., Aguado-Puente, P., Triscone, J.-M., Ghosez, Ph., et al. (2011), Ferroelectricity in ultrathin-film capacitors, Ch. 12, in *Oxide Ultrathin Films*, Science and Technology, Wiley.

47. Xu, G., Hiraka, H., Shirane, G., Li, J., Wang, J., and Viehland, D. (2005), Low symmetry phase in (001)$BiFeO_3$ epitaxial constrained thin films, *Appl. Phys. Lett.,* **86**, p. 182905.

48. Pertsev, N. A., Zembilgotov, A. G., and Tagantsev, A. K. (1998), Effect of mechanical boundary conditions on phase diagrams of epitaxial ferroelectric thin films, *Phys. Rev. Lett.*, **80**, p. 1988.

49. Kukhar, V. G., Pertsev, N. A., and Waser, R. (2001), Thermodynamic theory of epitaxial ferroelectric thin films with dense domain structure, *Phys. Rev. B*, **64**, p. 214103.

50. Noheda, B., Cox, D. E., Shirane, G., Gonzalo, J. A., Cross, L. E., and Park, S. E. (1999), A monoclinic ferroelectric phase in the $Pb(Zr_{1-x}Ti_x)O_3$ solid solution, *Appl. Phys. Lett.*, **74**, p. 2059.

51. Noheda, B., Wu, L., and Zhu, Y. (2002), Low-temperature superlattice in monoclinic $PbZr_{0.52}Ti_{0.48}O_3$, *Phys. Rev. B*, **66**, p. 060103.

52. Vanderbilt, D. and Cohen, M. H. (2001), Monoclinic and triclinic phases in higher-order Devonshire theory, *Phys. Rev. B.*, **63**, p. 094108.

53. Fu, H. and Cohen, R. E. (2000), Polarization rotation mechanism for ultrahigh electromechanical response in single-crystal piezoelectrics, *Nature*, **403**, p. 281.

54. Jang, H. W., Baek, S. H., Ortiz, D., Folkman, C. M., Das, R. R., Chu, Y. H., et al. (2008), Strain-induced polarization rotation in epitaxial (001) $BiFeO_3$ thin films, *Phys. Rev. Lett.*, **101**, p. 107602.

55. Bellaiche, L., García, A., and Vanderbilt, D. (2000), Finite-temperature properties of $Pb(Zr_{1-x}Ti_x)O_3$ alloys from first principles, *Phys. Rev. Lett.*, **84**, p. 5427.

56. Guo, R., Cross, L. E., Park, S. E., Noheda, B., Cox, D. E., Shirane, G. (2000), Origin of the High piezoelectric response in $PbZr_{1-x}Ti_xO_3$, *Phys. Rev. Lett.*, **84**, p. 5423.

57. Rossetti, G. A., Khachaturyan, A. G., Akcay, G., and Ni, Y. (2008), Ferroelectric solid solutions with morphotropic boundaries: Vanishing polarization anisotropy, adaptive, polar glass, and two-phase states, *J. Appl. Phys.*, **103**, p. 114113.

58. Choi, K. J., Biegalski, M., Li, Y. L., Sharan, A., Schubert, J., Uecker, R., et al. (2004), Enhancement of ferroelectricity in strained $BaTiO_3$ thin films, *Science*, **306**, p. 1005.

59. Haeni, J. H., Irvin, P., Chang, W., Uecker, R., Reiche, R., Li, Y. L., et al. (2004), Room-temperature ferroelectricity in strained $SrTiO_3$, *Nature*, **430**, p. 758.

60. Devonshire, A. F. (1954), Theory of ferroelectrics, *Adv. Phys.*, **3**, p. 85.

61. Uwe, H. and Sakudo, T. (1976), Stress-induced ferroelectricity and soft phonon modes in $SrTiO_3$, *Phys. Rev. B*, **13**, p. 271.

62. Béa, H., Dupé, B., Fusil, S., Mattana, R., Jacquet, E., Warot-Fonrose, B., et al. (2009), Evidence for room-temperature multiferroicity in a compound with a giant axial ratio, *Phys. Rev. Lett.,* **102**, p. 217603.

63. Zeches, R. J, Rossell, M, D., Zhang, J. X., Hatt, A. J., He, Q., Yang, C. H., et al. (2009), A strain-driven morphotropic phase boundary in $BiFeO_3$, *Science,* **326**, p. 977.

64. Wojdel, J. C. and Iñiguez, J. (2010), Ab initio indications for giant magnetoelectric effects driven by structural softness, *Phys. Rev. Lett.,* **105**, p. 037208.

65. Diéguez, O., González-Vázquez, O. E., and Wojdeł, Íñigue, J. C. (2011), First-principles predictions of low-energy phases of multiferroic $BiFeO_3$, *J. Phys. Rev. B,* **83**, p. 094105.

66. Damodaran, A. R., Liang, C. W., He, Q., Peng, C. Y., Chang, L., Chu, Y. H., et al. (2011), Nanoscale structure and mechanism for enhanced electromechanical response of highly strained $BiFeO_3$ thin films, *Adv. Mater.,* **23**, p. 3170.

67. Siemons, W., Chi, W. M., Howe, J., Biegalski, M. D., Balke, N., Maksymovych, P., et al. (2013), Phase transitions, phase coexistence, and piezoelectric switching behavior in highly strained $BiFeO_3$ films, *Adv. Mater.,* **25**, p. 5561.

68. Jin, Y. M., Wang, Y. U., Khachaturyan, A. G., Li, J. F., and Viehland, D. (2003), Conformal miniaturization of domains with low domain-wall energy: monoclinic ferroelectric states near the morphotropic phase boundaries, *Phys. Rev. Lett.,* **91**, p. 197601.

69. Bhattacharyy, S., Jinschek, J. R., Cao, H., Wang, Y. U., Li, J., and Viehland, D.(2008), Direct high-resolution transmission electron microscopy observation of tetragonal nanotwins within the monoclinic MC phase of $Pb(Mg_{1/3}Nb_{2/3})O_3$–$0.35PbTiO_3$ crystals, *Appl. Phys. Lett.,* **92**, p. 142904.

70. Beach, R. S., Borchers, J. A., Matheny, A., Erwin, R. W., Salamon, M. B., Everitt, B., et al. (1993), Enhanced Curie temperatures and magnetoelastic domains in Dy/Lu superlattices and films, *Phys. Rev. Lett.,* **70**, p. 3502.

71. Gan, Q., Rao, R. A., Eom, C. B., Garrett, J. L., and Lee, M. (1998), Direct measurement of strain effects on magnetic and electrical properties of epitaxial $SrRuO_3$ thin films, *Appl. Phys. Lett.,* **72**, p. 978.

72. Fuchs, D., Arac, E., Pinta, C., Schuppler, S., Schneider, R., and Löhneysen, H. V. (2008), Tuning the magnetic properties of $LaCoO_3$ thin films by epitaxial strain, *Phys. Rev. B,* **77**, p. 014434.

73. Daumont, C. J. M., Mannix, D., Venkatesan, S., Catalan, G., Rubi, D., Kooi, B. J., et al. (2009), Epitaxial $TbMnO_3$ thin films on $SrTiO_3$ substrates: a structural study, *J. Phys.: Condens. Matter,* **21**, p. 182001.

74. Marti, X., Skumryev, V., Ferrater, V., García-Cuenca, M. V., Varela, M., Sánchez, F., et al. (2010), Emergence of ferromagnetism in antiferromagnetic $TbMnO_3$ by epitaxial strain, *Appl. Phys. Lett.,* **96**, p. 222505.

75. Marti, X., Skumryev, V., Ferrater, V., García-Cuenca, M. V., Varela, M., Sánchez, F., et al. (2010), Strain-driven noncollinear magnetic ordering in orthorhombic epitaxial $YMnO_3$ thin films, *J. Appl. Phys.,* **108**, p. 123917.

76. Daumont, C. J. M., Farokhipoor, S., Ferri, A., Wojdeł, J. C., Íñiguez, J., Kooi, B. J., et al. (2010), Tuning the atomic and domain structure of epitaxial films of multiferroic $BiFeO_3$, *Phys. Rev. B,* **81**, p. 144115.

77. Farokhipoor, S., Magén, C., Venkatesan, S., Íñiguez, J., Daumont, C. J. M., Rubi, D., et al. (2014), Artificial chemical and magnetic structure at the domain walls of an epitaxial oxide, *Nature,* **515**, p. 379.

78. Farokhipoor, S. and Noheda, B. (2011), Conduction through 71° domain walls in $BiFeO_3$ thin films, *Phys. Rev. Lett.,* **107**, p. 12760.

79. Lee, J. H., Fang, L., Vlahos, E., Ke, X., Jung, Y. W., Kourkoutis, L. F., et al. (2010), A strong ferroelectric ferromagnet created by means of spin–lattice coupling, *Nature,* **466**, p. 954.

80. Feiming, B., Wang, J., Wuttig, M., Li, J., Wang, N., Pyatakov, A. P., et al. (2005), Destruction of spin cycloid in $(111)_c(111)_c$-oriented $BiFeO_3$ thin films by epitaxial constraint: Enhanced polarization and release of latent magnetization, *Appl. Phys. Lett.,* **86**, p. 32511.

81. Farokhipoor, S. and Noheda, B. (2014), Screening effects in ferroelectric resistive switching of $BiFeO_3$ thin films, *APL Mater.,* **2**, p. 056102.

82. Thomas, L. H. (1927), The calculation of atomic fields, *Proc. Cambridge Phil. Soc.,* **23**, p. 542.

83. P. Batra, P. and Silverman, B. D. (1972), Thermodynamic stability of thin ferroelectric films, *Solid State Commun.,* **11**, p. 291.

84. Massimiliano Stengel, M., Aguado-Puente, P., Spaldin, N. A., and Junquera, J. (2011), Band alignment at metal/ferroelectric interfaces: Insights and artifacts from first principles, *Phys. Rev. B,* **83**, p. 235112.

85. Catalan, G., Seidel, J., Ramesh, R., and Scott, J. F. (2012), Domain wall nanoelectronics, *Rev. Mod. Phys.,* **84**, p. 1.

86. Choi, T., Horibe, Y., Yi, H. T., Choi, Y. J., Wu, W., and Cheong, S. W. (2010), Insulating interlocked ferroelectric and structural antiphase domain walls in multiferroic $YMnO_3$, *Nat. Mater.,* **9**, p. 253.

87. Roitburd, A. (1976), Domain structure caused by internal stresses in heterophase solids, *Phys. Status Solidi. A,* **37**, p. 329.

88. Kittel, C. (1946), Theory of the structure of ferromagnetic domains in films and small particles, *Phys. Rev.,* **70**, p. 965.

89. Nesterov, O., Matzen, S., Magen, C., Vlooswijk, A. H. G., Catalan, G., and Noheda, B. (2013), Thickness scaling of ferroelastic domains in $PbTiO_3$ films on $DyScO_3$, *Appl. Phys. Lett.,* **103**, p. 142901.

90. Lubk, A., Gemming, S., and Spaldin, N. A. (2009), First-principles study of ferroelectric domain walls in multiferroic bismuth ferrite, *Phys. Rev. B,* **80**, p. 104110.

91. Streiffer, S. K., and Parker, C. B. (1998), Domain patterns in epitaxial rhombohedral ferroelectric films. I. Geometry and experiments, *J. Appl. Phys.,* **83**, p. 2742.

92. Nelson, C. T., Winchester, B., Zhang, Y., Kim, S.-J., Melville, A., Adamo, C., et al. (2011), Spontaneous vortex nanodomain arrays at ferroelectric heterointerfaces, *Nano Lett.,* **11**, p. 828.

93. Naumov, I. I., Bellaiche, L., and Fu, H. (2004), Unusual phase transitions in ferroelectric nanodisks and nanorods, *Nature,* **432**, p. 737.

94. Rodriguez, B. J., Gao, X. S., Liu, L. F., Lee, W., Naumov, I. I., Bratkovsky, A. M., et al. (2009), Vortex polarization states in nanoscale ferroelectric arrays, *Nano Lett.,* **9**, p. 1127.

95. Catalan, G., Béa, H., Fusil, S., Bibes, M., Paruch, P., Barthélémy, A., et al. (2008), Fractal dimension and size scaling of domains in thin films of multiferroic $BiFeO_3$, *Phys. Rev. Lett.,* **100**, p. 027602.

96. Balke, N., Choudhury, S., Jesse, S., Huijben, M., Chu, Y. H., Baddorf, A. P., et al. (2009), Deterministic control of ferroelastic switching in multiferroic materials, *Nat. Nanotechnol.,* **4**, p. 217.

97. Tenne, D. A., Bruchhausen, A., Lanzillotti-Kimura, N. D., Fainstein, A., Katiyar, R. S., Cantarero, A., et al. (2006), Probing nanoscale ferroelectricity by ultraviolet Raman spectroscopy, *Science,* **313**, p. 1614.

98. Li, Y. L., Hu, S. Y., Tenne, D., Soukiassian, A., Schlom, D. G., Xi, X. X., et al. (2007), Prediction of ferroelectricity in $BaTiO_3/SrTiO_3$ superlattices with domains, *Appl. Phys. Lett.,* **91**, p. 112914.

99. Balke, N., Winchester, B., Ren, W., Chu, Y. H., Morozovska, A. N., Eliseev, E. A., et al. (2011), Enhanced electric conductivity at ferroelectric vortex cores in $BiFeO_3$, *Nat. Phys.*, **8**, p. 81.

100. Kawasaki, M., Takahashi, K., Maeda, T., Tsuchiya, R., Shinohara, M., Ishiyama, O., et al. (1994), Atomic control of the $SrTiO_3$ crystal surface, *Science*, **266**, p. 1540.

101. Biswas, A., Rossen, P. B., Yang, C.-H., Siemons, W., Jung, M.-H., Yang, I. K., et al. (2011), Universal Ti-rich termination of atomically flat $SrTiO_3$ (001), (110), and (111) surfaces, *Appl. Phys. Lett*, **98**, p. 051904.

102. Chang, J., Park, Y. S., Kim, S. K. (2008), *Appl. Phys. Lett*, **92**, p. 152910.

103. Bachelet, R., Sanchez, F., Santiso, J., Munuera, C., Ocal, C., and Fontcuberta, J. (2009), Self-assembly of $SrTiO_3$(001) chemical-terminations: A route for oxide-nanostructure fabrication by selective growth. *Chem. Mater.*, **21**, p. 2494.

104. Park, J. W., Bogorin, D. F., Cen, C., Felker, D. A., Zhang, Y., Nelson, C. T., et al. (2010), Creation of a two-dimensional electron gas at an oxide interface on silicon, *Nat. Commun.*, **1**, p. 94.

105. Ohtomo, A. and Hwang, H. Y. (2004), A high-mobility electron gas at the $LaAlO_3/SrTiO_3$ heterointerface, *Nature*, **427**, p. 423.

106. Farokhipoor, S. and Noheda, B. (2014), Screening effects in ferroelectric resistive switching of BiFeO3 thin films, *APL. Mater.*, **2**, p. 056102.

107. Martin, L. M. and Schlom, D. G., (2012), Advanced synthesis techniques and routes to new single-phase multiferroics, *Curr. Opin. Solid State Mater. Sci.*, **16**, p. 199.

108. Houchmandzadeh, B., Lajzerowicz, J., and Salje, E. (1991), Order parameter coupling and chirality of domain walls, *J. Phys.: Condens. Matter*, **3**, p. 5163.

109. Donghwa Lee, D., Behera, R. K., Wu, P., Xu, H., Li, Y. L., Sinnott, S. B., et al. (2009), Mixed Bloch-Néel-Ising character of 180° ferroelectric domain walls, *Phys. Rev. B*, **80**, p. 06010.

110. Eliseev, E. A., Morozovska, A. N., Svechnikov, G. S., Gopalan, V., and Shur, V. Y. (2011), Static conductivity of charged domain walls in uniaxial ferroelectric semiconductors, *Phys. Rev. B*, **83**, p. 235313.

111. Jia, Ch-L., Mi, Sh-B., Urban, K., Vrejoiu, I., Alexe, M., and Hesse, D. (2008), Atomic-scale study of electric dipoles near charged and uncharged domain walls in ferroelectric films, *Nat. Mater.*, **7**, p. 57.

112. Pertsev, N. A. and Zembilgotov, A. G. (1995), Energetics and geometry of 90° domain structures in epitaxial ferroelectric and ferroelastic films, *J. Appl. Phys.*, **78**, p. 6170.

113. Zhirnov, V. A. (1959), A contribution to the theory of domain walls in ferroelectrics, *Sov. Phys. JETP,* **35**, p. 822.

114. Schilling, A., Adams, T. B., Bowman, R. M., Gregg, J. M., Catalan, G., and Scott, J. F. (2006), Scaling of domain periodicity with thickness measured in $BaTiO_3$ single crystal lamellae and comparison with other ferroics, *Phys. Rev. B*, **74**, p. 024115.

115. Lajzerowicz, J. and Niez, J. J. (1979), Phase transition in a domain wall, *J. Phys. Lett.,* **40**, p. 16.

116. Privrastka, J. and Janovec, V. (1999), Spontaneous polarization and/or magnetization in non-ferroelastic domain walls: symmetry predictions, *Ferroelectrics*, **222**, p. 23.

117. Salje, E. K. H. (2010), Domain boundaries as active memory devices: trajectories towards domain boundary engineering, *Chem. Phys. Chem.,* **11**, p. 940.

118. Seidel, J., Martin, L. W., He, Q., Zhan, Q., Chu, Y. H., Rother, A., et al. (2009), Conduction at domain walls in oxide multiferroics, *Nat. Mater.*, **8**, p. 229.

119. Seidel, J., Maksymovych, P., Batra, Y., Katan, A., Yang, S. Y., He, Q., et al. (2010), Domain wall conductivity in La-doped $BiFeO_3$, *Phys. Rev. Lett.*, **105**, p. 197603.

120. Chiu, Y. P., Chen, Y. T., Huang, B. Ch., Shih, M. Ch., Yang, J. Ch., et al. (2011), Atomic-scale evolution of local electronic structure across multiferroic domain walls, *Adv. Mater.,* **23**, p. 153.

121. Guyonnet, J., Gaponenko, I., Gariglio, S., and Paruch, P. (2011), Conduction at domain walls in insulating $Pb(Zr_{0.2}Ti_{0.8})O_3$ thin films, *Adv. Mater.,* **23**, p. 5377.

122. Meier, D., Seidel, J., Cano, A., Delaney, K., Kumagai, Y., Mostovoy, M., et al. (2012), Anisotropic conductance at improper ferroelectric domain walls, *Nat. Mater.,* **11**, p. 284.

123. Wu, W., Horibe, Y., Lee, N., Cheong, S. W., and Guest, J. R. (2012), Conduction of topologically protected charged ferroelectric domain walls, *Phys. Rev. Lett.,* **108**, p. 077203.

124. Rama, K., Vasudevan, R. K., Wu, W., Guest, J. R., Baddorf, A. P., Morozovska, A. N., et al. (2013), Domain wall conduction and polarization-mediated transport in ferroelectrics, *Adv. Funct. Mater.,* **23**, p. 2592.

125. Lane, W., Martin, L. M., Chu, Y. H., Holcomb, M. B., Huijben, M., Yu, P., et al. (2008), Nanoscale control of exchange bias with $BiFeO_3$ thin films, *Nano Lett.*, **8**, p. 2050.

126. Yang, S. Y., Seidel, J., Byrnes, S. J., Shafer, P., Yang, C. H., Rossell, M. D., et al. (2010), Above-bandgap voltages from ferroelectric photovoltaic devices, *Nat. Nanotech.*, **5**, p. 143.

127. Seidel, J., Fu, D., Yang, S. Y., Alarcón-Lladó, E., Wu, J., Ramesh, R., et al. (2011), Efficient photovoltaic current generation at ferroelectric domain walls, *Phys. Rev. Lett.*, **107**, 126805.

128. He, Q., Yeh, C. H., Yang, J. C., Singh-Bhalla, G., Liang, C. W., Chiu, P. W., et al. (2012), Magnetotransport at domain walls in $BiFeO_3$, *Phys. Rev. Lett.*, **108**, p. 067203.

129. Catalan, G. (2012), On the link between octahedral rotations and conductivity in the domain walls of $BiFeO_3$, *Ferroelectrics*, **433**, p. 65.

130. Salje, E. K. H. and Zhang, H. (2009), Domain boundary engineering, *Phase Transitions*, **82**, p. 452.

131. Schmid. H. (2008), Some symmetry aspects of ferroics and single phase multiferroics, *J. Phys.: Condens. Matter*, **20, p.** 43420.

132. Kumagai, Y. and Spaldin, N. A. (2013), Structural domain walls in polar hexagonal manganites, *Nat. Commun.*, **4**, p. 1540.

133. Artyukhin, S., Delaney, K. T., Spaldin, N. A., and Mostovoy, M. (2014), Landau theory of topological defects in multiferroic hexagonal manganites, *Nat. Mater.*, **13**, p. 42.

134. Fiebig, M., Goltsev, A. V., Lottermoser, T., and Pisarev, R. V. (2004), Structure and interaction of domain walls in $YMnO_3$, *J. Magn. Magn. Mater.*, **353**, p. 272.

135. Cano, A. and Levanyuk, A. P. (2010), Pseudo proper ferroelectricity in thin films, *Phys. Rev. B*, **81**, p. 172105.

136. Wada, S., Yako, K., Yokoo, K., Kakemoto, H., and Tsurumi, T. (2006), Domain wall engineering in barium titanate single crystals for enhanced piezoelectric properties, *Ferroelectrics*, **334**, p. 17.

137. Sluka, T., Tagantsev, A. K., Bednyakov, P., and Setter, N. (2013), Free-electron gas at charged domain walls in insulating $BaTiO_3$, *Nat. Commun.*, **4**, p. 1808.

138. Geng, Y., Lee, N., Choi, Y. J., Cheong, S. W., and Wu, W. (2012), Collective magnetism at multiferroic vortex domain walls, *Nano. Lett.*, **12**, p. 6055.

139. Zhang, Q. M., Wang, H., Kim, N., and Cross, L. E. (1994), Direct evaluation of domain-wall and intrinsic contributions to the dielectric and piezoelectric response and their temperature dependence on lead zirconate-titanate ceramics, *J. Appl. Phys.*, **75**, p. 454.

140. Yan, W., Phillips, L. C., Barbone, M., Hämäläinen, S. J., Lombardo, A., Ghidini, M., et al. (2016), Long spin diffusion length in few-layer Graphene flakes, *Phys. Rev. Lett.*, **114**, p. 147201.

Chapter 6

Spintronic Functionalities in Multiferroic Oxide-Based Heterostructures

I. Fina[a,b] and X. Martí[c]
[a]*Institut de Ciència de Materials de Barcelona (CSIC), Campus de Bellaterra, 08193 Bellaterra, Spain*
[b]*Departament Física Aplicada i Òptica, Universitat de Barcelona, Martí i Franquès 1, 08028 Barcelona, Spain.*
[c]*Institute of Physics, Academy of Sciences of the Czech Republic, v.v.i., CZ–16253 Praha 6, Czech Republic*
ignasifinamartinez@gmail.com

6.1 Introduction

Spintronics [1] is the area that merges electronics with the spin functionalities, and it is a rapidly developing area that has already delivered commercial devices, viz. non-volatile magnetic random access memory (MRAM) [2]. Spintronics foresees future storage memories, and other information technologies (with enhanced properties) based on spin. Currently, spintronics relies on ferromagnetic materials. In ferromagnets, the magnetic moment can be modified by the application of an external magnetic field, and therefore "0"s and "1"s can be written (Fig. 6.1). The two magnetic

Oxide Spintronics
Edited by Tamalika Banerjee

ISBN 978-981-4774-99-4 (Hardcover), 978-0-429-46819-3 (eBook)
www.jennystanford.com

memory states are antiparallel, and these can be read by means of an adjacent magnetic layer (separated by a non-magnetic one) with fixed magnetization. The use of this architecture leads to the occurrence of giant magnetoresistance (GMR, Nobel Prize 2007 [3]). GMR can be used to electrically read two magnetic states. Tunnel magnetoresistance (TMR) can be also used to read magnetic information. For TMR, a magnetic tunnel junction architecture, made of two magnetic layers (one fixed and the other containing the magnetic information as in GMR) separated by a thin insulating layer, is used. The TMR in magnetic tunnel junctions is a consequence of spin-dependent tunneling, which means that electron tunneling (i.e., device resistance) depends on its spin state and the differences in the probability of their tunneling while crossing the thin insulating layer. Thus, the difference in the measured resistance defined by the two different magnetic states is larger than in GMR devices. Older and well-known phenomena are magnetoresistance (MR), and anisotropic magnetoresistance (AMR). MR consists in any variation of the resistance state in a magnetic (or non-magnetic) material under the application of an external magnetic field and its origin can be diverse, as it will be discussed later in detail. AMR is a function of the microscopic magnetic moment vector; it is the direction of the spin-axis rather than the direction of the macroscopic magnetization that determines the effect. Therefore, AMR can be used to read perpendicular, instead of antiparallel magnetic states.

Recent reports from SIA and SRC, both semiconductor industry associations, point out that the continuous increase of data storage density is reaching its saturation [4]. Hence, the semiconductor industry finds that an even more important requirement than data density increase is to improve the energy efficiency. Chappert et al. emphasized in their review entitled *The Emergence of Spin Electronics in Data Storage* that "writing is the problem" [1]. In longitudinal recording systems, the magnetization of the recorded bit lies in the plane of the disk. An inductive write element records the data in horizontal magnetization patterns. Alternatives, to simplify and/or decrease the power consumption of the longitudinal recording systems, have been explored. For hard disk device, see Ref. [5] as a complete review.

The mentioned spintronic effects (TMR, GMR, AMR, and MR), which allow the control of electron flow by magnetization, have their reciprocal effect. Thus, the magnetization can be controlled by current injection. This is a recently developed writing technique that does not require the presence of any external applied magnetic field, and it is called the *spin-transfer torque* [6]. Although direct writing by electric current presents the convincing advantages of confinement of the switching area and large reduction of consumed energy compared with techniques where an applied magnetic field is required, the amount of consumed energy is still large, and the increase of energy required to keep the temperature of the cell caused by dissipated energy by Joule effect is an issue [7]. For these reasons, there is still a growing interest in finding alternative procedures to process magnetic information and write on spintronic devices using electric fields [8, 9], and fully insulating structures completely avoiding the presence of electric currents. Studies have reported that the current can be replaced by electric fields, which can allow saving a significant amount of energy in magnetic tunnel junctions [10]. However, this method cannot avoid the presence of high current due to the rather low resistance of magnetic tunnel junction (MTJ).

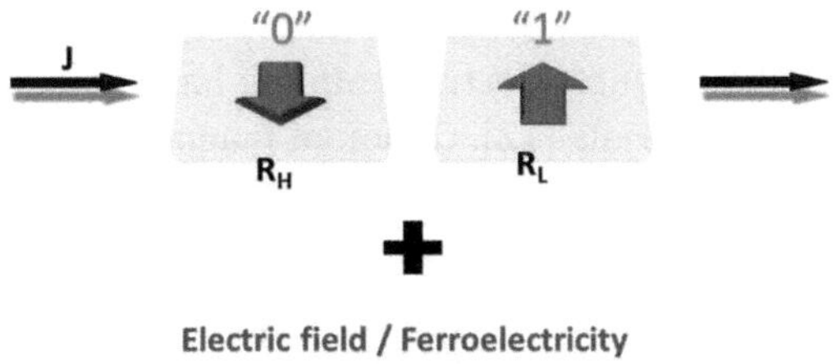

Figure 6.1 In spintronic-based memory devices, different magnetic states correspond to different memory states. The read-out of the different magnetization states is easy owing to the so-called spintronic functionalities (i.e., magnetoresistance, anisotropic magnetoresistance, giant magnetoresistance, or tunnel magnetoresistance). Therefore, two different memory states correspond to different resistive states (RH and RL). The manipulation of magnetization in traditional spintronic devices is achieved by means of the generation of an external magnetic field, which is costly in terms of power consumption and dissipation. In some multiferroic materials, where ferromagnetism and ferroelectricity are combined, the electric field can be used to modify the magnetic state, avoiding the requirement of large power consumption and dissipation. Multiferroic materials can be read-out by the same spintronic functionalities as in pure ferromagnetic materials.

Multiferroics are an interesting alternative to pure ferromagnets. Multiferroic materials are those materials where one can find the coexistence of more than one ferroic order. Of technological relevance is the fact that one can find coexistence of both ferromagnetic (switchable net magnetic moment under application of magnetic field) and ferroelectric (spontaneous surface charge switchable by an electric field) orders. If coupling between them exists, one can envisage the control of surface charge by the magnetic field (so-called *direct magnetoelectric effect*), or the control of magnetization by an electric field (so-called *converse magnetoelectric effect*). As far as the ferroelectric nature of the envisaged material would guarantee its insulating nature, electric currents through the material would be avoided. The absence of currents can lead to an important decrease in the power consumption of the writing procedure. Moreover, the Joule heating issue inherent to the presence of currents will be also avoided, resulting in an important decrease of the refrigeration demand of a multiferroic-based electronic element compared with the ferromagnetic-based ones.

Important milestones in the control of magnetic order by an electric field have been achieved without the use of oxide compounds [11]. However, scarce literature is available on the use of materials that are not oxides, signaling the relevant role of oxides in this field.

In oxides, the magnetoelectric coupling has been demonstrated in single-phase materials [12]. Cycloidal magnets are an important example of single-phase magnetoelectric materials; however, these are not proper multiferroic materials, because they are antiferromagnetic. Moreover, these only show large effects at low temperatures, making them, in principle, uninteresting for applications. The fact is that single-phase multiferroic/magnetoelectric materials at room temperature are scarce makes composite materials an interesting alternative to them. Composite materials are a combination of ferromagnetic and ferroelectric materials at room temperature; therefore the material resulting from their combination must be multiferroic also at room temperature (if structural properties are preserved). Usually, in most of the studied systems, both or one of them is an oxide. In composite materials, the magnetoelectric coupling is always mediated at the interface and several effects can make it possible [13, 14].

If multiferroic/magnetoelectric materials are technologically exported, they will make use of a spintronic functionality as a read-out technique. Spintronic probing techniques have been broadly used to probe magnetoelectric coupling. Therefore, it is very relevant to analyze (as we will do in the present chapter) the results obtained up to now on systems where spintronic functionalities are characterized on multiferroic systems. Here we will not only focus on systems that show magnetoelectric coupling but also on these that, even being multiferroic, do not show coupling. In the present work, we will classify important results published up to now in the participating spintronic functionality:

(a) Tunnel magnetoresistance
(b) Giant magnetoresistance
(c) Anisotropic magnetoresistance
(d) Magnetoresistance or resistance manipulation by an electric field

These four spintronic functionalities can be combined with multiferroicity in different manners. The following classification helps to visualize the level of integration between the spintronic functionality and the multiferroic material:

(a) Integrated: The multiferroic material is, at the same time, a part of the spintronic device, therefore the resistive state would change accordingly with the magnetic state of the multiferroic material [Fig. 6.2(a), top panel].

(b) Coupled: A ferroelectric or multiferroic material changes its electric or magnetic state, and it is somehow coupled with a spintronic device producing also a change in resistance resulting from a change in the magnetic state [Fig. 6.2(a) bottom panel].

From this, the latter group (coupled spintronic-multiferroic structure) is the envisaged oxide-based magnetoelectric random access memories drawn by M. Bibes and A. Barthelemy in Ref. [15] in 2008, Fig. 6.2(b). In Ref. [15], $BiFeO_3$ was proposed as an active material, and the tunneling device was a passive structure used to read the change in the magnetic state of $BiFeO_3$ induced by an electric field. Strictly speaking, $BiFeO_3$ is not a multiferroic material. Like the aforementioned cycloidal magnets, $BiFeO_3$ is a robust ferroelectric but it is antiferromagnetic at room temperature. In the envisaged structure, the coupling between the antiferromagnetic and electric order in the material is transferred to the tunneling device through magnetic exchange coupling. We will recall $BiFeO_3$ in this chapter due to its relevant role in the field of multiferroics.

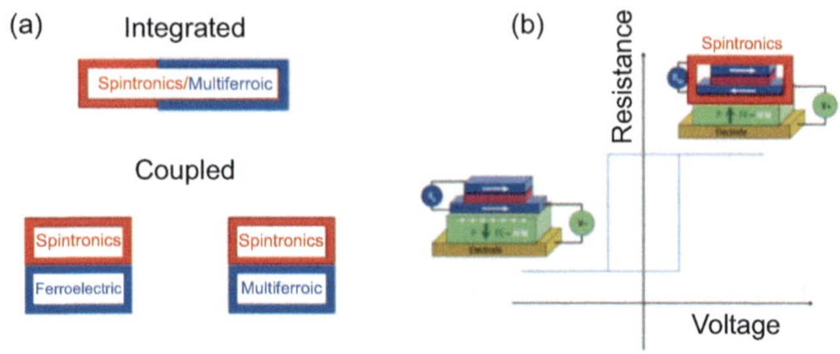

Figure 6.2 (a) Classification of multiferroic/spintronic device according to the integration of the multiferroic material in the spintronic device. The upper panel shows the integrated multiferroic-spintronic structure where the multiferroic material is part of the spintronic device. The bottom panels show coupled multiferroic-spintronic structures, where the multiferroic or ferroelectric material is used to manipulate the magnetic state in a spintronic structure. (b) Sketch of a possible coupled multiferroic-spintronic device. The binary information is stored by the magnetization direction of the bottom ferromagnetic layer (blue) part of a magnetic tunnel junction, which can be electrically read-out through tunnel magnetoresistance. The two parallel (left) or antiparallel (right) states of the magnetic junction can be modified by the voltage applied to the multiferroic material ($BiFeO_3$) underneath. The magnetization of the bottom ferromagnetic layer is coupled to the spins in the multiferroic (small white arrows). In the presence of large magnetoelectric coupling in the multiferroic material, the magnetic state of the interface can be modified reversing the ferroelectric polarization, and concomitantly the magnetic and resistive state of the magnetic tunnel junction on top. Reprinted from Ref. [15], with permission from Springer Nature, Copyright 2008.

In the following sections, we will describe the works on multiferroic oxides where tunnel magnetoresistance, giant magnetoresistance, anisotropic magnetoresistance, and magneto-resistance are characterized. We will also describe the few works done on spintronic characterization at multiferroic domain walls, which belongs to a particular class of spintronic functionality in a multiferroic system. We arrange the sections from TMR, the most appealing spintronic functionality, to MR, in principle the less attractive one. Therefore, the reader will see that the degree of complexity throughout sections decreases. In this work, we do not include results obtained by other spintronic functionalities such as spin Hall effect, or anomalous spin Hall effect, since the work on these topics using multiferroic, reported up to the moment this chapter was written, was scarce.

6.2 Tunnel Magnetoresistance

In magnetic tunnel junctions, metallic ferromagnetic materials acting as emitting, and receiving electrodes are spaced by a very thin layer of an insulating non-magnetic material. TMR results from the spin-dependent tunneling, whose origin can be found in the splitting of electronic bands for up and down spin states [band scheme of Fig. 6.3(a)]. Therefore, the population of carriers for each spin state is different, and the probability to tunnel from the emitting to the receiving electrode in MTJ is different if the magnetic state of the electrodes is antiparallel or parallel [Fig. 6.3(a)]. TMR is defined as TMR = $(R_{AP} - R_P)/R_P$, where R_{AP} is the resistance state for the magnetic configuration where both electrodes have antiparallel magnetization, and R_P is the resistance state for the magnetic configuration where both electrodes have parallel magnetization. The simplest description of the different resistive states is given by Julliere's model that relates the magnitude of TMR with the spin polarization [16].

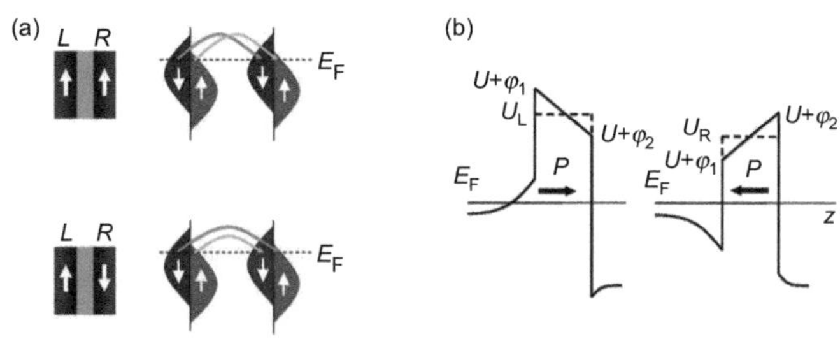

Figure 6.3 (a) Schematic representation of the two possible resistive states in a magnetic tunnel junction. In the top panel, both electrodes are in parallel configuration and the probability of an electron to tunnel from one to the other is different from the case when their magnetic state is antiparallel, as shown in the bottom panel. Adapted from Ref. [17]. (b) Schematic representation of the potential profile in a Metal-FE-Metal junction for polarization pointing to the left and right, where δ/ε ratio is smaller for the metal on the left than for the one on the right. ε is the metal permittivity and δ is its screening length. The dashed lines show the average potential for tunneling electrons across the ferroelectric barrier. The horizontal solid line denotes the Fermi energy, E_F. Adapted from Ref. [18] and corrected as in Ref. [19].

Non-equivalent to TMR, but with the similar consequences, is the tunneling electroresistance effect (TER). In a tunnel junction where the insulating spacer is ferroelectric and the two electrodes are not necessarily ferromagnetic but with necessarily different δ/ε ratio (where ε is the metal permittivity and δ is its screening length), the probability of an electron to tunnel is dictated by the shape of the potential barrier. As sketched in [Fig. 6.3(b)], depending on the direction of the polarization of the ferroelectric materials, the barrier profile is different. Therefore, an electron will have more probability to tunnel if the polarization is pointing to the left than if it is pointing to the right. The scenario and the consequence depicted in [Fig. 6.3(b)] correspond to the situation where the left electrode has a smaller δ/ε ratio than the right one. TER is defined as TER = $(R_\uparrow - R_\downarrow)/R_\downarrow$, where $R_\uparrow$, $R_\downarrow$ correspond to the polarization states for up (right) and down (left) polarization, respectively, although one can also find results in the literature where TER is defined as TER = $(R_H - R_L)/R_L$, where R_H, R_L correspond to the high and low resistive states, respectively. Even though the concept behind TER effect was proposed time ago [20], it was not until the observation of FE order for very thin ferroelectric films [21–23] that the feasibility of a tunnel ferroelectric barrier was demonstrated.

In the mentioned scenario, where the electrostatic effect is governing the change in tunneling current, the correspondence between the resistance state (high or low), and the polarization (up or down) is one to one. Therefore, for the case depicted in Fig. 6.3(b), the high resistive state corresponds to polarization pointing to the right, and the low resistive state corresponds to polarization pointing to the left. However, one can find that literature is not always coherent. This is because the presented scenario is not the only applicable one and other effects might be more predominant or coexisting. The first alternative results from changes in the bonding between atoms upon switching of ferroelectric polarization that can have a deep impact on the nature of the electronic configuration at the interface. Thus, the effective work function (φ) changes, resulting in a change of the electrostatic potential, and a concomitant change of the tunneling current. The second alternative is the presence of piezoelectricity. Piezoelectricity* is inherent to ferroelectric

*Piezoelectricity is defined as the change of electric polarization by strain in the direct effect and the change of strain by an electric field in the converse one, occurring in some non-centrosymmetric crystals.

materials and it can also result in a change of the actual insulating layer thickness, because of changes in the polar state. However, here if one neglects the presence of electric fields, externally imposed, or built-in (as it is the ferroelectric imprint field), the two resistance states would correspond to $P = \pm Ps$ and $P = 0$, since strain depends on the P absolute value [24, 25]. In the presence of an external electric field (easy to find in a FETJ where electrodes with different work-functions are used), the piezoelectric hysteresis loop shifts along the voltage axis resulting in two different strain states at electric remanence. Finally, one must take into account that ionic conduction can also be responsible for important changes in resistance without or partially without any important role of ferroelectric polarization and tunneling current [26].

Having established the TMR and TER effects, the combination of ferroelectric, and ferromagnetic materials, or by using a multiferroic material in a tunnel junction architecture results in the so-called multiferroic tunnel junctions, where TER and TMR effects can exist, and in the presence of magnetoelectric coupling, can cross-talk. Multiferroic tunnel junctions can be divided into two different big groups:

(a) **Single-phase**, where the insulating spacer is multiferroic itself and one or both electrodes are ferromagnetic.

(b) **Composite**, where the insulating spacer is a ferroelectric or a multiferroic and one or both of the electrodes are ferromagnetic.

In some cases, the spacer is a multiferroic, but only its ferroelectric character plays a role in the tunneling effect. In this case, we consider that the junction belongs to the second group.

6.2.1 Single Phase

The studies on magnetic tunnel junctions where the spacer is a single-phase multiferroic material are very limited due to the usually inherent leaky character of multiferroic materials. Leakage current can easily hide the presence of tunneling current. In fact, $BiMnO_3$ is the only material that has shown the coexistence of TMR and TER [27]. $BiMnO_3$ is multiferroic, but with ferromagnetic T_C occurring at very low temperature, which limits applications. In Fig. 6.4(a), it is shown the TMR curve for a FM/FM-I/non-magnetic tunnel junction (in this case the FM-I is also ferroelectric, LSMO/$BiMnO_3$/Au) after applying appropriate electric prepoling pulses. For prepoling pulses of opposite sign, polarization switches and the resistance changes (as a result of TER), and the overall TMR curve shifts. Moreover, there is a slight difference in the TMR value [27% (after −2 V) and 35% (after +2 V)] indicating some coupling between electric and magnetic states. This effect will be discussed in more detail in the section devoted to TMR on composite multiferroics, where the found effects are larger. Therefore, neglecting the small coupling observed, the tunnel magnetoresistive experiments in $BiMnO_3$ have shown the ability of this material to show four resistive states [two-electric and two-magnetic, Fig. 6.4(b)], becoming the first demonstration of a multiferroic memory.

Single-phase four states multiferroic memory at room temperature might have some interest for some niche of applications. Recent works have shown that there is a short list of available single-phase multiferroic materials at room temperature. In some cases with observed magnetoelectric coupling, ε-Fe_2O_3 [28, 29], $Ga_{2-x}Fe_xO_3$ [30, 31], $(Ga,Fe)_2O_3$ [32], $Pb(Zr,Ti)O_3$–$Pb(Fe,Ta)O_3$ [33–35], $Pb(Fe, M)_x(Zr,Ti)_{(1-x)}O_3$ [M = Ta, Nb] [36], $(1-x)BiTi_{(1-y)/2}Fe_yMg_{(1-y)/2}O_3$–$(x)CaTiO_3$ [37]. However, reproducing the TMR and TER experiments on $BiMnO_3$ at room temperature remains elusive.

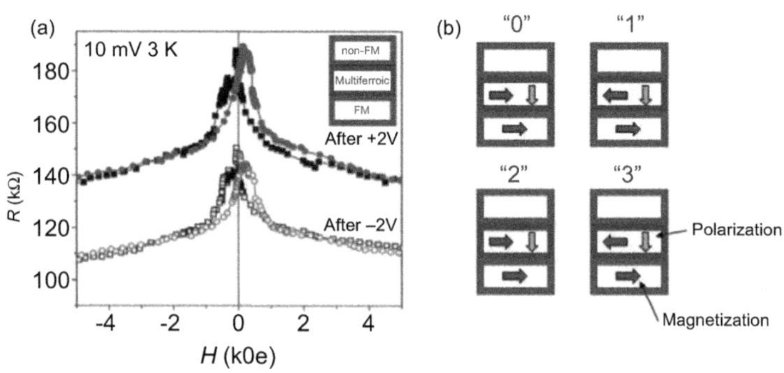

Figure 6.4 Single-phase multiferroic tunnel junction. (a) Tunnel magnetoresistance curves at 3 K and 10 mV for V = +2 V (solid symbols), and −2 V (open symbols) prepoling voltage for LSMO/$BiMnO_3$/Au. Reprinted from Ref. [27], with permission from Springer Nature, Copyright 2007. (b) Schematics of the four resistive states (two-magnetic and two-electric) that can be obtained in a multiferroic tunnel junction, where the multiferroic material is the spacer, as in LSMO/$BiMnO_3$/Au junction case.

6.2.2 Composite Multiferroics

Now, we recall the device sketched in Fig. 6.2(b) where TMR is controlled by an electric field. The experimental realization of the mentioned device remains elusive; however, alternative tunnel junctions where multiferroicity is present have been realized, and the obtained results are described as follows.

The simplest multiferroic tunnel junction is that where the spacer is ferroelectric and the emitting and the collecting electrodes are ferromagnetic [sketch in Fig. 6.5(a)]. In this type of devices, magnetoelectric coupling (if present) takes place at the interface via a list of phenomena: elastic coupling between ferromagnetic/magnetostrictive material and a ferroelectric/piezoelectric one, modification of charge doping by an electric field and the concomitant modification in the magnetic ordering, and orbital reconstruction upon ferroelectric switching [13, 14]. In fact, it must be stated that in most of the reported systems, all these phenomena must coexist together. Distinguishing between the predominant one might be difficult and in some cases impossible, which superimposes to the different phenomena that can result in TER as mentioned in the introduction. However, the experimental facts that probe magnetoelectric coupling in these systems are irrefutable.

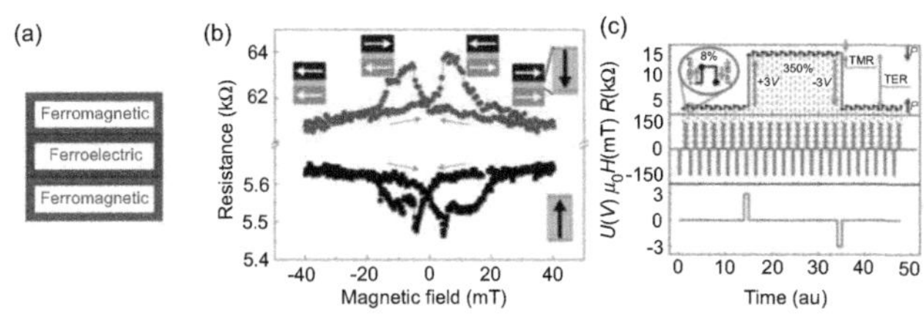

Figure 6.5 (a) Results on ferromagnetic/ferroelectric/ferromagnetic tunnel junctions. (b) TMR for polarization pointing backwards Co (top) and towards Co (bottom) in an LSMO/PZT/Co junction. Reprinted from Ref. [38], with permission from Springer Nature, Copyright 2012. (c) Sequentially, modified four resistive states (magnetic and electric) (top panel) measured at remanence with magnetic (middle panel), and electric pulses (bottom panel). Adapted from Ref. [39], with permission from Springer Nature, Copyright 2015. In the figure, all experiments were performed at a low temperature.

Of interest for us is that the spin polarization (the one that determines the TMR) can be modified by the electrically induced changes of magnetic order. Therefore, TMR value varies with an electric field, this is what we have called *modulation of TMR* (also applicable for GMR, AMR, and MR), or tunnel electro-magneto-resistance (TEMR), defined as the relative variation of TMR upon polarization switching TEMR = $(TMR_H - TMR_L)/TMR_L$, where TMR_H, TMR_L correspond to the TMR values for high and low values, respectively. In LSMO/FE/Fe [40, 41] or Co [38, 39] structures, this fact has been shown. In the LSMO/FE/Fe, large negative TMR was found for polarization pointing to Fe, and smaller when pointing away from it. In further investigations [41], it was shown that the change in the spin polarization was finding its origin on the ferromagnetic-like character of $BaTiO_3$, thus providing an interface possessing both magnetic and electric orders. Remarkably, when replacing Fe by Co [38], the TMR value (smaller in amplitude) was also negative for polarization pointing to Co, but positive when pointing away from it, thus meaning that the major spin population in one of the electrodes was changing its sign upon ferroelectric switching [Fig. 6.5(b)]. A combination of the ferromagnetic-like character antiferromagnetically coupled to one of the ferromagnetic electrodes (Co, in Co/PTO/LSMO structures) was found to be at the origin of large exchange bias effect in multiferroic tunnel junction without requiring any antiferromagnetic, or hard magnetic layer, thus producing a multiferroic "spin valve" [Fig. 6.5(c)] [39]. A similar spin-valve effect was reported using an antiferromagnetic layer in LSMO/BTO/Co/IrMn, in this latter case with modulation of TMR [42]. The last example of a multiferroic junction where exchange bias is present is the reported LSMO/BFO/LSMO junction, but here the exchange bias effect is low due to the symmetric configuration of the junction [43]. Nonetheless, TMR can also be modified by the electric field in a multiferroic tunnel junction by modifying the electric properties of the insulating spacer. Hambe et al. [44], showed a reversible modulation of TMR from 61 to 69 % depending on the sign of the prepoling voltage. However, in this case, it is argued that ionic displacements are at the origin of the observed modulation of TMR without requiring the presence of magnetoelectric coupling.

Strongly correlated systems, where charge, magnetism, and strain are intimately coupled are very attractive from a fundamental and applications point of view [45]. The electric field effect on magnetic properties was a rapidly emerging field, achieving large effects in a relatively short period of time [46, 47]. In recent works, the large electrically induced changes in the magnetic/electric properties of manganites have been used to modulate TMR/TER [48, 49].

Also by modulating the magnetic properties, and subsequently the transport properties, of one of the ferroelectric/ferromagnetic interfaces in a multiferroic tunnel junction, large pure TER can be obtained. This can be done by modulating the electronic phase in a manganite depending on the ferroelectric polarization. Therefore, the manganite layer $La_{2/3}Sr_{1/3}MnO_3$ can be either metallic or insulating, thus decreasing or increasing the width of the tunneling barrier. This might also result in modulation of TMR [48, 49] by an electric field, similar to [40], but with different origin. Ferroelectric polarization switch can result also in the modulation of the transport properties in the ferroelectric itself, resulting also in the modulation of the tunneling width. In LSMO/BTO/Pt large values for TER (up to 3×10^4 %) can be obtained [50]. In Co/PTO/LSMO, the modulation of the tunneling thickness due to metallization of the last layers of the ferroelectric material at the PTO/Co interface was at the origin of large TER (≥230 %) [51].

Now, we focus on a particular example of coupled multiferroic magnetic tunnel junctions. It is well-known that in granular magnetic materials, the grain boundaries can act as insulating domain walls, resulting in the observation of tunneling current (at very low temperature) in single films of manganite magnetic materials [52–54]. In grainy manganite/FE bilayers, it is a natural argument that the switching of the ferroelectric polarization that results in accumulation or depletion of carriers in the manganite layer can modulate not only its conductivity and/or magnetic state; but also the thickness of the insulating granular boundary thickness. Therefore, an electric modulation of the tunneling barrier height and/or width at the grain boundaries takes place. Thus, electroresistance up to near 1000% can be observed while T_C is being modulated by 16 K [55].

As we will also show the case of other spintronic functionalities, the characterization of magnetic tunnel junctions grown on top of a piezoelectric material, have recently given some interesting results obtaining a larger modulation of TMR by an electric field at room temperature [56, 57].

Finally, it must be stressed that the device sketched in Fig. 6.2(b) should not be restricted to the use of $BiFeO_3$. Large magnetoelectric coupling between antiferromagnetic and ferroelectric order has been demonstrated in other compounds such as $TbMnO_3$ [12], $TbMn_2O_5$ [58], and other $REMnO_3$ (RE = rare earth) compounds [59–62], among other more complex and more recently studied oxides [63]. In these materials, the coupling is intrinsic thus the ferroelectric polarization appears thanks to a particular magnetic order, a cycloidal one. However, we find two reasons that prohibit their application. First is that the cycloidal order results from magnetic frustration and therefore it can only appear at very low temperature. Second is that there is no clear procedure reported to exploit this magnetoelectric coupling in a structure where the cycloidal order is somehow coupled to a ferromagnetic order in an adjacent ferromagnetic layer.

6.3 Giant Magnetoresistance

If in a magnetic tunnel junction, one replaces the insulating spacer by a non-magnetic conductive material, the new device architecture is that of a GMR. In the GMR architecture, it is important to distinguish between two subgeometries: (a) transversal and (b) longitudinal. In the transversal (or current perpendicular to the plane, CPP), the current is perpendicular to the spacer plane. Therefore, if the magnetic alignment of the emitting and receiving electrode is parallel, the resistance is low, and if it is antiparallel, the resistance is high (in the archetypical case of positive GMR). As most of the current is scattered at the interface between layers in the longitudinal geometry (or current in plane, CIP), GMR is also present; however, as far as the current is parallel to the interface, the efficiency is lower. The former has the disadvantage of being more difficult to grow; this is the reason that in all the multiferroic structures showing GMR, the latter architecture has been used. As in TMR, GMR is defined as $GMR = (R_{AP} - R_P)/R_P$, where R_{AP} is the resistance state for the magnetic configuration where both metals have antiparallel magnetization, and R_P is the resistance state for the magnetic configuration where both metals have parallel magnetization.

In the GMR architecture, there is no insulating layer involved. This makes impossible to have a GMR where the multiferroic material is part of it since ferroelectricity only exists in insulating materials. Therefore, GMR architecture cannot be an integrated multiferroic-spintronic device. However, the rapid developmental studies on metallic or semiconducting ferroelectrics might allow it, at least from a fundamental point of view [64]. Therefore, the study of GMR in multiferroic devices is limited to the electric manipulation of GMR in a coupled spintronic-multiferroic structure, similar to that one introduced in the previous section and introduced in Ref. [15], replacing the magnetic tunnel junction with a GMR structure.

In $BiFeO_3$/FM multiferroic structures, the most relevant results have been obtained. The fact that $BiFeO_3$ is not ferromagnetic makes the participation of a coupled ferromagnetic material and magnetoelectric coupling necessary factors for the envisaged device. In the presence of these necessary factors, GMR is an excellent option to probe changes in magnetic order [Fig. 6.6(a)]. In early works, it was already demonstrated the presence of coupling between magnetism and applied electric field in equivalent systems [65, 66]. Afterwards, it was shown that the cycloidal order superimposed to the collinear antiferromagnetic order and the ferroelectric order was coupled [67] [Fig. 6.6(b)] being that the fundamental mechanism for the earlier found coupling. Owing to the presence of this type of coupling, it was demonstrated that the magnetic order in an adjacent to $BiFeO_3$ ferromagnetic layer could be rotated by the application of an electric field in $BiFeO_3$ [68] [Fig. 6.6(c)]. In previous experiments, AMR [69] was used to probe the magnetic order, and therefore the experiments will be discussed in more details in the pertinent section. Remarkably, it was reported that the observed electrically stimulated GMR contrast, although small ($\approx$1.5%), was almost the same that the one obtained while sweeping the magnetic field [compare Fig. 6.6(c and d)] and thus reversing the magnetization by 180°. The small pure GMR was expected due to the used GMR device was a longitudinal one. However, the fact that both, electrically and magnetically induced changes in resistance were similar indicated that coupling was very efficient. This was argued to be owing to the particular ferroelectric switching procedure used.

Other systems where the magnetoelectric coupling is purely strain mediated or is argued to be strain mediated have been studied, since the early publication on PZT/spin-valve GMR structure [70]. The fact that all the characterization in piezoelectric/GMR systems was performed in CIP configuration, limited the obtained variation of GMR induced by an electric field to small numbers (<1%) [71–73].

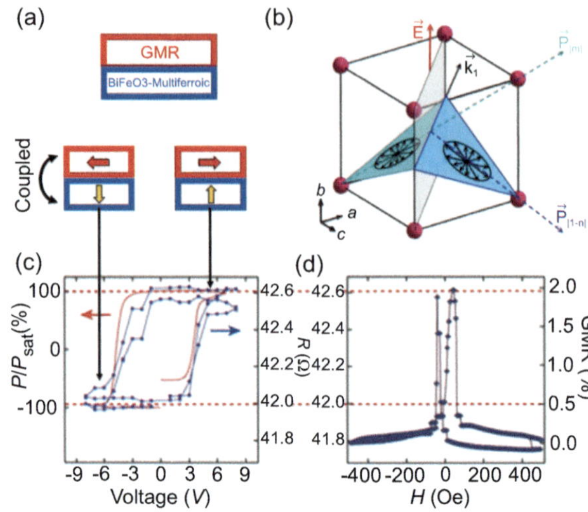

Figure 6.6 (a) The magnetic state of the free layer of the GMR structure (red) is coupled to the polarization of $BiFeO_3$, therefore if polarization on $BiFeO_3$ is manipulated the magnetic state changes and the resistive state in the GMR structure is modified accordingly. (b) Schematics of the rotation of the cycloidal plane accompanying the change of the polarization direction. Reprinted from Ref. [67], with permission from American Physical Society, Copyright 2008. (c) Change of polarization (red line) and resistance (blue points) as a function of the applied electric field in a BFO/CoFe/Cu/Co structure. (d) GMR in the CoFe/Cu/Co structure grown on top of BFO. Reprinted from Ref. [68], with permission from Springer Nature, Copyright 2014.

6.4 Anisotropic Magnetoresistance

Anisotropic magnetoresistance (AMR) was discovered by Lord Kelvin more than 100 years ago in a piece of iron [74]. The AMR results from the change of the conductivity of a ferromagnetic material depending on the direction of magnetization with respect to the injected current used to measure the resistance. It results that the resistivity (ρ) depends on the angle (θ) between the current and the applied magnetic field as $\rho(\theta) = \rho_\perp + (\rho_{||} - \rho_\perp) \cos 2\theta$ [75]; thus, AMR is an even function of the applied magnetic field and it can not be used to distinguish between antiparallel states, but only between perpendicular ones. The AMR amplitude is small in front of the amplitude that one can get using TMR or GMR architectures; however, it presents the enormous advantage of being a very simple technique. Similar to GMR, AMR cannot exist in multiferroic materials because of its insulating nature, and therefore we will focus on coupled spintronic-multiferroic structures, as shown in the top the sketches of Fig. 6.7.

Seminal work on the modulation of AMR by an electric field was shown by the characterization of Pt/$YMnO_3$/Py sandwich structures [76]. In this work, the electric control of the antiferromagnetic order in $YMnO_3$ (due to domain wall coupling between ferroelectric and antiferromagnetic domain walls as predicted by Goltsev et al. [77]), and consequently the exchange bias effect between Py and $YMnO_3$ allowed to strongly tune the AMR of the Py layer [Fig. 6.7(a)].

In manganite/ferroelectric structures [45, 77], modulation of the AMR was shown, as depicted in Fig. 6.7(b). In this latter case, the field induces a change in the doping state of the manganite films, resulting in an anisotropy change that modifies the observable AMR.

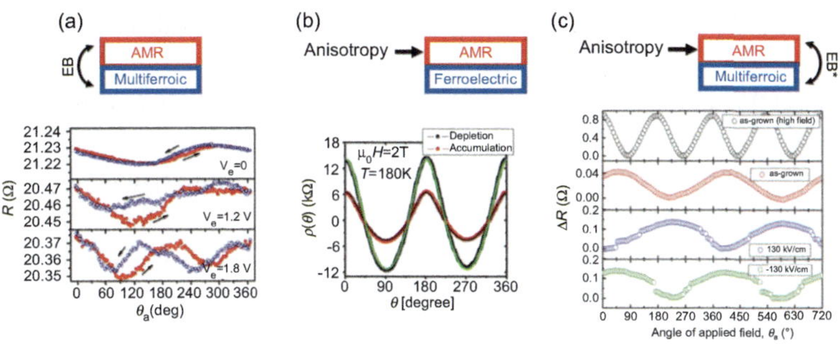

Figure 6.7 (a) Electric control of AMR in $YMnO_3$/Py structures, in which coupling is mediated by magnetic exchange coupling. Reprinted from Ref. [76], with permission from American Physical Society, Copyright 2006. (b) Electric control of transversal AMR in $La_{0.825}Sr_{0.125}MnO_3$/PZT heterostructures where, upon ferroelectric switching, the magnetic anisotropy of the manganite changes. Reprinted from Ref. [78], with permission from American Physical Society, Copyright 2014. (c) Electric control of AMR in BFO/Co structures. Reprinted from Ref. [69], with permission from American Physical Society, Copyright 2011.

Important changes of AMR by an electric field are those obtained by J. T. Heron et al. in BFO/Co structures [68]. In BFO/Co, as shown in Fig. 6.7(c), the AMR was shifting by 180° at room temperature. In the experiment, AMR was used to probe the alignment of the magnetic moment, and AMR measurements are performed at a very low magnetic field. The observed $\cos 2\theta$ can become $\cos\theta$ due to pinning of the magnetic moment that only allows the magnetization to slightly move around its pinned position. Therefore, the opposite magnetic states can be distinguished by a shift of 180° in the AMR data, as shown in Fig. 6.8(c), meaning that the modulation was −100%. It was argued the 180° magnetic reversal took place thanks to the coupling between the canted BFO moment and the ferroelectric polarization, which at the same time was coupled to Co.

As in the case of tunnel magnetoresistance, results on the electric modulation of AMR have been reported with the use of a piezoelectric material of large piezoelectric coefficient and magnetoelectric coupling mediated by strain. In Ref. [79], Hong et al. showed a sizeable, but small modulation of AMR in a $Ni/SiO_2/Ti/$(011)–PMN–PT heterostructure.

6.5 Magnetoresistance or Electric-Field Modulation of Resistance and Magnetism

Magnetoresistance is defined as the change of the resistivity of a material by the application of an external magnetic field. Magnetoresistance itself can have several different origins. The first one results from the action of Lorentz force and it takes place in any magnetic or non-magnetic metallic material. In ferromagnetic materials, its origin can also be due to grainy magnetoresistance or anisotropic magnetoresistance. In grainy magnetoresistance, the conductivity of the material changes according to the number of interfaces with different magnetic moment (domain walls) where stronger scattering exists. Therefore, one can find the maximum variation of resistance in the coercive field, where the number of domains and domain walls are maximum. In AMR magnetoresistance, the change in resistance results from the misalignment of some or all magnetic domains. Therefore, the angle between the injected current and the magnetic moment is an average non-zero, and the conductivity increases or decreases accordingly. Finally, magnetoresistance can also appear owing to the presence of spin–orbit coupling, the conductivity in this case is modified by any change in the magnetic moment or magnetic order.

Therefore, transport measurements are directly related to the magnetization. This has also allowed to infer changes in transition temperature in multiferroic structures. In ferroelectric/

ferromagnetic semiconductor devices, where the coupling is mediated by an electric field (ferroelectric field effect devices), T_C has been tuned by 5–10 K degree. Fully oxide structures, with a strongly correlated material (SCM) [80] acting as ferromagnetic showed the largest modulation on the magnetic properties induced by ferroelectric polarization reversal [47, 81]. The observed shift of the Curie temperature (T_C) for the ferroelectrically gated SCM is much larger (by 40 K) than the one measured for an FM-SC channel (compare, for instance, Refs. [81] and [82]). The effects are also large (TC shifts for 16K) in polycrystalline SCM/ferroelectric structures grown on Si [55].

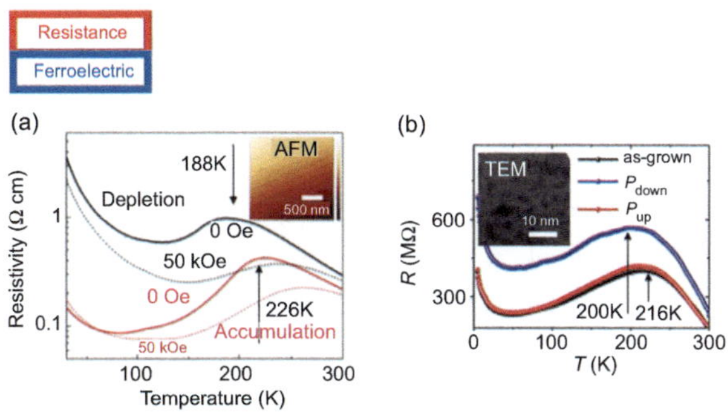

Figure 6.8 (a) Resistance versus temperature measured upon polarization switching inducing the accumulation/depletion of the carrier at the interface in a $PbZr_{0.2}Ti_{0.8}O_3/La_{0.8}Sr_{0.2}MnO_3/SrTiO_3(001)$ fully epitaxial heterostructure structure. Good quality of the film is shown by the AFM image included as an inset. Reprinted from Ref. [47], with permission from AIP Publishing, Copyright 2011. (b) Equivalent measurement than in panel (a), in a fully polycrystalline $PbZr_{0.2}Ti_{0.8}O_3/La_{0.825}Sr_{0.175}MnO_3$ structure grown on Si. In the inset, it can be observed the polycrystalline character of the sample revealed by a TEM image. Adapted from Ref. [55], with permission from Springer Nature, Copyright 2015.

In FeRh/piezoelectric structures, the measurement of the resistance was used to probe that owing to strain mediated coupling complete modulation of the antiferromagnetic-to-ferromagnetic transition was achieved by an electric field when using $BaTiO_3$ as a piezoelectric material [83], instead of the partial one obtained in the case that the used piezoelectric material was PMNPT [84–86].

In pioneering multiferroic structures, a standard measurement of resistance has been used to measure important variations of magnetization [87] where the magnetoelectric coupling was mediated by a field effect. The fact is that the measurement of the resistive state versus electric field can be used as a tool to probe magnetism and give an interesting insight into the physical mechanism triggering magnetoelectric coupling [88]. In Fig. 6.9(a), it is shown that the resistance in a $La_{0.7}Sr_{0.3}MnO_3$/PZT structure is a hysteretic function of the applied electric field at remanence, thus reproducing the typical ferroelectric loop like the shape of the ferroelectric underneath. Therefore, in this case, the coupling is mediated by field effect induced by the ferroelectric that modifies the charge density in the ferromagnet, and concomitantly its transport and magnetic properties. Instead, in Fig. 6.9(b), similar structure $La_{0.8}Ca_{0.2}MnO_3$/PZT shows butterfly shape piezoelectric loop. Thus, in this latter case, the coupling is elastic owing to the piezoelectric nature of the ferroelectric and the magnetostrictive effect of the ferromagnet.

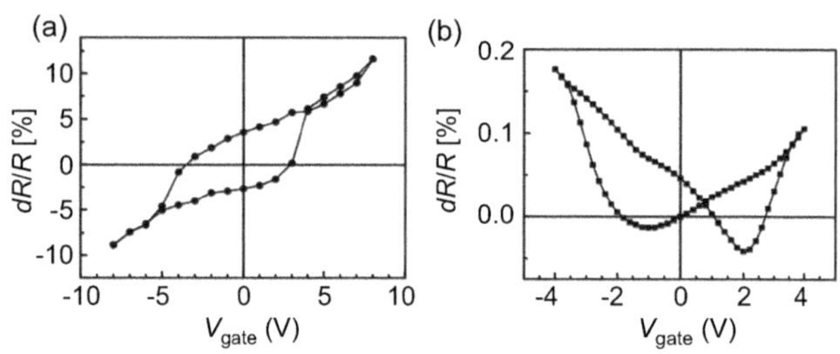

Figure 6.9 (a) Resistance modulation *dR/R* versus piezoelectric voltage in a $La_{0.8}Ca_{0.2}MnO_3$(10 nm)/PZT structure, with dominating field effect. (b) The same experiment than in Fig. 6.9(a), in $La_{0.7}Sr_{0.3}MnO_3$/PZT, where strain effect dominates. Reprinted from Ref. [88], with permission from John Wiley and Sons, Copyright 2006.

Direct experiments focused on the electric control of magnetoresistance are scarce [89].

6.6 Spintronic Functionalities at $BiFeO_3$ Domain Walls

Enhanced conductivity at BiFeO3 domain walls has been reported using proximity probe techniques [90]. Li doping in BFO allowed large improvement of insulating properties of ferroelectric domain walls in $BiFeO_3$. Therefore, allowing the characterization of transport properties of ferroelectric domain walls, dependence on temperature using macrocontacts. In Ref. [91], AMR at the domain walls revealed that a sizeable response can be observed, indicating their ferromagnetic character. Interestingly, the result showed a large hysteresis depending on the increase or decrease of the angle of the current with respect to the magnetic field. This was argued to result from the magnetic exchange produced between the ferromagnetic domain wall and the antiferromagnetic domains of $BiFeO_3$ [Fig. 6.10(a)]. The AMR was already observed in Ref. [92] at room temperature, but in this latter case using as-grown in-plane domain walls present at $BiFeO_3$ and in-plane contact geometry [Fig. 6.10(b)].

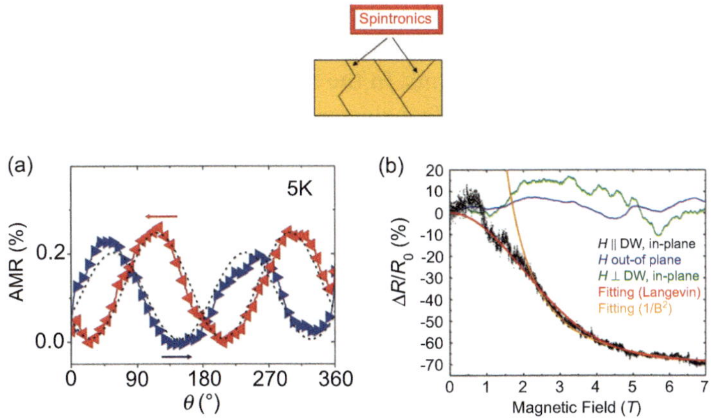

Figure 6.10 (a) AMR measured at $BiFeO_3$ domain walls. Reprinted from Ref. [91], with permission from John Wiley and Sons, Copyright 2014. (b) MR for applied magnetic fields along the different direction with respect to the as-grown BiFeO3 domain walls, showing the presence of AMR. Reprinted from Ref. [92], with permission from American Physical Society, Copyright 2012.

6.7 Conclusions and Perspectives

The tremendous effort done by the scientific community on the characterization of multiferroic materials has led to several results, important from the fundamental and technological points of view.

In Table 6.1, the results obtained in multiferroic systems where any spintronic functionality (TMR, AMR, GMR, MR) has been reported are summarized. Regarding TMR, we can conclude that there is a lack of work performed on systems that can work at room temperature. This is mainly because the multiferroic materials or structures employed so far lose their multiferroic character at room temperature, and electric or magnetic state–dependent tunneling only appears at low temperature. The exceptions are structures where the magnetic tunnel junction is not a multiferroic itself, in particular, the MTJ/piezoelectric structures, where electric control of TMR has been observed. Unlike the lack of results at room temperature reported for TMR, interesting results have been obtained in BFO-based structures for GMR and AMR. The absolute variations of resistivity in these latter systems are always low due to the low GMR values used in planar configuration and the low values of AMR. In addition, modulation of GMR and AMR at room temperature has been achieved in GMR/piezoelectric and AMR/piezoelectric structures similar to TMR/piezoelectric systems. In the spintronic/piezoelectric systems studied up to now, the piezoelectric material is a single crystal with the obvious disadvantage for applications. Regarding MR, because of the aforementioned reasons, the works are also limited, which make it difficult to extract any conclusion.

Therefore, two are the open venues regarding the electric control of spintronic functionalities:

(a) To expand the knowledge on magnetoelectric coupling in BFO-based or similar structure to achieve electric control of magnetic order that produces larger resistance variations in a coupled spintronic-multiferroic.
(b) To study the viability and the applicability of spintronic/piezoelectric systems, in particular, how the observed effects can be also preserved after nanostructuration, as required for miniaturized devices.

Finally, one should remark that the fact that AMR and MR were observed in BFO domain walls opens the door to new unknown applications, where the electric manipulation of spintronic nano-objects is necessary.

Table 6.1 Survey of the TER, ER, TMR, GMR, AMR, MR results obtained in several multiferroic systems based on oxide materials. The modulation is the variation of the spintronic functionality (*SF* = TMR, GMR, AMR, MR) upon the application of an electric field. Modulation = $(SF_{high} - SF_{low})/SF_{low}$)

Material	TER	ER	Modulation	TMR	GMR	AMR	MR	Temperature	Ref.
LSMO/BMO/Au	27%		30%	30%				3 K	[27]
LSMO/BFO/LSMO	45%		13%	69%				80.0	[44]
LSMO/BTO/Fe	16%		467%	45%				4.2 K	[40]
LSMO/PZT/Co	1000%		–3693%	10%				<200 K	[38]
LSMO/BTO/LCMO	8000%		∞	85%				<100 K	[48]
LSMO/PTO/Co	350%		33%	8%				<140 K	[39]
LSMO/BTO/Co/IrMn	10000%		100%	20%				10 K	[42]
LSMO/BFO/LSMO	15%		18%	3%				10 K	[43]
PMNPT/CoFeB/AlO/CoFeB		15%	#	40%				RT	[56]
PMNPT/Ta/CofeB/MgO/CoFeB/Ta		100%	#	100%				RT	[57]
PZT/Ta/IrMn/FeCo/Cu/NiFe/Ta		$	15%		2%			RT	[70]
BFO/CoFeB/Co		$	40%		80%			RT	[71]*
PZT/Pt/IrMn/Cu/Co/Cu/CoFeB/MgO		0.02%	≈0%		0.4%			RT	[72]

(*Continued*)

Table 6.1 (*Continued*)

Material	TER	ER	Modulation	TMR	GMR	AMR	MR	Temperature	Ref.
BFO/CoFe/Cu/CoFe		1.5%	$		2%			RT	[65]
BTO/Fe/Cu/Co		$	25%		2.5			RT	[73]
YMO/Py		100%	275%			0.1%		5 K	[76]
BFO/CoFeB		−100%	−100%			0.1%		RT	[69]
LSMO/PZT		50%	50%			1%		<250 K	[78]
PMNPT/Ti/SiO2/Ni		$	0.15%			1%		RT	[79]
PZT/LSMO(poly)		1000%	33.0%				50%	<200 K	[55]*
PMNPT/Co		0.15	≈0%				0.4%	RT	[89]
BFO domain walls		400%	≈0%			0.2%		5 K	[91]
BFO domain walls		$	$			$		10 K	[92]

TMR, AMR, and MR highest value if several values are reported
#Important shape modulation without change in GMR amplitude
*Irreversible
$Data not available
RT: room temperature

Acknowledgments

We acknowledge financial support from the Spanish MINECO (MAT2015-73839-JIN). We also acknowledge support from the ERC Advanced Grant No. 268066, from the Ministry of Education of the Czech Republic Grant No. LM2015087, from the Grant Agency of the Czech Republic Grant No. 14-37427. ICMAB-CSIC authors acknowledge financial support from the Spanish Ministry of Economy and Competitiveness, through the "Severo Ochoa" Programme for Centers of Excellence in R&D (SEV-2015-0496). I.F. acknowledges Beatriu de Pinós postdoctoral scholarship (2011 BP-A_2 00014) from AGAUR-Generalitat de Catalunya and Juan de la Cierva-Incorporación postdoctoral fellowship (IJCI-2014-19102) from the Spanish Ministry of Economy and Competitiveness of Spanish Government. I. F. acknowledges Ramón y Cajal contract RYC-2017-22531. J. Sort and M. Qian are acknowledged for critical reading.

References

1. C. Chappert, A. Fert, and F. N. Van Dau, *Nat. Mater.*, **6**, p. 813, 2007.
2. J. Daughton, Magnetoresistive Random Access Memory (MRAM), http://www.nve.com/Downloads/mram.pdf, 2000.
3. *The Nobel Prize in Physics 2007 — Advanced Information*, Nobelprize.org., Nobel Media AB 2014, http://www.nobelprize.org/nobel_prizes/physics/laureates/2007/advanced.html, Accessed date: 14/10/2016.
4. *Rebooting the IT Revolution: A Call to Action (SIA/SRC, 2015)*, available at http://go.nature.com/urvkhw, Accessed date: 10/10/2016.
5. A. Moser, K. Takano, D. T. Margulies, M. Albrecht, Y. Sonobe, Y. Ikeda, S. Sun, and E. E. Fullerton, *J. Phys. D: Appl. Phys.*, **35**, p. R157, 2002.
6. J. C. Slonczewski, *J. Magn. Magn. Mater.*, **159**, p. L1, 1996.
7. E. Kültürsay, M. Kandemir, A. Sivasubramaniam, and O. Mutlu, in *Performance Analysis of Systems and Software (ISPASS), IEEE International Symposium on* (IEEE, 2013), p. 256, 2013.
8. D. E. Nikonov and I. A. Young, *J. Mater. Res.*, **29**, p. 2109, 2014.
9. S. Manipatruni, D. E. Nikonov, and I. A. Young, *Nat. Phys.*, **14**, p. 338, 2018.

10. Y. Shiota, T. Nozaki, F. Bonell, S. Murakami, T. Shinjo, and Y. Suzuki, *Nat. Mater.,* **11**, p. 39, 2012.

11. I. Stolichnov, et al., *Nat. Mater.,* **7**, p. 464, 2008.

12. T. Kimura, T. Goto, H. Shintani, K. Ishizaka, T. Arima, and Y. Tokura, *Nature,* **426**, p. 55, 2003.

13. J. Ma, J. Hu, Z. Li, and C. W. Nan, *Adv. Mater.,* **23**, p. 1062, 2011.

14. C. A. F. Vaz, *J. Phys.: Condens. Matter,* **24**, p. 333201, 2012.

15. M. Bibes and A. Barthélémy, *Nat. Mater.,* **7**, p. 425, 2008.

16. M. Julliere, *Phys. Lett. A,* **54**, p. 225, 1975.

17. E. Tsymbal, 'Tunnel magnetoresistance', available at https://unlcms.unl.edu/cas/physics/tsymbal/reference/spin-dependent_tunneling.shtml. Accessed date: 21/09/2016.

18. M. Y. Zhuravlev, R. F. Sabirianov, S. Jaswal, and E. Y. Tsymbal, *Phys. Rev. Lett.,* **94**, p. 246802, 2005.

19. M. Y. Zhuravlev, R. F. Sabirianov, S. S. Jaswal, and E. Y. Tsymbal, *Phys. Rev. Lett.,* **102**, p. 169901, 2009.

20. L. Esaki, R. Laibowitz, and P. Stiles, 'IBM Technical Disclosure Bulletin', **13**, p. 114, 1971.

21. T. Tybell, C. Ahn, and J. M. Triscone, *Appl. Phys. Lett.,* **75**, p. 856, 1999.

22. C. Lichtensteiger, J. M. Triscone, J. Junquera, and P. Ghosez, *Phys. Rev. Lett.,* **94**, p. 047603, 2005.

23. M. Stengel, D. Vanderbilt, and N. A. Spaldin, *Nat. Mater.,* **8**, p. 392, 2009.

24. C. Newcomb and I. Flinn, *Electron. Lett.,* **11**, p. 442, 1982.

25. M. G. Cain, *Characterisation of Ferroelectric Bulk Materials and Thin Films,* Springer, 2014.

26. M. Li, et al., *Adv. Electron. Mater.,* **1,** 2015.

27. M. Gajek, M. Bibes, S. Fusil, K. Bouzehouane, J. Fontcuberta, A. Barthelemy, and A. Fert, *Nat. Mater.,* **6**, p. 296, 2007.

28. M. Gich, I. Fina, A. Morelli, F. Sánchez, M. Alexe, J. Gàzquez, J. Fontcuberta, and A. Roig, *Adv. Mater.,* **26**, p. 4645, 2014.

29. C. Kadlec, et al., *Phys. Rev. B,* **88**, p. 104301, 2013.

30. A. Thomasson, S. Cherifi, C. Lefevre, F. Roulland, B. Gautier, D. Albertini, C. Meny, and N. Viart, *J. Appl. Phys.,* **113**, p. 214101, 2013.

31. K. I. Kobayashi, T. Kimura, H. Sawada, K. Terakura, and Y. Tokura, *Nature,* **395**, p. 677, 1998.

32. S. H. Oh, J. H. Lee, R. H. Shin, Y. Shin, C. Meny, and W. Jo, *Appl. Phys. Lett.,* **106**, p. 142902, 2015.
33. D. Evans, A. Schilling, A. Kumar, D. Sanchez, N. Ortega, M. Arredondo, R. Katiyar, J. Gregg, and J. Scott, *Nat. Commun.,* **4**, p. 1534, 2013.
34. D. Evans, A. Schilling, A. Kumar, D. Sanchez, N. Ortega, R. Katiyar, J. Scott, and J. Gregg, *Philos. Trans. R. Soc. London, A* **372**, p. 20120450, 2014.
35. D. M. Evans, M. Alexe, A. Schilling, A. Kumar, D. Sanchez, N. Ortega, R. S. Katiyar, J. F. Scott, and J. M. Gregg, *Adv. Mater.,* **27**, p. 6068, 2015.
36. D. A. Sanchez, et al., *J. Appl. Phys.,* **113**, p. 074105, 2013.
37. P. Mandal, M. Pitcher, J. Alaria, H. Niu, P. Borisov, P. Stamenov, J. Claridge, and M. Rosseinsky, *Nature,* **525**, p. 363, 2015.
38. D. Pantel, S. Goetze, D. Hesse, and M. Alexe, *Nat. Mater.,* **11**, p. 289, 2012.
39. A. Quindeau, I. Fina, X. Marti, G. Apachitei, P. Ferrer, C. Nicklin, E. Pippel, D. Hesse, and M. Alexe, *Sci. Rep.*, **5**, p. 9749, 2015.
40. V. Garcia, et al., *Science,* **327**, p. 1106, 2010.
41. S. Valencia, et al., *Nat. Mater.,* **10**, p. 753, 2011.
42. H. Mao, et al., *J. Appl. Phys.,* **116**, p. 053703, 2014.
43. Y. Liu, Y. Yin, S. Dong, S. Yang, T. Jiang, and X. Li, *Appl. Phys. Lett.,* **104**, p. 043507, 2014.
44. M. Hambe, A. Petraru, N. A. Pertsev, P. Munroe, V. Nagarajan, and H. Kohlstedt, *Adv. Funct. Mater.*, **20**, p. 2436, 2010.
45. J. Hoffman, X. Pan, J. W. Reiner, F. J. Walker, J. P. Han, C. H. Ahn, and T. P. Ma, *Adv. Mater.,* **22**, p. 2957, 2010.
46. X. Hong, J. B. Yau, J. Hoffman, C. Ahn, Y. Bason, and L. Klein, *Phys. Rev. B,* **74**, p. 174406, 2006.
47. C. Vaz, J. Hoffman, Y. Segal, J. Reiner, R. Grober, Z. Zhang, C. Ahn, and F. Walker, *Phys. Rev. Lett.,* **104**, p. 127202, 2010.
48. Y. Yin, et al., *Nat. Mater.,* **12**, p. 397, 2013.
49. Y. Yin, et al., *J. Appl. Phys.,* **117**, p. 172601, 2015.
50. G. Radaelli, D. Gutiérrez, F. Sánchez, R. Bertacco, M. Stengel, and J. Fontcuberta, *Adv. Mater.,* **27**, p. 2602, 2015.
51. A. Quindeau, V. Borisov, I. Fina, S. Ostanin, E. Pippel, I. Mertig, D. Hesse, and M. Alexe, *Phys. Rev. B,* **92**, p. 035130, 2015.
52. R. Gross, et al., *J. Magn. Magn. Mater.,* **211**, p. 150, 2000.

53. K. Steenbeck, T. Eick, K. Kirsch, H. G. Schmidt, and E. Steinbeib, *Appl. Phys. Lett.,* **73**, p. 2506, 1998.

54. I. B. Shim, Y. J. Oh, and S. Y. Choi, *J. Korean Phys. Soc.*, **37**, p. 425, 2000.

55. I. Fina, G. Apachitei, D. Preziosi, H. Deniz, D. Kriegner, X. Marti, and M. Alexe, *Sci. Rep.*, **5**, p. 14367, 2015.

56. P. Li, et al., *Adv. Mater.,* **26**, p. 4320, 2014.

57. Z. Zhao, M. Jamali, N. D'Souza, D. Zhang, S. Bandyopadhyay, J. Atulasimha, and J.-P. Wang, *Appl. Phys. Lett.,* **109**, p. 092403, 2016.

58. N. Hur, S. Park, P. A. Sharma, J. S. Ahn, S. Guha, and S. W. Cheong, *Nature,* **429**, p. 392, 2004.

59. T. Kimura, G. Lawes, T. Goto, Y. Tokura, and A. Ramirez, *Phys. Rev. B,* **71**, p. 224425, 2005.

60. I. Fina, L. Fabrega, X. Marti, F. Sanchez, and J. Fontcuberta, *Appl. Phys. Lett.,* **97,** 2010.

61. I. Fina, L. Fabrega, X. Marti, F. Sanchez, and J. Fontcuberta, *Phys. Rev. Lett.,* **107,** 2011.

62. I. Fina, V. Skumryev, D. O'Flynn, G. Balakrishnan, and J. Fontcuberta, *Phys. Rev. B,* **88**, p. 100403, 2013.

63. T. Kimura, *Annu. Rev. Condens. Matter Phys.,* **3**, p. 93, 2012.

64. A. Filippetti, V. Fiorentini, F. Ricci, P. Delugas, and J. Íñiguez, *Nat. Commun.,* **7**, p. 11211, 2016.

65. Y.-H. Chu, et al., *Nat. Mater.,* **7**, p. 478, 2008.

66. S. Lee, W. Ratcliff, S. W. Cheong, and V. Kiryukhin, *Appl. Phys. Lett.,* **92**, p. 2906, 2008.

67. D. Lebeugle, D. Colson, A. Forget, M. Viret, A. Bataille, and A. Gukasov, *Phys. Rev. Lett.,* **100**, p. 227602, 2008.

68. J. Heron, et al., *Nature,* **516**, p. 370, 2014.

69. J. T. Heron, et al., *Phys. Rev. Lett.,* **107**, p. 217202, 2011.

70. D. H. Giang, V. Thuc, and N. Duc, *J. Magn. Magn. Mater.,* **324**, p. 2019, 2012.

71. J. Allibe, S. Fusil, K. Bouzehouane, C. Daumont, D. Sando, E. Jacquet, C. Deranlot, M. Bibes, and A. Barthelemy, *Nano Lett.,* **12**, p. 1141, 2012.

72. N. Lei, et al., *Nat. Commun.,* **4**, p. 1378, 2013.

73. S. S. Pillai, H. Kojima, M. Itoh, and T. Taniyama, *Appl. Phys. Lett.,* **107**, p. 072903, 2015.

74. W. Thomson, *Proc. R. Soc. London,* **8**, p. 546, 1856.

75. I. Campbell and A. Fert, *Ferromagnetic Materials*, EP Wolfarth, North-Holland, Amsterdam, 1982.
76. V. Laukhin, et al., *Phys. Rev. Lett.*, **97**, p. 227201, 2006.
77. A. Goltsev, R. Pisarev, T. Lottermoser, and M. Fiebig, *Phys. Rev. Lett.*, **90**, p. 177204, 2003.
78. D. Preziosi, I. Fina, E. Pippel, D. Hesse, X. Marti, F. Bern, M. Ziese, and M. Alexe, *Phys. Rev. B*, **90**, p. 125155, 2014.
79. B. Hong, et al., *Mater. Lett.*, **169**, p. 110, 2016.
80. C. Ahn, J. M. Triscone, and J. Mannhart, *Nature*, **424**, p. 1015, 2003.
81. H. J. Molegraaf, J. Hoffman, C. A. Vaz, S. Gariglio, D. Van Der Marel, C. H. Ahn, and J. M. Triscone, *Adv. Mater.*, **21**, p. 3470, 2009.
82. I. Stolichnov, et al., *Phys. Rev. B*, **83**, p. 115203, 2011.
83. Z. Liu, et al., *Phys. Rev. Lett.*, **116**, p. 097203, 2016.
84. Y. Lee, et al., *Nat. Commun.*, **6**, p. 5959, 2015.
85. J. D. Clarkson, et al., *Sci. Rep.*, **1**, p. 15460, 2017.
86. I. Fina, et al., *ACS Appl. Mater. Interfaces*, **9**(18), p. 15577, 2007.
87. T. Zhao, S. B. Ogale, S. R. Shinde, R. Ramesh, R. Droopad, J. Yu, K. Eisenbeiser, and J. Misewich, *Appl. Phys. Lett.*, **84**, p. 750, 2004.
88. K. Dörr and C. Thiele, *Phys. Status Solidi B*, **243**, p. 21, 2006.
89. C. Zhou, C. Zhang, J. Yao, and C. Jiang, *Appl. Phys. Lett.*, **109**, p. 112404, 2016.
90. J. Seidel, et al., *Nat. Mater.*, **8**, p. 229, 2009.
91. J. H. Lee, I. Fina, X. Marti, Y. H. Kim, D. Hesse, and M. Alexe, *Adv. Mater.*, **26**, p. 7078, 2014.
92. Q. He, et al., *Phys. Rev. Lett.*, **108**, p. 067203, 2012.

Chapter 7

Novel Functionalities in Oxide Magnetic Tunnel Junctions: Spin Filtering by Interface-Induced Magnetism

Zouhair Sefrioui,[a,b] **Yaohua Liu,**[c] **Carlos Leon,**[a,b]
Suzanne G. E. te Velthuis,[d] **Manuel Bibes,**[e]
Agnes Barthelemy,[e] **and Jacobo Santamaria**[f]

[a]*GFMC, Departamento de Física de Materiales, Universidad Complutense de Madrid, 28040 Madrid, Spain.*
[b]*Instituto de Magnetismo Aplicado, Universidad Complutense de Madrid, 28040 Madrid, Spain.*
[c]*Quantum Materials Division, Oak Ridge National Laboratory, Oak Ridge, TN 37831, USA*
[d]*Materials Science Division, Argonne National Laboratory, Argonne, Illinois 60439, USA.*
[e]*Unité Mixte de Physique CNRS/Thales, 1 Avenue A. Fresnel, 91767 Palaiseau, France and Université Paris-Sud, 91405 Orsay, France.*
[f]*Unité Mixte de Physique CNRS/Thales, 1 Avenue A. Fresnel, 91767 Palaiseau, France and Université Paris-Sud, 91405 Orsay, France*
jacsan@fis.ucm.es

Oxide Spintronics
Edited by Tamalika Banerjee

ISBN 978-981-4774-99-4 (Hardcover), 978-0-429-46819-3 (eBook)
www.jennystanford.com

7.1 Introduction

The discovery of the conducting state at the interface between $LaAlO_3$ and $SrTiO_3$, two-band insulators [1, 2] have generated a lot of interest aimed at exploring new electronic states at oxide interfaces but also driven to functionalize them into novel device concepts. The technological opportunities can be dramatically expanded if instead of the (in principle) relative inert-band insulators, one combines transition metal oxides with correlated electrons. Complex transition metal oxides are a wide family of materials, which contain elements with incomplete *d* shells. The localized *d* bands poorly screen the Coulomb interaction and give rise to electron correlations responsible for a strong entanglement between the various electronic and lattice degrees of freedom. This is at the origin of the rich phase diagrams of these compounds where the different electronic states with similar characteristic energies compete fiercely. Materials of this family can be found in almost every electronic or lattice ground state of the solid matter, including superconductivity, ferromagnetism, antiferromagnetism, ferroelectricity, multiferroicity, etc. [3–5], and thus display a wide variety of properties and responses.

Many complex oxides share a common perovskite structure with similar lattice parameters, which allow the growth of heterostructures that grow cube on cube preserving a high degree of crystalline perfection. Oxygen octahedra surrounding the transition metal ion arranged in the cubic sites are the basic building block of the structure. Distortions and rotations of the octahedra determine the degree of hybridization between 3*d* and oxygen bands and modify the hierarchy of the crystal field e_g and t_{2g} levels, which result from the Coulomb interaction with the oxygen ions. Moreover, distortions can be artificially manipulated through strain engineering, opening a path for the design of the electronic states happening at the interfaces [6, 7].

The fabrication techniques have reached a level of control comparable to the semiconductor technology, and interfaces can be engineered with atomic precision allowing combining the lattices of dissimilar materials with very good crystalline matching. In analogy to the (relative inert) semiconductor interfaces where interesting phenomena and even novel states of matter have been found, the delicate entanglement between the various degrees of freedom across the oxide interfaces, along with the possibility of coupling different order parameters is entitled to be the source of emergent electronic states with exciting properties. Orbital reconstruction resulting from structural (lattice distortions) and electronic processes occurring at interfaces stabilize novel electronic phases. Emergent electronic states may nucleate at the interface between dissimilar oxides [8, 9] as the result of the broken symmetry and of the discontinuity in important quantities such as charge density n, repulsion energy U, and bandwidth W, which govern the correlated electronic states [4, 10, 11]. In addition, charge is known to leak across the interfaces due to the difference in the electrochemical potentials [12], that in the case of correlated electron systems may have deeper consequences for the electronic states than in the case of conventional semiconductors which admit a description in terms of single electron schemes [13, 14]. New forms of couplings between different long-range orders with high technological potential for new device concepts [8, 9, 13, 14] thus appear. This is the arena of the emerging "oxide electronics," an actively pursued strategy based on exploiting the technological opportunities offered by devices based on correlated oxides [8, 9, 13–15]. This is in fact one of the alternative routes being sought to enable a higher density of information storage and more efficient computation speed than the current 2-bits-based von Neumann architectures, which is having difficulties to maintain Moore's law pace of increase of the computing speed. However, the challenge of oxide electronics has so far not been fulfilled, partly due to the incomplete understanding of the complex physics involved, but also to the lack of simple device concepts with externally tunable responses. In addition, the complexity of the electronic interactions has prevented the possibility of designing oxide interfaces with the same ease as for semiconductor interfaces [13, 14].

The electronic and orbital reconstructions occurring at oxide interfaces offer highly efficient avenues to tailor their magnetic states. In perovskite oxides with the orbital moment quenched by the octahedral crystal field, magnetism is largely determined by the spin–spin interaction [1, 2]. Spin interaction is governed by charge transfer processes, which can be real as in double-exchange or virtual as in superexchange. The sign (ferromagnetic or antiferromagnetic) of the superexchange interaction is captured by the Goodenough–

Kanamori rule [16, 17] as a function of filling and overlap in a way determined by the Pauli exclusion principle. Thus, charge transfer processes, structural distortions (Jahn–Teller), and rotations of the oxygen octahedra, with a direct effect on band filling and orbital polarization, provide direct access to the magnetic interactions [18, 19]. Furthermore, the bond reconstruction occurring at the atomically sharp oxide interfaces may draw new superexchange paths (B–O–B′) between different B-site cations at an $ABO_3/A'B'O_3$ interface, which may trigger magnetic interactions not existing in either of the original compounds. This magnetic interaction can in turn be manipulated through the control of the bond angle by strain engineering.

In this paper, we offer a series of examples of our recent work on how interfacially induced magnetism at oxide interfaces can be used to tailor novel functionalities in magnetic tunnel junctions. In particular, we will show that the spin reconstruction at the interfaces drives them into a novel magnetic state that acts as a spin filter. Spin filtering has been observed when current is driven through a ferromagnetically insulating (FI) barrier and results from different barrier heights for both spin orientations. However, ferromagnetic insulators are scarce, and their use as tunneling barriers requires a high degree of structural and chemical compatibility with the rest of the layers, which is usually difficult to attain.

We will explicitly distinguish between two related scenarios. First, we will consider cases in which the origin of the induced magnetism is the superexchange interaction between B-site cations at both sides of the interface resulting from the bond reconstruction. This is the case of the cuprate/manganite $La_{0.7}Ca_{0.3}MnO_3/YBa_2Cu_3O_7$ (LCMO/YBCO), or the ferrite/manganite $La_{0.7}Sr_{0.3}MnO_3/LaFeO_3$ interfaces, which will be described where respectively Mn–O–Cu and Mn–O–Fe superexchange paths induce (ferro) magnetism at the barrier interfaces. Doped cuprates do not show long-range magnetic order. Although parent compounds of cuprates are antiferromagnets, long-range order dies rapidly at low doping levels, yet strong antiferromagnetic correlations persist in the lightly doped compounds. $LaFeO_3$, on the other side, is an antiferromagnetic Mott insulator and the induced magnetism results essentially from the canting of a previously existing antiferromagnetic Fe lattice. On the other hand, we will also consider a second scenario consisting of induced magnetism due to phase nucleation at manganite/manganite $La_{0.7}Ca_{0.3}MnO_3/La_{0.3}Ca_{0.7}MnO_3$ interfaces with different doping levels. Here the electronic reconstruction driven by charge transfer together with the symmetry breaking of the interface is responsible for the nucleation of a magnetic state which is not present in the bulk phase diagram.

The new form of spin filtering occurring at the oxide interface, which we discuss in this paper, may draw a new direction for the design of novel spin filters. Moreover, spin-filtering barriers can be used in combination with ferromagnetic electrodes to increase the tunneling magnetoresistance (TMR) of magnetic tunnel junctions. We will further show that the interfacial magnetism offers an interesting alternative route for the electric field control of the magnetism of the electrodes in a magnetic tunnel junction, a functionality, which is being actively pursued for the development of low-dissipation spintronics [20].

7.2 Induced Magnetism at Oxide Interfaces

Induced magnetism at an oxide interface was first reported by B. Keimer's group at the Max Planck Institute [21, 22], for interfaces between LCMO ferromagnetic (half-metallic) manganite and YBCO high T_C superconductor cuprate. X-ray magnetic circular dichroism experiments showed a dichroic signal at the Cu edge evidencing a Cu magnetic moment aligned antiparallel to Mn moments at the interface. The antiferromagnetic superexchange interaction through the interface results from an orbital reconstruction by which holes, otherwise located in the x^2–y^2 orbital, are transferred to the $3z^2$–r^2 orbital shifted in energy due to hybridization. The interface $3z^2$–r^2 electrons experience and antiferromagnetic superexchange with the e_g electrons of the manganite, which is transmitted to the $d(x^2$–$y^2)$ electrons by the strong Hund's coupling interaction of the cuprate. As a result, there is an induced negative spin polarization at the cuprate caused by the canting of the antiferromagnetic Cu lattice to attain the 0.2 μ_B/Cu atom measured by the dichroism experiment [21, 22].

The interfaces between ferromagnetic (half-metallic) manganites and cuprates have focused a lot of interest in recent years. Both families have perovskite-related structures with very similar lattice parameters, and it is possible to grow them epitaxially one on top of the other with atomically sharp interfaces, bringing the two antagonistic electronic orders to direct contact. Interesting physics result from their interplay, which to large extent governed by the induced magnetic moment in the Cu sublattice at the interface. Interface structure and composition examined by scanning transmission electron microscopy (STEM) show good epitaxial properties and coherent growth. A net transfer of electrons occurs from the manganite to the cuprate, being the charge profile influenced by the mismatch of the electrochemical potential and the details of bonding, and hybridization at the interface. This electron transfer is responsible for a hole doping of the manganite, which causes additional suppression of the LCMO layer magnetization as measured by polarized neutron reflectometry [23, 24]. Electron energy loss spectroscopy (EELS) spectrum images acquired in LCMO/YBCO/LCMO trilayers to investigate the interface structure evidence that CuO chains of YBCO are missing at both interfaces, being the BaO atomic plane termination for both the top and bottom interfaces of the cuprate, and MnO_2 of the manganite at both interfaces. This symmetrical interface termination ensures an Mn–O–Cu superexchange path yielding Cu net magnetism at both interfaces of an ferromagnet/insulator/ferromagnet (F/S/F) trilayer [25, 26]. The effect of the induced magnetism at the interface is to create a spin polarization in the electron gas of the cuprate over a short distance from the interface, which has a drastic effect in suppressing superconductivity in ultrathin layers. The long-range suppression of the superconductivity in LCMO/YBCO epitaxial heterostructures over length scales much larger (1–2 orders of magnitude) than the coherence length of the cuprate [27, 28], or the inverse proximity effect in F/S/F trilayers, where superconductivity is favored for parallel alignment of the magnetic moments of the ferromagnets [29–35] have been recently shown to result from the Cu magnetism [36, 37]. Furthermore, it may also be at the origin of the triplet superconducting correlations, which propagate over tens of nanometers deep into the ferromagnet [38, 39].

Interestingly, this class of low-dimensional magnetism appearing at the interface between magnetic and non-magnetic oxides originating at the new superexchange path occurring at the boundary between both materials is quite ubiquitous, and in fact, has been observed experimentally also at manganite/titanate [40–42], manganite/ferrite [43], and nickelate/titanate interfaces [44]. The case of Ti^{4+} titanates requires a mechanism of charge transfer into the otherwise empty conduction band in Ti^{4+} compounds. Ti^{3+} with a d^1 structure may result from the presence of oxygen vacancies or from the doping due to anomalous interface terminations. For example, we have shown that an extra LaO plane at $La_{0.7}Sr_{0.3}MnO_3$/$SrTiO_3$ interface has a donor character and makes the LSMO/STO interface to resemble a $La_{0.7}Sr_{0.3}MnO_3/LaTiO_3$ one. In titanates, the small overlap of t_{2g} orbitals mediated by weak O π-bonds that are not directed along [100] bond directions give rise to the small crystal field splitting and the weak JT coupling which yield a larger tendency to orbital degeneracy and may thus display orbital moments and low-temperature ferromagnetism. Interestingly, while X-ray magnetic circular dichroism (XMCD) spectra show that the Mn orbital moment is essentially quenched as expected, there is a significant orbital component to the magnetic moment of the Ti. The AF alignment of the Ti moment with respect to the Mn moment found experimentally [21, 40] can be explained according to Goodenough–Kanamori rules in terms of the dominant superexchange process between the Ti t_{2g} electrons and the half-filled Mn t_{2g} bands.

Induced magnetism has also been found at $La_{0.7}Sr_{0.3}MnO_3$/$BiFeO_3$ (LSMO/BFO) interfaces by R. Ramesh's group [43]. At this interface, there is a strong orbital reconstruction originating a $d(x^2-y^2)$ orbital polarization at the interface. This destabilizes the ferromagnetic interaction along the z-direction due to the disruption of the double-exchange path perpendicular to the interface. In this system, Fe in a 3+ oxidation state is in a d^5 electronic high spin state and the down spin $d(x^2-y^2)$ state is located far above the gap, and thus charge transfer is not allowed. According to the Goodenough–Kanamori rule, the superexchange coupling between Fe^{3+} and Mn^{3+} (with in plane $d(x^2-y^2)$ orbital ordering), or Mn^{4+} is expected to be strongly ferromagnetic. On the other hand, the interfacial Mn layer with modified $d(x^2-y^2)$ orbital ordering naturally leads to the antiferromagnetic coupling with its neighboring Mn layer. Interfacial Mn changes its interplane coupling to the next Mn in the second layer, giving rise to an interface Mn layer coupled AF to the second Mn layer. As a result, the interfacial Fe spins and the Mn spins in the bulk LSMO region are coupled antiferromagnetically.

Another interesting scenario is the induced Ti magnetism in the $BaTiO_3$/Fe interface [45–48], which suggests a new path towards artificial multiferroics. Evidence of a Ti ferromagnetic state has been obtained from X-ray resonant magnetic scattering hysteresis loops at Ti $L_{2,3}$ and O K edges. Combined interface imaging by aberration-corrected STEM EELS and modeling suggests that in this case, Ti is in a 4+ state. Ti atoms are spin polarized due to the hybridization with Fe and spin-polarized oxygen atoms suggesting that the mechanism at play may be different from that outlined previously in manganite/titanate heterojunctions. Not completely understood also is the magnetic response found in the LAO/STO system. The first reports of magnetism by Brinkman and colleagues [49] based on hysteretic magnetotransport measurement [50], were followed by claims of room temperature ferromagnetism based on SQUID magnetometry [51], or on torque magnetometry by Li and colleagues [52]. Bert and colleagues [53], using a scanning Hall probe contact microscopy have found evidence of micron size magnetic patches with a large associated dipole field consistent with magnetism generated by localized carriers. On the other hand, neutron scattering experiments on LAO/STO superlattices with conducting and non-conducting interfaces do not detect any magnetic state, suggesting that possibly the integrated moment is small and below the detection limit of the technique [54]. This may be related to the very recent small Ti-localized moments detected by dichroism experiments [55] in amounts way below the concentration necessary to yield magnetic coupling between neighbors. Authors suggest the presence of localized moments at oxygen vacancies, which are coupled ferromagnetically by the free carriers of the 2D electron gas via indirect exchange, as found in the ferromagnetic 2D electron gases found in titanates [56, 57].

7.3 Spin Filtering by Barrier-Induced Magnetism: Generalities

Spin filtering was first introduced by Esaki [58] in junctions with normal electrodes and an EuS barrier. Later the concept was explored by Moodera et al., in hybrid junctions containing spin-filtering barriers and superconducting electrodes by the so-called Tedrow–

Meservey experiments [59]. There the spin-split density of state of the superconductor was used to analyze the spin polarization of the current (see Fig. 7.1a). Using this method, large spin polarization (85%) was measured in Au/EuS/Al superconducting tunnel junctions [60, 61]. EuS is a ferromagnetic semiconductor, which can produce highly spin-polarized currents due to the spin filtering by the exchange split conduction band. Utilized as a tunnel barrier, this spin splitting produces different barrier heights for spin-up and spin-down electrons, which result in different tunneling currents (depending exponentially on barrier height [62]) for both spin channels See sketch in Fig. 1(a). As a result, carriers tunneling from the Au layer became spin polarized due to the different transmission probabilities for spin-up and spin-down electrons (see below). The spin polarization was analyzed by the spin-split density of states of the superconducting Al at low temperatures (0.4 K). In other experiments with EuSe barriers (EuSe is an insulating ferromagnet under a large applied magnetic field), Moodera et al. obtained nearly 100% spin polarization [63].

The transmission coefficient for an electron with its spin parallel to the magnetization of the FI layer (subindex P in Eq. 7.1) will be increased due to a reduced barrier height. Using the Wenzel–Kramer–Brillouin (WKB) approximation, the transmission coefficient can be written as [64]:

$$T_{\mathrm{P,\,AP}} = T_0 e^{-\kappa_{P,\,AP} d} \text{ with } \kappa_{\mathrm{AP,P}} = \left(2m(U_0 - E_{\mathrm{F}} \pm \frac{\Delta E_{\mathrm{ex}}}{2})/\hbar^2 \right)^{1/2} \quad (7.1)$$

where U_0 is the conduction band edge in absence of magnetism, E_{F} is the Fermi level, and ΔE_{ex} measures the exchange splitting between the spin-up and spin-down bands of the ferromagnet. In Eq. 7.1, we have not considered energy-dependent barrier heights, which have to be considered for example in presence of strong applied electric fields.

In that case, the transmission coefficient takes the form:

$$T_{\mathrm{P,AP}} = T_0 \exp\left\{ -\int_0^d \left[2mU_0(x) - E_{\mathrm{F}} \pm \frac{\Delta E_{\mathrm{ex}}}{2})/\hbar^2 \right]^{1/2} dx \right\} \quad (7.2)$$

where the integral is computed over the width of the barrier, d.

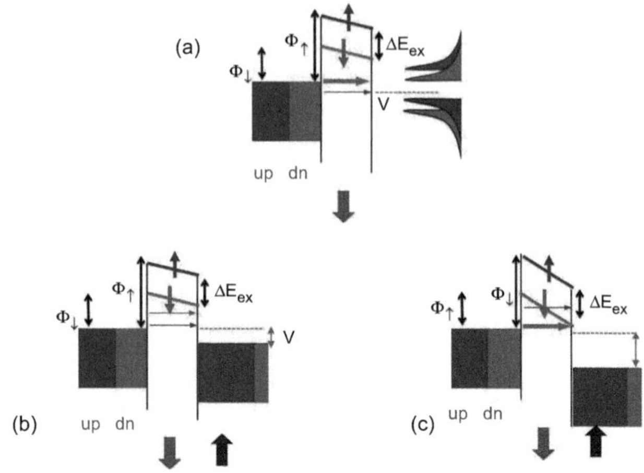

Figure 7.1 (a) Spin filtering in a normal metal/ferromagnetic insulator/superconductor (N/FI/S) junction. The current injected from a normal electrode across a ferromagnetic insulator is spin-polarized due to different barrier heights for spin-up and spin-down electrons. Spin polarization is analyzed with the spin-split density of states of a superconductor. (b and c) Fowler–Nordheim tunneling in an a normal metal/ferromagnetic insulator/ferromagnetic metal (N/FI/FM) junction. At low bias (panel b), there is direct tunneling for both spin channels. At high-bias (panel c) Fowler–Nordheim tunneling for down spins increases the transmission probability.

In spin filters, MR is potentially very large due to the so-called Fowler–Nordheim mechanism [65]. In conventional magnetic tunnel junctions, TMR is known to decrease monotonically with increasing bias and temperature due to the excitation of magnons [66] that cause spin mixing, however, in spin filters, TMR does not depend monotonically with bias showing a pronounced increase at the onset of the Fowler–Nordheim (FN) mechanism [65] for the majority spin channel. Due to the modulation of the tunneling barrier with the applied voltage, (tunneling) electron wave vectors can become real in parts of the barrier, which reduces the decay of the transmission probability. An increase of the spin-up current occurs when the applied voltage exceeds, say, the spin-up barrier, which results from the preferred transmission of spin-up electrons relative to spin-down spins, yielding an increase of MR [see Fig. 7.1(b and c)]. This excess TMR disappears when on further increasing the bias, the applied voltage exceeds the spin-down barrier height and the FN tunneling contributes to both spin channels. The hallmark of the spin-filtering phenomenon is thus a highly anomalous bias dependence of the TMR: an abrupt increase of MR at intermediate bias was observed at the onset of the FN tunneling regime [67]. Also, the preferred transmission of spin-up electrons in FN tunneling results in a decrease of the junction resistance when the temperature is decreased below the Curie point of the FI barrier as the result of the reduced barrier height for electrons with spins parallel to the barrier moment [68].

An interesting concept is that of double–spin filters, which are based on the independent switching of two FI barriers and do not need the superconducting electrode. Double–spin filtering by a multilayered FI barrier in tunneling devices with non-magnetic metal (NM) electrodes was theoretically introduced by Worledge and Geballe [64]. This concept basically exploits the spin polarization of the current across an NM/FI sandwich and its modulation with a composite barrier FI_1/FI_2 made of two ferromagnetic insulators which can be switched independently. Notice that there is no need of the spin-split density of states of the superconductor to analyze the polarization of the current across the spin filter. For the composite barrier, a reduction of the barrier height will occur for parallel alignment of the magnetizations of the ferromagnets as compared to the antiferromagnetic configuration where the flipped moment yields an increase of the barrier height for these electrons (subindex *AP* in Eq. 7.1). Notice that the height of the tunnel barrier is reduced in ΔE_{ex} for spins parallel to the magnetization of the barrier, and accordingly, the transmission coefficient is enhanced exponentially. The conductance can be expressed as the (parallel) sum of the contributions of the two spin channels, and in general, it will be shunted by the electrons with spin parallel to the magnetization of the barrier. This concept has been experimentally realized recently in Moodera's group [69] in structures containing $EuS/Al_2O_3/EuS$ hybrid tunnel barriers. EuS orders ferromagnetically with a bulk Curie temperature (T_C) of 16.6 K. An abrupt increase of MR is found at intermediate bias voltage exceeding the barrier height for majority electrons. As a result, there is preferential FN tunneling of spin-up (majority) electrons for parallel alignment of the magnetizations of the insulating layers. This gives rise to the pronounced peaks in the bias-dependent TMR for both bias polarities shown in Ref. [69]. Asymmetries in MR peak height may result from differences in the thickness of the ferromagnetic layers.

Alternatively, spin filtering has also been experimentally observed in quasi-magnetic tunnel junctions, which also do not need the analyzing superconducting electrode. Instead, a single ferromagnetic electrode used to analyze the tunneling current polarized by the NM/FI junction by measuring the resistance change when the alignment between barrier's and electrode's ferromagnetic moments is changed from parallel (low resistance) to antiparallel (large resistance) [69–75] [see Fig. 7.1(b and c)].

In view of the large spin polarization of the tunneling current in spin filters, the combination of spin- filtering barriers with ferromagnetic electrodes was suggested as a way to increase the spin polarization of the tunneling current in a magnetic tunnel junction. The first report of the use of a spin-filtering tunnel barrier to enhance the TMR in magnetic tunnel junctions was authored by Le Clair [76–78]. In this report, the modest increase of the TMR of ferromagnet/insulator/ferromagnet (F/I/F) tunnel with ferromagnetic (F) electrodes was observed due to spin filtering by a Gd_2O_3 barrier, which in analogy to the case of EuSe [76–78], is proposed to become ferromagnetically insulating under strong magnetic fields.

7.4 Spin Filtering by Cuprate Barriers in Manganite Tunnel Junctions

Here we describe the use of the induced moment at the interface between cuprates and manganites to manipulate tunneling transport in magnetic tunnel junctions. Tunnel structures were fabricated with ferromagnetic metallic electrodes made of LCMO manganite separated by a cuprate barrier made of insulating $PrBa_2Cu_3O_7$ (PBCO) [79, 80]. This cuprate is isostructural to the YBCO and has essentially the same growth properties. The interfaces are symmetrical with BaO planes of the cuprate facing MnO_2 planes of the manganite and missing CuO chains at the interface [25, 26]. This ensures the Mn–O–Cu superexchange path across the interface, which we discuss later [81, 82]. We will show here that the induced magnetism at the barrier interfaces is at the bottom of an interesting spin-filtering functionality. We will furthermore show that it can be responsible for a magnetoelectric coupling at the interface, which allows switching electrode magnetization with the sole action of an electric field.

A first important question regards the evolution of the spin polarization of the manganites with temperature. As shown by photoemission experiments, [78]La/Ca manganites are half-metals, i.e., ferromagnetic, which at zero temperature show a full spin polarization at the Fermi level. This is enabled by only one spin band crossing the Fermi level, while an energy gap is observed for the other spin orientation. This condition does not last when the temperature is increased. At finite temperature, the spin polarization decreases due to the shrinking of the gap in the minority spins band, which eventually closes completely at the Curie temperature. At low temperatures, due to the locking of the minority spin channel, magnetization reversal is expected to occur by spin-lattice coupling through anisotropy fluctuations driven by spin–orbit interaction, which is, in fact, a slow process [84]. However, minority electron excitations are possible due to the (virtual) magnon excitations [85]. On the other hand, when the temperature is increased, fluctuations of the magnetization due to activated magnetic excitations (spin waves) mix the two spin channels [85–87]. Moreover, transport measurements of magnetic tunnel junctions have shown that the spin polarization at manganite interface decays faster than the bulk magnetization, suggesting the presence of an effective mechanism to reverse spins at the interfaces [88].

Samples described here are LCMO (8 nm)/$PrBa_2Cu_3O_7$(3–8 nm)/ LCMO (25–50 nm) trilayers grown on $SrTiO_3$ substrates using (high-pressure) pure oxygen by RF sputtering at elevated temperatures [81]. Scanning transmission electron microscopy (STEM) images and electron energy loss spectroscopy (EELS) profiles in Fig. 7.2 show that, despite occasional interface steps of one unit cell high, the barrier is continuous and atomically flat over large lateral distances, and the Mn oxidation state measured by EELS is close to its nominal 3.3 value. The magnetic behavior evaluated using polarized neutron reflectometry showed that while bottom layer had biaxial [110] easy axes, the top layer with larger coercivity had easy axes directed along [100] directions [81]. PNR experiments by analyzing both the non-spin-flip and the spin-flip cross-sections allow examining non-collinear magnetization states: magnetic fields applied in plane in the [110] direction in a hysteresis loop sequence showed that bottom layer magnetization switched abruptly with the nucleation of large domains while the top layer exhibited substantial magnetization rotation with the nucleation of small domains, i.e., magnetization can be independently switched in bottom and top layers due to their different magnetic anisotropy.

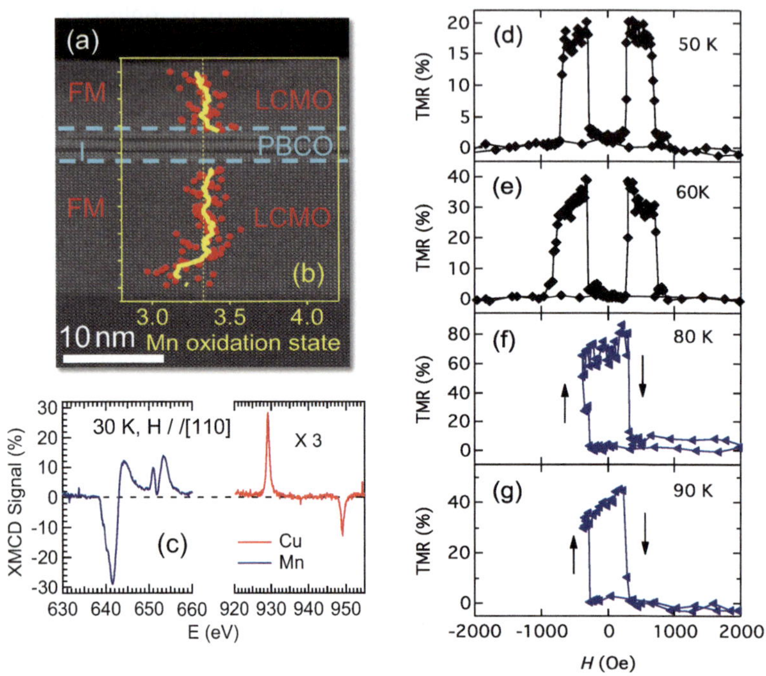

Figure 7.2 (a) Low-magnification STEM image shows the good structural quality of the PBCO barrier as well as its continuity over long distances. (b) Yellow inset: Mn oxidation state from the EELS O K edge fine structure. (c) TEY XMCD $L_{2,3}$ spectra of Mn (blue) and Cu (red). Notice the antiparallel alignment between Cu and Mn moments. (d–g) Transport data from a junction with 6 nm PBCO barrier. TMR versus magnetic field plots at different temperatures, black curves are full hysteresis loops and blue curves are minor loops. Notice the TMR suppression at low temperatures. Figures adapted from Refs. [81, 82].

X-ray absorption experiments with polarization analysis were conducted at the $L_{2,3}$ edges of Mn and Cu. Measurements performed in the total electron yield (TEY) collection mode displayed a strong dichroic signal at the Cu edge, clearly indicating that Cu gets magnetized at the interface (see Fig. 7.2c). Measurements as functions of temperature and field showed that Cu net moment closely follows the evolution of Mn moment, supporting that the origin of Cu magnetism is the superexchange interaction with Mn.

LCMO/PBCO/LCMO trilayers patterned into 16–100 μm^2 square-shaped magnetic tunnel junctions (MTJs) using standard lithography and Ar ion milling techniques showed large TMR. Sweeping in-plane magnetic field applied in the [110] direction produced abrupt switches between parallel and antiparallel (or misaligned) resistance states [see Fig. 7.2(d–g)]. The temperature dependence of the TMR exhibited an anomalous non-monotonic decrease when the temperature was reduced below 90 K (see Fig. 7.3a). This is an unexpected behavior since typically TMR would increase due to the increase of spin polarization when the temperature is reduced. Moreover, since the PNR experiments showed full magnetization for both layers at low temperatures, it could not be ascribed to depolarization due to inhomogeneous magnetization.

Here we argue that this non-monotonic behavior of the TMR with temperature is in fact one of the signatures of the "negative" spin-filtering tunnel barrier. The induced moment in the insulating barrier gives rise to spin splitting of its valence band of the PBCO, due to which, there are different barrier heights for spin-up and spin-down electrons.

A short comment is in order regarding the complexity added by the combination of half-metallic electrodes in a magnetic tunnel junction with the spin-filtering property of the barrier interfaces. In magnetic tunnel junctions, the tunneling probability is determined by the spin polarization, p, of the electrodes as described by Julliere's formula [89], which for symmetrical electrodes can be written as:

$$TMR = \frac{2p^2}{1-p^2} \tag{7.3}$$

The TMR closely tracks the spin polarization mainly of the electrode at its interface with the tunneling barrier. Equation 7.3 has to be modified if there is spin selectivity due to the barrier, for an example, occurs in the case of symmetry filtering of the wave function by some crystalline tunnel barriers which preferentially allow the passage of electrons with a certain Bloch state symmetry

[90, 91], yielding much larger TMR values than expected from the spin polarization of the electrodes. In spin filters, the spin selectivity is due to the difference in barrier height for both spin orientations can be turned on and off by playing with the relative alignment of the magnetizations of the FI layers.

A spin polarization of the tunneling current may thus result either from the injection of carriers from a ferromagnetic electrode with spin polarization p defined in terms of the density of states at the Fermi level for both spin bands $p = \frac{n_{\uparrow} - n_{\downarrow}}{n_{\uparrow} + n_{\downarrow}}$ (as in magnetic tunnel junctions with a non-magnetic barrier), or from the different transmission coefficients for spin-up and spin-down electrons $p_{\mathrm{T}} = \frac{T_{\uparrow} - T_{\downarrow}}{T_{\uparrow} + T_{\downarrow}}$ (as in the case of a FI barrier with non-magnetic electrodes). In general, the spin polarization of the tunneling current with both ferromagnetic electrodes and possible spin filtering by the barrier can be written as:

$$p_{\mathrm{eff}} = \frac{n_{\uparrow}T_{\uparrow} - n_{\downarrow}T_{\downarrow}}{n_{\uparrow}T_{\uparrow} + n_{\downarrow}T_{\downarrow}} \tag{7.4}$$

or alternatively in terms of the spin polarization p,

$$p_{\mathrm{eff}} = \frac{(1+p)T_{\uparrow} - (1-p)T_{\downarrow}}{(1+p)T_{\uparrow} + (1-p)T_{\downarrow}} \tag{7.5}$$

Notice that in Eq. 7.4, the transmission coefficients $T_{\uparrow,\downarrow}$ depend on the magnetic alignment of the injector electrode moment with the magnetization of the barrier according to Eq. 7.1 .

Since Cu moment is antiferromagnetically aligned to Mn moment at the interface, there will be a larger barrier for majority electrons than for minority electrons. This will depolarize the tunneling current in a negative spin-filtering effect triggered by the different transmission probabilities of spin-up and spin-down electrons. However, notice that the interfacial spin filter is not switchable since it is coupled to the electrode by the very strong superexchange interaction. Its effect will then be reduced the effective spin polarization of the tunneling current.

The spin polarization and the exchange splitting are directly correlated since spin polarization is proportional to magnetization and magnetization scales directly with exchange splitting. Therefore, we have assumed that exchange splitting and spin polarization have similar dependences on temperature. Notice that they have opposite effects on the effective spin polarization of the tunneling current. When the temperature is decreased, the increase of spin polarization is accompanied by an increase of the exchange splitting which enhances the tunnel barrier for minority electrons relative to the majority electrons, thus reducing their transmission probability. As a result, the effective polarization can be reduced and even its sign may be reversed, as shown in Fig. 7.3b. It is precisely the decrease of the effective spin polarization of the tunneling current, the responsible of the non-monotonic decreases the TMR when the temperature is decreased. The experimental situation can thus be modeled within the WKB approximation by assuming the same temperature dependence for the spin polarization and the exchange splitting, and parameterizing their zero temperature values. We have shown that for a given value of the spin polarization, increasing exchange splitting may result in a reduction or even in a sign change of the TMR qualitatively reproducing the obtained results. Further details can be found in reference [81].

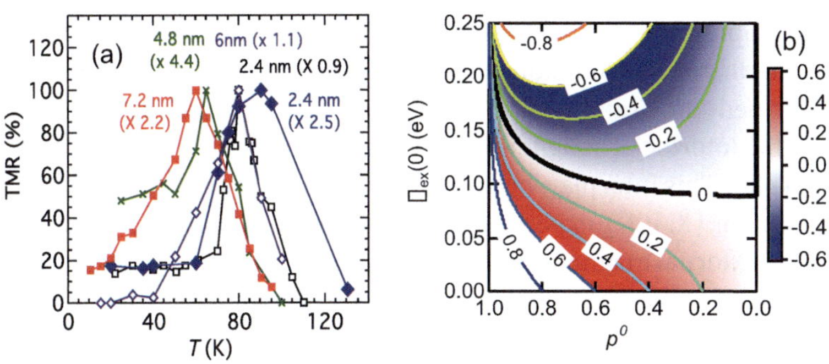

Figure 7.3 (a) Temperature dependence of the TMR of various magnetic tunnel junctions with LCMO electrodes and different $PrBa_2Cu_3O_7$ barriers thicknesses (TMR scales are different for the different junctions). Notice the anomalous low-temperature suppression of the TMR. (b) Effective spin polarization according to Eq. 7.5 for different values of the spin polarization of the electrode (horizontal axis) and of the zero-temperature exchange splitting in the interfacial PBCO region (vertical axis). Notice that the effective spin polarization can be reversed in a parameter range. Reprinted from Ref. [81], with permission from American Physical Society, Copyright 2013.

In summary, we have discussed recent evidence on the spin filtering by induced magnetism in an insulating $PrBa_2Cu_3O_7$ barrier. Due to the antiparallel alignment of the induced Cu moment relative to Mn moment, the spin filtering is negative, i.e., the spin-filtering effect triggers a reduction instead of an increase of the TMR. However, it becomes clear that the induced magnetism at barrier interfaces can be explored as a new route to artificial spin filters, which may increase the TMR of oxide tunnel junctions. An interesting aspect, which needs to be addressed is the tunability of the effect with external electric fields, a subject which will be addressed in the following paragraphs.

7.4.1 Electric Field Control of Magnetism

Once having shown that the modified bonding at the interface between magnetic (manganite) and non-magnetic (cuprates) oxides produces an interesting example of low-dimensional magnetism, which is capable of filtering spins. Now, we describe how the interfacial magnetic Cu state can be manipulated by an electric field. We will discuss a magnetoelectric effect, which can be used to electrically switch the magnetization of the magnetic layers without an applied magnetic field.

Samples are the LCMO(8 nm)/PBCO(4 nm)/LCMO(25–50 nm) magnetic tunnel junctions with an ultrathin PBCO barrier [81] described previously, but now the barrier will be thin enough for the interfacial magnetic states at the respective interfaces to interplay. This interplay was evidenced from X-ray magnetic circular dichroism (XMCD) experiments performed on trilayer samples in which external magnetic fields were used to switch between parallel and antiparallel alignment of the electrode moments [82]. Due to the small probing depth, the TEY Mn $L_{2,3}$ signal comes only from the top manganite layer, as shown by the similar Mn $L_{2,3}$ signal in the parallel and antiparallel alignment of the magnetic layers. On the other hand, there is a very significant suppression (by more than a factor of 2) of the Cu $L_{2,3}$ signal in the antiparallel configuration [see Fig. 7.4(a and b)] which provides strong evidence that the Cu moment at the top interface (being probed) senses the Mn moment at the bottom interface. Thus, the PBCO layer is exchange-coupled to both LCMO layers and develops an *exchange-spring state* for non-collinear magnetic configurations of the manganite layers (see the sketches in Fig. 7.4c). As described by Salafranca and Okamoto [36], the strong exchange field at the interface couples Cu and Mn moments antiferromagnetically. This coupling decays exponentially in the bulk of the barrier over a length scale of 3 uc. For a thin barrier such as ours, both exponential tails overlap, resulting in a magnetic coupling throughout the whole Cu oxide.

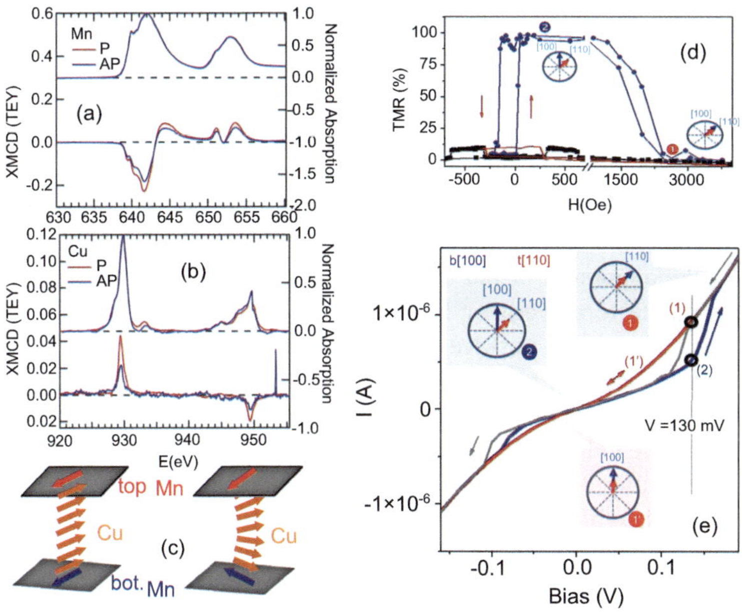

Figure 7.4 Total electron yield (TEY) $L_{2,3}$ spectra of Mn (a) and Cu (b) XAS (upper panel) and XMCD (lower panel) at 30 K in the remnant state (*P*), and at *H* = 60 Oe (AP). (c) Sketch illustrating magnetic coupling through the induced Cu moment at the interface. Both Mn magnetizations point in different directions due to their different easy axis. (b) MR versus field at low 60 mV (blue) and high 250 mV bias. The high-resistance state stable at zero magnetic field is the misaligned state 2 (see text). Notice that at large applied bias, the parallel state is stable in zero magnetic field (black data points or red minor loop). (c) Sketches indicate the magnetic alignment between the top and bottom layers. (d) IV curves (up and down sweeps) for the high-resistance state 2 (blue line) and low-resistance state 1 (red line) in zero magnetic field measured at 95 K. Notice that the IV curve starting at state 2 is hysteretic. Reprinted from Ref. [82], with permission from Springer Nature, Copyright 2014.

The antiferromagnetic coupling at the cuprate–manganite interface originates from the orbital reconstruction and charge transfer [36, 81, 82]. Because of the strong hybridization of $3z^2$–r^2 orbitals at the cuprate–manganite interface, the partially filled Cu x^2–y^2 orbitals experience the Hund's coupling interaction with the down spin hybrid $3z^2$–r^2 orbitals. However, the size of the AF Cu moment will depend on the filling of the x^2–y^2 orbital (it would be maximum for half-filling and zero for a complete filling of the orbital). The filling of this level is controlled by the charge transfer, which can be electrically controlled as in a p–n junction. Increasing bias will increase the filling at one interface and decrease it at the other, increasing the coupling strength.

The MR of the junctions closely tracks the misalignment between the top and the bottom layer moments due to their different easy axis directions. When a magnetic field is applied along the [110] direction, the saturation (low resistance) level corresponds to a parallel alignment of the electrode magnetizations along the easy axis of the top layer. For low-applied bias, the resistance sharply increases still at a positive field up to a high resistance level (state 2) corresponding to misaligned magnetizations, with the bottom layer magnetization now pointing along its [100] easy axis (see blue TMR curve in Fig. 7.4d). On the other hand, for the higher bias level, the transition into the misaligned state 2 is absent and the parallel state 1 is stable in zero field (see black TMR curve and red minor loop in Fig. 7.4d). Notice that in their respective bias ranges, high- and low-resistance states are stable in zero applied magnetic field. This evidences that increasing electric field has stabilized a collinear magnetization state along [110], indicating that the strength of the ferromagnetic coupling mechanism is electric field dependent.

To explore the effect of the electric field on switching magnetization, Fig. 7.4e shows *I*–*V* curves measured at zero magnetic field in state 2 (blue). For comparison, the red curve highlights the effect of saturating magnetization of both layers in the [100] direction, i.e., in state 1 (red). Notice that by sweeping the electric field in the zero field curve, there is a transition between a high-resistance state (at low electric fields), and a low-resistance state (at high electric fields) due to the switching of the magnetization of the softer bottom layer from [110] to [100] resulting from the enhanced ferromagnetic coupling at large bias. This transition is hysteretic that allows electric selection of the magnetic states at the same voltage.

We have thus shown electric field control of the relative alignment of the magnetic moments of the electrodes of an all-oxide magnetic tunnel junction with manganite electrodes and an insulating cuprate barrier. The superexchange induced moment in the Cu sublattice results in a ferromagnetic coupling between electrode moments for sufficiently thin cuprate barriers, which can be controlled by an external electric field. This enables the electric control of the two magnetization states and their corresponding resistance states. This opens a route that takes advantage of novel electronic phenomenon emerging at interfaces between correlated oxides and demonstrates reversible voltage-driven magnetization switching without magnetic field. This discovery enables a new avenue towards electric field control of magnetism, a great challenge in low-dissipation spintronics.

7.5 Spin Filtering by Imprinting Ferromagnetism in an Antiferromagnetic Barrier

Now we discuss the effect of induced magnetism at epitaxial interfaces between in antiferromagnetic $LaFeO_3$ (LFO) and ferromagnetic $La_{0.7}Sr_{0.3}MnO_3$ (LSMO). This is an interesting case of a new type of magnetism induced in a layer with a pre-existing magnetic order. $LaFeO_3$ is an antiferromagnetic charge transfer insulator with a Néel temperature of 740 K [92] and $La_{0.7}Sr_{0.3}MnO_3$ (LSMO), a half-metallic ferromagnet [93, 94] with a high Curie temperature T_C = 350 K. A net magnetic moment is induced at the interface with LSMO such that the magnetic domain structure of the manganite is replicated in the first atomic planes of the antiferromagnetic LFO. We show that these interfacial magnetic states yield an emerging spin-filtering functionality, which deeply affects the transport of spins across the interface, providing an additional example of the spin-filtering functionality discussed in this chapter.

The samples of this study were good quality epitaxial heterostructures combining LSMO and LFO [95] as evidenced by scanning transmission electron microscopy (STEM) where high-angle annular dark field (HAADF) images showed that the interface had high crystalline quality at atomic level. Chemical structure of the interface was examined by acquiring elemental maps using electron energy loss spectroscopy (EELS). The simultaneously acquired annular dark field (ADF) image as well as the elemental maps corresponding to the Mn $L_{2,3}$, Fe $L_{3,2}$, and La $M_{4,5}$ absorption edges and upper panels (Fig. 7.5a) show coherent growth and atomically sharp interfaces. X-ray absorption spectroscopy (XAS) (fluorescence yield) measured on an $LFO_{1.2}/LSMO_{3.5}$ bilayer at Mn $L_{3,2}$ and Fe $L_{3,2}$ edges confirmed the $Mn^{3+/4+}$ mixed valence in LSMO and dominant Fe^{3+} character in LFO. The XMCD experiments measured at the Fe $L_{3,2}$ edge showed a non-zero signal, indicating the presence of a net magnetic moment in the LFO layer. The induced moment is aligned antiparallel to that of Mn and can be estimated using sum rules to be close to 0.1 μ_B/Fe (if one assumes that only the FeO_2 plane closest to the interface acquires a magnetic moment). In view of the antiferromagnetic character of the LFO layer, this moment can be thought to result from a canting of the AF lattice.

Magnetic tunnel junctions fabricated with the layer sequence [$LSMO_{35}/LFO_{3.5}/LSMO_8$] where the thickness is expressed in nanometers showed moderate TMR = $(R_{high} - R_{low})/R_{low} \sim 30\%$ when the magnetic field was swept between parallel and antiparallel alignment of the magnetization of the LSMO electrodes. TMR values for this system were lower than the largest values found in all-oxide tunnel junctions [88, 96, 97]. The domain structures at the bottom and top LFO_{bot} and LFO_{top} interfaces in MTJs were examined by means of XMCD-PEEM measurements. Experiments were conducted on [$LSMO_{35}/LFO_{1.2}$] samples and [$LSMO_{35}/LFO_{3.5}/LSMO_8$] 8.5 × 2.8 μm^2 mesa structures, respectively (see sketch in Fig. 7.5b), due to the small probing depth of the technique provided separated access to both interfaces. XMCD-PEEM images obtained at both Mn and Fe L_3 edges provide direct access to the magnetic domain structure of both interfaces. Magnetic domains are clearly resolved at the LFO interfaces, which closely replicate the domain structure of the LSMO but with antiparallel alignment as highlighted by the red and blue colors in Fig. 7.5b. This demonstrated that the LSMO domains are magnetically imprinted into the LFO layers through the Fe–Mn superexchange interaction as concluded from XMCD data.

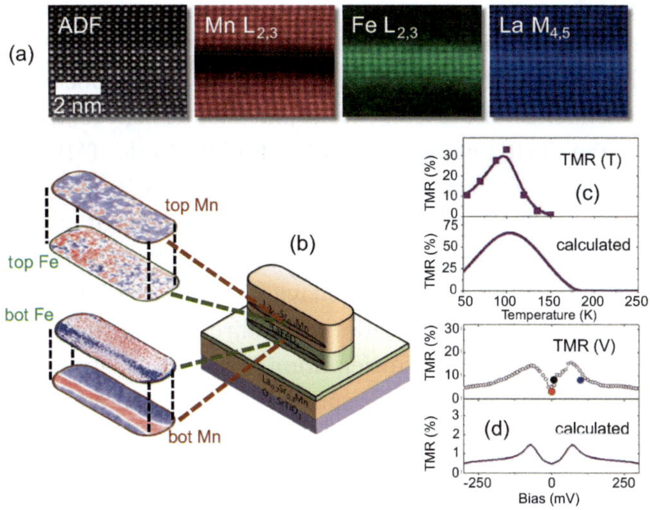

Figure 7.5 (a) Annular dark field (ADF) signal acquired simultaneously with EEL spectrum image and atomic column resolution elemental maps. Atomic resolution elemental maps corresponding to (from left to right) Mn $L_{3,2}$, Fe $L_{3,2}$ and La $M_{4,5}$ edges. (b) XMCD–PEEM images at E = 642.2 eV (Mn $L_{3,2}$ edge) and E = 709.7 eV (Fe $L_{3,2}$ edge). The schematics are of a patterned (LSMO35/LFO3.5/LSMO8) heterostructure where the Mn images were obtained and of a patterned (LSMO35/LFO1.2) heterostructure where the Fe images were obtained. The XMCD color scale is normalized to magnetic moment pointing along the long axis of the junction at 0° (blue) and red 180° (red). Notice the imprint of the ferromagnetic manganite domains into the ferrite. (c) Temperature dependence of the TMR: upper panel shows experimental data and the bottom panel shows a simulation. (d) Bias dependence of the TMR measured at 50 K. Upper panel shows experimental data and the bottom panel of a simulation. Reprinted from Ref. [95], with permission from Springer Nature, Copyright 2015.

Transport experiments conducted on the LSMO/LFO/LSMO magnetic tunnel junctions exhibited the two hallmarks of spin filtering, namely non-monotonic behavior of TMR with temperature and with bias. Looking first at the temperature dependence of the TMR, we found that upon decreasing temperature, TMR rose from about ~1% at 150 K to ~33% at 100 K, but it decreased to about 8% only at 25 K (see Fig. 7.5c). This non-monotonic dependence contrasts with the monotonic increase of TMR with decreasing temperature usually found in manganite tunnel junctions, resulting from the increase of spin polarization [88]. Also, resistance exhibited the non-monotonic behavior characteristic of spin filters, where the decrease results from the reduced barrier height for the majority spins in the parallel configuration [98]. Concerning the bias dependence of the TMR, we found that at high temperatures, close to the Curie temperature of the top electrode, TMR followed the usual monotonic decrease when the bias was increased. At low temperatures, TMR increases with increasing voltage up to about 70 mV and then decreases as bias increases further (see Fig. 7.5d). As mentioned previously, this behavior is expected for spin filters [67, 96, 99] due to the transition from direct tunneling to Fowler–Nordheim tunneling for one orientation of spins. Tunnel transmission is then strongly favored for one spin direction compared to the other and TMR reaches a maximum value [69, 99] followed by the normal TMR decrease with bias when FN regime occurs for both spin channels.

As discussed before, the induced magnetic moment at the interface exchange splits the band structure of the LFO, in a way that the interface planes behave as ferromagnetic insulators. This results in different transmission coefficients for spin-up and spin-down electrons across the tunnel barrier in magnetic tunnel junctions. Due to the antiparallel alignment of the induced moment relative to that of the electrodes, its effect is increasing the barrier for the majority spin channel that can reduce the effective spin polarization of the tunneling current or even reverse it. The temperature dependence of the TMR has been modeled [81, 94] in terms of the temperature dependence of the barrier exchange splitting (which will follow the spin polarization of the electrodes at the interface) [88, 93]. The results of this simulation, which can be found in Ref. [95], reproduce qualitatively the observed behavior although with larger values of the MR. This can be understood in terms of magnon excitations by the tunneling electrons [100–102] (not taken into account by simulation) with a strong depolarizing effect when the temperature is lowered below 100 K.

In summary, the spin reconstruction occurring at the interfaces of LSMO/$LaFeO_3$/LSMO tunnel junctions deeply modifies their spintronic response. The spin-filtering functionality due to the interface induced magnetism hinders the transport of majority spins for the parallel alignment of the magnetic moments of the electrodes.

This results in a reduced spin polarization of the tunneling current that modifies the temperature dependence of the TMR and causes an anomalous increase of the TMR with bias.

7.6 Spin Filtering by Emerging Ferromagnetism in Phase-Separated Barriers

A different source of interface magnetism at oxide interfaces is phase nucleation. It is known that as the result of electron correlations there are competing interactions with similar characteristic energies. This has been frequently observed in cuprates and also in manganites, systems both prone to phase separation. An interesting question is whether there is a possibility of controlling phase separation with external variables, and in particular, with the symmetry breaking at interfaces, which has been proposed to be the source of emergent electronic states. Here we discuss [96] heterostructures combining $La_{0.7}Ca_{0.3}MnO_3$ (LC3MO), a highly spin-polarized FM–M with a Curie temperature (T_C) of 270 K, with $La_{0.3}Ca_{0.7}MnO_3$(LC7MO), a charge ordered insulator below 260 K, becoming AFM below 150 K [103]. In particular, we describe the use of ultrathin $La_{0.3}Ca_{0.7}MnO_3$ as tunnel barriers in magnetic tunnel junctions with $La_{0.7}Ca_{0.3}MnO_3$ (LC3MO) electrodes. The good lattice matching between these materials allows the growth of heterostructures with sharp interfaces, not only structurally, but also chemically, i.e., despite the similar composition at both sides of the interface, the chemical integrity is preserved with little or no interdiffusion as measured from STEM EELS with atomic column resolution [see Fig. 7.6 (a–c)]. Polarized neutron reflectometry experiments, a technique that due to its depth resolution is ideally suited to explore magnetism of buried layers, highlighted the presence of substantial magnetic signal coming from the barrier at low temperatures. Fits of neutron reflectivity spectra provide an estimate or the barrier moment that can be as large as 1 μ_B/Mn, which is highly surprising in an antiferromagnetically insulating barrier. Understanding the origin of this magnetic moment in the barrier is in this case more complicated than the simple superexchange scenario discussed previously and phase nucleation has to be considered. Especially due to the symmetrical doping at both sides of the interface, the leveling of the chemical potential requires a half-doping (x = 0.5) at the interface, a composition corresponding to the boundary between ferromagnetic/metallic and antiferromagnetic/insulating phases [104]. Moreover, we expect that the existence of a wide gap insulating phase for $La_{0.5}Ca_{0.5}MnO_3$(LC5MO) may lead to pinning of the electron density at 0.5 in the ultrathin layer [105]. We have examined the nucleation of a ferromagnetic phase in the half-doped interface due to the close presence of the ferromagnetic metallic electrodes. Self-consistent calculations using a model [105–108] that includes the kinetic energy (double-exchange), the Jahn–Teller phonons, the AFM super-exchange, and the Coulomb interaction were performed for LC3MO/$La_{0.5}Ca_{0.5}MnO_3$/LC3MO heterostructures. $La_{1-x}Ca_xMnO_3$ at x = 0.5 (LC5MO) shows a similar kind of CE-type AFM ordering as LC7MO [108]. We have found that an LC5MO monolayer between two LC3MO FM electrodes acquires a finite FM moment and is insulating within a parameter range that would correspond to CE order in the bulk, i.e., a different phase is stabilized due to the close proximity of the ferromagnetic electrodes. The density of states projected on the LC3MO FM–M layers and on the LC5MO layer with a net magnetic moment due to the canted AFM order corresponding to 10% magnetization shows a gap as displayed in Fig. 7.3. On the other hand, it is clear that the finite value of the density of states projected to electrodes shows that they remain conductive. The ultrathin LC5MO layer gains kinetic energy by acquiring some ferromagnetic moment and balance Jahn–Teller coupling and superexchange interaction with a tendency to lower moment. The emergence of the ferromagnetic state may in fact be an experimental realization of the theoretically proposed tendency to ferromagnetism of antiferromagnetic manganites in proximity to ferromagnets [104, 107, 109, 110]. This model calculation provides a theoretical footing for the existence of an FM insulating state in terms of canting of the CE phase at the interface although from the experimental data at hand it is not possible to confirm it.

In the following text, we show that the induced magnetic state in the barrier acts as a spin filter. We describe the (perpendicular) magnetotransport properties of LC3MO (35 nm)/LC7MO (1.2 nm)/LC3MO (8 nm) magnetic tunnel junctions patterned in the form of micron size pillars. A very large TMR in excess of 500% (see Fig. 7.6c) was obtained at low temperatures, a value among the largest TMR reported in oxide tunnel junctions. The TMR disappeared at temperatures of 180 K close to the Curie point of the bottom electrode. The depressed T_C at the interface suggests that possible reduction of the spin polarization. Yet, the large values of the TMR may constitute the first indication of spin filtering by the induced ferromagnetic states at the barrier interfaces as proposed previously [65, 76–78], which repolarizes the tunneling current by limiting the transport of minority spins. Further confirmation for the spin-filtering scenario is obtained from the bias and temperature dependences of the TMR. Looking at the bias dependence of the TMR at low temperatures, it first decreases as expected for low bias values, but it displays a maximum if bias is further increased (see Fig. 7.6d). The non-monotonic behavior of the TMR disappears when the temperature is increased above 100 K, way before the TMR vanishes completely.

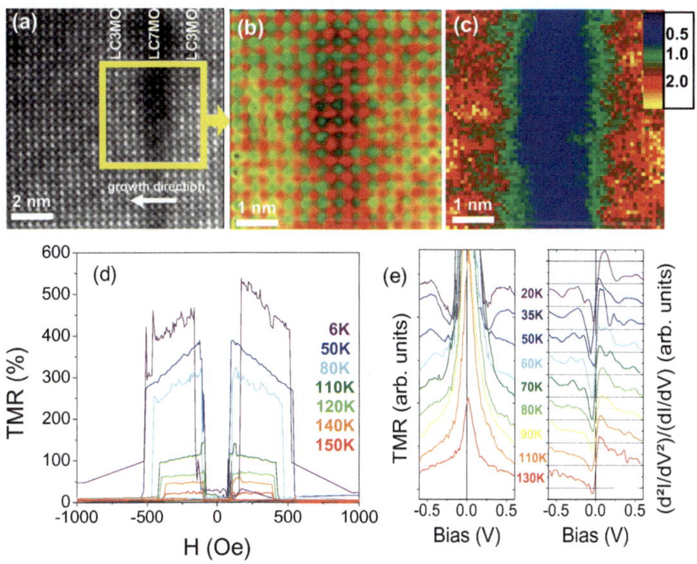

Figure 7.6 (a–c) Atomic-scale structural and chemical characterization. (a) Raw HAADF ultra STEM image of an LC3MO/LC7MO/LC3MO interface area. (b) EELS chemical map with atomic column resolution: La $M_{4,5}$ (green) and Mn $L_{2,3}$ (red) maps. (c) La/Ca concentration ratio from the same spectrum image. Notice that the La/Ca ratio drops from 2 to 0.5 over one unit cell when passing from electrode to barrier. (d) TMR curves of an LC3MO(8 nm)/LC7MO(1.2 nm)/LC3MO(50 nm) MTJ at 10 mV and different temperatures. (d) TMR(V) curve obtained from (both) I(V) and R(H) curves. (e) Bias dependence of derivative plots $(d^2I/dV^2)/(dI/dV)$ at various temperatures to illustrate the non-monotonic bias dependence of the TMR below the Curie point of the barrier. Reprinted from Ref. [96], with permission from John Wiley and Sons, Copyright 2010.

The situation may in fact be of a double-spin filter by FI states at both interfaces each coupled ferromagnetically to the neighboring electrode. In this scenario, a spin filtering of the tunneling current occurs for parallel alignment of the electrodes, which favors the tunneling of majority electrons due to the enhanced barrier for minority electrons. In the antiparallel configuration, the effect is (partially) cancelled by the antiparallel alignment of the magnetic moments of the interfaces.

We have thus demonstrated spin filtering by interface magnetism induced by phase nucleation at correlated oxide interfaces. This spin-filtering functionality is comparable to the one found using chalcogenides [60], or other oxides and may inspire a strategy to the scarcity of spin-filtering barriers [74, 76]. Moreover, the phase nucleation concept put forward in this study may draw a new path towards the functionalization of phase separation for oxide electronic and spintronic devices.

7.7 Conclusions

We have reviewed recent work showing that the new bond appearing at epitaxial interfaces between magnetic transition metal oxides is the source of interesting forms of induced magnetism driven either by the superexchange interaction across the new bond or the tendency to phase nucleation triggered by correlations. Of particular interest is the interface between a ferromagnetic and a non-magnetic oxide where magnetism can be transferred to the non-ferromagnetic compound.

We have considered cases in which the non-ferromagnetic oxide has a magnetic moment and no magnetic order such as doped titanates, and cases where the non-ferromagnetic oxide possess a ground state with a long-range magnetic order, such as $La_{0.7}Ca_{0.3}MnO_3/PrBa_2Cu_3O_7$, $La_{0.7}Sr_{0.3}MnO_3/LaFeO_3$, or $La_{0.7}Ca_{0.3}MnO_3/La_{0.3}Ca_{0.7}MnO_3$ with a ferromagnetic manganite, is put in contact with an antiferromagnet. A net magnetic moment is induced in the non-ferromagnetic material at the interface, which in the latter cases (in fact, the main focus of this chapter) can be understood in terms of canting of the antiferromagnetic sublattices. Surprisingly the induced moment can be surprisingly large reaching values in the range 0.1–0.2 μ_B per B-site atom. Quite interestingly, this leads to ground states in specific cases which do not exist in the bulk phase diagrams of the individual compounds, for example, $LaFeO_3$ with ferromagnetic domains of the manganite imprinted in its spin structure or $La_{0.3}Ca_{0.7}MnO_3$ where the immediate vicinity of a ferromagnetic phase with different chemical potential favors the nucleation of a ferromagnetic CE phase.

We have discussed that the induced magnetism has a spin-filtering functionality in oxide magnetic tunnel junctions. In all the cases analyzed here wherein the induced moment is driven by superexchange interaction, the magnetic alignment between the electrode and induced moments was antiferromagnetic. As a consequence, spin filtering is negative, i.e., there is an increased barrier for the majority spin channel as compared to the minority. This gives rise to an anomalous non-monotonic dependence of the TMR with bias, which increases (with bias) in a voltage range due to the onset of the Fowler–Nordheim mechanism. On the other hand, in the phase nucleation case, the induced moment is parallel to the moment of the electrodes, which is responsible for very large values of the TMR.

In summary, we have shown a novel spin-filtering functionality at oxide interfaces, which could offer design opportunities and alternatives to the scarcity of the canonical insulating ferromagnetic spin filters. The cases discussed here represent examples of the functionalization of oxide interfaces into new device concepts and may signal new avenues towards a future oxide electronics.

Acknowledgments

JS thanks support by CNRS and Bourse D'Alembert of the Universite Paris Saclay. Work at UCM was supported by Spanish MINECO through grant MAT2014-52405-C02-01 and by CAM through grant CAM S2013/MIT-2740. Work at Argonne National Laboratory and use of the Advanced Photon Source was supported by the U.S. Department of Energy, Office of Basic Energy Sciences, under contract no. DE-AC02-06CH11357. We acknowledge financial support from the European Research Council (ERC) Advanced Grant FEMMES, No. 267579 and the Labex NanoSaclay project FIRET. The ALICE project is supported by the BMBF contract No. 05K10PC2. The research leading to these results has received funding from the European Community's Seventh Framework Programme (FP7/2007–2013), under grant agreement no. 226716.

References

1. A. Ohtomo, D. A. Muller, J. L. Grazul, and H. Y. Hwang, Artificial charge-modulation in atomic-scale perovskite titanate superlattices, *Nature,* 419, pp. 378–380 (2002).
2. A. Ohtomo and H. Hwang, A high-mobility electron gas at the $LaAlO_3$/$SrTiO_3$ heterointerface, *Nature,* 427, pp. 423–426 (2004).
3. M. Imada, A. Fujimori, and Y. Tokura, Metal-insulator transitions, *Rev. Mod. Phys.,* 70, p. 1039 (1998).
4. E. Dagotto and Y. Tokura, Strongly correlated materials: present and future, *MRS Bull.,* 33, p. 1037 (2008).
5. D. Khomskii, *Transition Metal Compounds,* Cambridge University Press (2014).
6. Y. Tokura and N. Nagaosa, Orbital physics in transition-metal oxides, *Science,* 288, pp. 462–468 (2000).
7. J. M. Rondinelli and N. A. Spaldin, Structure and properties of functional oxide thin films: insights from electronic-structure calculations, *Adv. Mater.,* 23, p. 3363 (2011).
8. J. Mannhart and D. G. Schlom, Oxide interfaces an opportunity for electronics, *Science,* 327, p. 1607 (2010).
9. H. Y. Hwang, et al., Emergent phenomena at oxide interfaces, *Nat. Mater.,* 11, pp. 103–113 (2012).
10. C. H. Ahn, et al., Electrostatic modification of novel materials, *Rev. Mod. Phys.,* 78, pp. 1185–1212 (2006).
11. P. Zubko, S. Gariglio, M. Gabay, P. Ghosez, and J. Triscone, Interface physics in complex oxide heterostructures, *Annu. Rev. Condens. Matter Phys.,* 2, pp. 141–165 (2011).
12. S. Okamoto and A. Millis, Electronic reconstruction at an interface between a Mott insulator and a band insulator, *Nature,* 428, pp. 630–633 (2004).
13. Editorial, The interface is still the device, *Nat. Mater.,* 11, pp. 91 (2012).
14. J. Chakhalian, A. J. Millis, and J. Rondinelli, Whither the oxide interface, *Nat. Mater.,* 11, p. 92–94 (2012).

15. J. H. Ngai, F. J. Walker, and C. H. Ahn, Correlated oxide physics and electronics, *Annu. Rev. Mater. Res.,* 44, pp. 1–17 (2014).

16. J. B. Goodenough, Theory of the role of covalence in the perovskite-type manganites [La, M(II)]MnO_3, *Phys. Rev. B,* 100, pp. 564–573 (1955).

17. J. Kanamori, Superexchange interaction and symmetry properties of electron orbitals, *J. Phys. Chem. Solids,* 10, pp. 87–98 (1959).

18. A. Bhattacharya and S. J. May, Magnetic oxide heterostructures, *Annu. Rev. Mater. Res.,* 44, pp. 65–90 (2014).

19. J. M. D. Coey, Ariando, and W. E. Pickett, Magnetism at the edge: new phenomena at oxide interfaces, *MRS Bull.,* 38, pp. 1040–1047 (2013).

20. S. Fusil, V. Garcia, A. Barth´el´emy, and M. Bibes, Magnetoelectric devices for spintronics, *Annu. Rev. Mater. Res.,* 44, pp. 91–116 (2014).

21. J. Chakhalian, et al., Magnetism at the interface between ferromagnetic and superconducting oxides, *Nat. Phys.,* 2, p. 244 (2006).

22. J. Chakhalian, et al., Orbital reconstruction and covalent bonding at an oxide interface, *Science,* 318, pp. 114–1117 (2007).

23. A. Hoffmann, S. G. E. te Velthuis, Z. Sefrioui, J. Santamaria, S. Park, and M. Fitzsimmons, Suppressed magnetization in LCMO/YBCO superlattices, *Phys. Rev. B,* 72, R140407 (2005).

24. J. W. Freeland, J. J. Kavich, K. E. Gray, L. Ozyuzer, H. Zheng, J. F. Mitchell , M P Warusawithana, P Ryan, X Zhai, R H Kodama, and J N Eckstein, *J. Phys. Condens. Matter.,* 19, 315210 (2007).

25. C. Visani, et al., Symmetrical interfacial reconstruction and magnetism in $La_{0.7}Ca_{0.3}MnO_3/YBa_2Cu_3O_7/La_{0.7}Ca_{0.3}MnO_3$ heterostructures, *Phys. Rev. B,* 84, 060405 (2011).

26. Juan Salafranca, et al., Competition between covalent bonding and charge transfer at complex-oxide interfaces, *Phys. Rev. Lett.,* 112, 186902 (2014).

27. Z. Sefrioui, et al., Ferromagnetic/Superconducting proximity effect in $La_{0.7}Ca_{0.3}MnO_3/YBa_2Cu_3O_{7-x}$ superlattices, *Phys. Rev. B,* 67, 214511 (2003).

28. V. Peña, et al., Coupling of superconductors through a half-metallic ferromagnet: evidence for a long-range proximity effect, *Phys. Rev. B,* 69, 224502 (2004).

29. V. Peña, Z. Sefrioui, D. Arias, C. Leon, J. L. Martinez, S. G. E. te Velthuis, A. Hoffmann, and J. Santamaria, Giant magnetoresistance in ferromagnet/ superconductor structures, *Phys. Rev. Lett.,* 94, 57002 (2005).

30. V. Peña, C. Visani, J. Gacia-Barriocanal, Z. Sefrioui, D. Arias, C. Leon, and J. Santamaria, Spin diffusion vs. proximity effect at ferromagnet/superconductor YBCO/LCMO interfaces, *Phys. Rev. B,* 73, 104513 (2006).

31. C. Visani, et al., Spin-dependent magnetoresistance at ferromagnet/superconductor/ferromagnet $La_{0.7}Ca_{0.3}MnO_3/YBa_2Cu_3O_7/La_{0.7}Ca_{0.3}MnO_3$ trilayers, *Phys. Rev. B,* 75, 54501 (2007).

32. N. M. Nemes, et al., Origin of the inverse spin-switch behavior in manganite/cuprate/manganite trilayers, *Phys. Rev. B,* 78, 094515 (2008).

33. N. M. Nemes, C. Visani, Z. Sefrioui, C. Leon, J. Santamaria, M. Iglesias, F. Mompean, and M. Garcia-Hernandez, Exchange-bias-modulated inverse superconducting spin switch in CoO/Co/YBCO/LCMO thin film hybrids, *Phys. Rev. B,* 81, 024512 (2010).

34. C. Visani, et al., Directionally controlled superconductivity in ferromagnet/superconductor/ferromagnet trilayers with biaxial easy axes, *Phys. Rev. B,* 81, 094512 (2010).

35. N. M. Nemes, et al., Magnetic memory based on $La_{0.7}Ca_{0.3}MnO_3/YBa_2Cu_3O_7/La_{0.7}Ca_{0.3}MnO_3$ ferromagnet/superconductor hybrid structures, *App. Phys. Lett.,* 97, 032501 (2010).

36. J. Salafranca and S. Okamoto, *Phys. Rev. Lett.,* 105, 256804 (2010).

37. Y. Liu, et al., Effect of interface-induced exchange fields on cuprate-manganite spin switches, *Phys. Rev. Lett.,* 108, 207205 (2012).

38. T. Hu, H. Xiao, C. Visani, Z. Sefrioui, J. Santamaria, C. Alamsan, Evidence from magnetoresistance measurements for an induced triplet superconducting state on $La_{0.7}Ca_{0.3}MnO_3/YBa_2Cu_3O_{7-\delta}$ multilayers, *Phys. Rev. B,* 80, R060506 (2009).

39. C. Visani, et al., Villegas, Equal-spin Andreev reflection and long-range coherent transport in high-temperature superconductor/half-metallic ferromagnet junctions, *Nat. Phys.,* 8, p. 539 (2012).

40. J. Garcia-Barriocanal, et al., Spin and orbital Ti magnetism at $LaMnO_3/SrTiO_3$ interfaces, *Nat. Commun.,* 1(82), doi: 10.1038/ncomms1080 (2010).

41. A. Tebano, et al., Evidence of Orbital Reconstruction at Interfaces in Ultrathin $La_{0.67}Sr_{0.33}MnO_3$ Films, *Phys. Rev. Lett.,* 100, 137401 (2008).

42. F. Y. Bruno, et al., Electronic and magnetic reconstructions in $La_{0.7}Sr_{0.3}MnO_3/SrTiO_3$ heterostructures: a case of enhanced interlayer coupling controlled by the interface, *Phys. Rev. Lett.,* 106, 147205 (2011).

43. P. Yu, et al., Interface ferromagnetism and orbital reconstruction in BiFeO–La0.7Sr0.3MnO3 heterostructures, *Phys. Rev. Lett.*, 105, 027201 (2010).

44. M. N. Grisolia, et al., Hybridization-controlled charge transfer and induced magnetism at correlated oxide interfaces, *Nat. Phys.*, 12, p. 484 (2016).

45. S. Valencia, et al., Interface-induced room-temperature multiferroicity in BaTiO3, *Nat. Mater.*, 10, p. 753 (2011).

46. C. G. Duan, S. S. Jaswal, and E. Y. Tsymbal, Predicted magnetoelectric effect in Fe/BaTiO3 multilayers: ferroelectric control of magnetism. *Phys. Rev. Lett.*, 97, 047201 (2006).

47. M. Fechner, I. V. Maznichenko, S. Ostanin, A. Ernst, J. Henk, P.Bruno, and I. Mertig, Magnetic phase transition in two-phase multiferroics predicted from first principles, *Phys. Rev. B*, 78, 212406 (2008).

48. M. Fechner, S. Ostanin, and I. Mertig, Effect of oxidation of the ultrathin Fe electrode material on the strength of magnetoelectric coupling in composite multiferroics, *Phys. Rev. B*, 80, 094405 (2009).

49. A. Brinkman, et al., Magnetic effects at the interface between non-magnetic oxides, *Nat. Mater.*, 6, p. 493 (2007).

50. D. A. Dikin, et al., Coexistence of superconductivity and ferromagnetism in two dimensions, *Phys. Rev. Lett.*, 107, 056802 (2011).

51. Ariando, et al., Electronic phase separation at the LaAlO3 /SrTiO3 interface, *Nat. Commun.*, 2, p. 188 (2011).

52. L. Li, C. Richter, J. Mannhart, and R. C. Ashoori, Coexistence of magnetic order and two-dimensional superconductivity at $LaAlO_3/SrTiO_3$ interfaces, *Nat. Phys.*, 7, p. 762 (2011).

53. J. A. Bert, B. Kalisky, C. Bell, M. Kim, Y. Hikita, H. Y. Hwang, and K. A. Moler, Direct imaging of the coexistence of ferromagnetism and superconductivity at the $LaAlO_3/SrTiO_3$ interface, *Nat. Phys.*, 7, p. 767 (2011).

54. M. R. Fitzsimmons, et al., Upper limit to magnetism in $LaAlO_3/SrTiO_3$ heterostructures, *Phys. Rev. Lett.*, 107, 217201 (2011).

55. M. Salluzzo, et al., Origin of interface magnetism in $BiMnO_3/SrTiO_3$ and $LaAlO_3/SrTiO_3$ heterostructures, *Phys.Rev. Lett.*, 111, 87204 (2013).

56. P. Moetakef, J. R. Williams, D. G. Ouellette, A. P. Kajdos, D. Goldhaber-Gordon, S. James Allen, and Susanne Stemmer, Carrier-controlled ferromagnetism in $SrTiO_3$, *Phys. Rev. X*, 2, 21014 (2012).

57. S. Stemmer and S. J. Allen, Two-dimensional electron gasses at complex oxide interfaces, *Annu. Rev. Mater. Res.*, 44, pp. 151–171 (2014).

58. L. Esaki, P. J. Stiles, and S. von Molnar, Magnetointernal field emission in junctions of magnetic insulators, *Phys. Rev. Lett.,* 19, p. 852 (1967).

59. R. Meservey and P. M. Tedrow, Spin-polarized electron tunneling, *Phys. Rep.,* 238, p. 173 (1994).

60. J. S. Moodera, X. Hao, G. A. Gibson, and R. Meservey, Electron-spin polarization in tunnel junctions in zero applied field with ferromagnetic EuS barriers, *Phys. Rev. Lett.,* 61, p. 637 (1988).

61. X. Hao, J. S. Moodera, and R. Meservey, Spin-filter effect of ferromagnetic europium sulfide tunnel barriers, *Phys. Rev. B,* 42, p. 8235 (1990).

62. W. F. Brinkman, R. C. Dynes, and J. M. Rowell, Tunneling conductance of asymmetrical barriers, *J. Appl. Phys.,* 41, p. 1915 (1970).

63. J. S. Moodera, R. Meservey, and X. Hao, Variation of the electron-spin polarization in EuSe tunnel junctions from zero to near 100% in a magnetic field, *Phys. Rev. Lett.,* 70, p. 853 (1993).

64. D. C. Worledge and T. H. Geballe, Magnetoresistive double-spin filter tunnel junction, *J. Appl. Phys.,* 88, p. 5277 (2000).

65. R. H. Fowler and L. Nordheim, Electron emission in intense electric fields, *Proc. R. Soc. A,* 119, p. 173 (1928).

66. J. S. Moodera, J. Nowak, and R. J. M. van de Veerdonk, Interface magnetism and spin wave scattering in ferromagnet-insulator-ferromagnet tunnel junctions, *Phys. Rev. Lett.,* 80, p. 2941 (1998).

67. T. Nagahama, T. S. Santos, and J. S. Moodera, Enhanced magneto transport at high bias in quasi-magnetic tunnel junctions with EuS spin-filter barriers, *Phys. Rev. Lett.,* 99, 16602 (2007).

68. T. Santos, et al., Determining exchange splitting in a magnetic semiconductor by spin-filter tunneling, *Phys. Rev. Lett.,* 101, 147201 (2008).

69. G.-X. Miao, M. Muller, and J. S. Moodera, Magnetoresistance in double-spin filter tunnel junctions with non-magnetic electrodes and its unconventional bias dependence, *Phys. Rev. Lett.,* 102, 76601 (2009).

70. P. LeClair, et al., Magnetic and structural properties of EuS for magnetic tunnel junction barriers, *Appl. Phys. Lett.,* 80, p. 625 (2002).

71. U. Luders, et al., Spin filtering through ferrimagnetic $NiFe_2O_4NiFe_2O_4$ tunnel barriers, *Appl. Phys. Lett.,* 88, 082505 (2006).

72. U. Luders, et al., Bias dependence of tunnel magnetoresistance in spin filtering tunnel junctions: experiment and theory, *Phys. Rev. B,* 76, 134412 (2007).

73. A. V. Ramos, et al., $CoFe_2O_4$/buffer layer ultrathin heterostructures on Si(001), *Appl. Phys. Lett.,* 91, 122107 (2007).

74. M. Gajek, et al., Room temperature spin filtering in epitaxial cobalt-ferrite tunnel barriers, *Phys. Rev. B,* 72, 020406(R) (2005).

75. M. Gajek, et al., Tunnel junctions with multiferroic barriers, *Nat. Mater.,* 6, p. 296 (2007).

76. P. LeClair, J. S. Moodera, and R. Meservey, Ferromagnetic-ferromagnetic tunneling and th spin filter effect, *J. Appl. Phys.,* 76, p. 6546 (1994).

77. C. R. Chang and S. P. Chen, Spin polarized tunneling through a ferromagnetic barrier, *Chin. J. Phys.* (Taipei), 36, p. 85 (1998).

78. C. R. Chang and S. P. Chen, Enhancement of tunneling magnetoresistance through a magnetic barrier, *J. Magn. Magn. Mater.,* 209, p. 61 (2000).

79. R. Fehrenbacher and T. M. Rice, Unusual electronic structure of $PrBa_2Cu_3O_7$, *Phys. Rev. Lett.,* 70, p. 3471 (1993).

80. U. Kabasawa, et al., Size effect on variable-range-hopping transport in $PrBa_2Cu_3O_{7-x}$, *Rev. Lett.,* 70, p. 1700 (1993).

81. Y. Liu, F. A. Cuellar, Z. Sefrioui, J. W. Freeland, M. R. Fitzsimmons, C. Leon, et al., Emergent spin-filter at the interface between ferromagnetic and insulating layered oxides, *Phys. Rev. Lett.,* 111, 247203 (2013).

82. F. A. Cuellar, Y. H. Liu, J. Salafranca, N. Nemes, E. Iborra, G. Sanchez-Santolino, et al., Reversible electric-field control of magnetization at oxide interfaces, *Nat. Commun.,* 5, p. 4215 (2014).

83. J. H. Park, E. Vescovo, H. J. Kim, C. Kwon, R. Ramesh, and T. Venkatesan, Direct evidence for a half-metallic ferromagnet, *Nature,* 392, pp. 794–796 (1998).

84. G. M. Müller, J. Walowski, M. Djordjevic, G.-X. Miao, A. Gupta, A. V. Ramos, et al., Spin polarization in half-metals probed by femtosecond spin excitation, *Nat. Mater,* 8, p. 56 (2009).

85. L. Chioncel, et al., Non-quasiparticle states in Co2MnSi evidenced through magnetic tunnel junction spectroscopy measurements, *Phys. Rev. Lett.,* 100, 086402 (2008).

86. P. A. Dowben and R. Skomski, Finite-temperature spin polarization in half-metallic ferromagnets, *J. Appl. Phys.,* 93, pp. 7948–7950 (2003).

87. M. Lezaic, Ph. Mavropoulos, J. Enkovaara, G. Bihlmayer, and S. Blügel, Thermal collapse of spin polarization in half-metallic ferromagnets, *Phys. Rev. Lett.,* 97, 026404 (2006).

88. V. Garcia, M. Bibes, A. Barthélémy, M. Bowen, E. Jacquet, J. P. Contour, and A. Fert, Temperature dependence of the interfacial spin polarization of $LaSrMnO_3$, *Phys. Rev. B,* 69, 052403 (2004).

89. M. Julliere, Tunneling between ferromagnetic films, *Phys. Lett. A,* 54, p. 225 (1975).

90. W. H. Butler, X. G. Zhang, T. C. Schulthess, and J. M. MacLaren, Spin-dependent tunneling conductance of Fe|MgO|Fe sandwiches, *Phys. Rev. B,* **63**, 054416 (2001).

91. J. Mathon and A. Umerski, Theory of tunneling magnetoresistance of an epitaxial Fe/MgO/Fe(001) junction, *Phys. Rev. B,* 63, 220403(R) (2001).

92. J. W. Seo, et al., Antiferromagnetic $LaFeO_3$ thin films and their effect on exchange bias, *J. Phys.: Condens. Matter,* 20, 264014 (2008).

93. J. Park, et al., Magnetic properties at surface boundary of a half-metallic ferromagnet $La_{0.7}Sr_{0.3}MnO_3$, *Phys. Rev. Lett.,* 81, pp. 1953–1956 (1998).

94. M. Bibes, J. E. Villegas, and A. Barthélémy, Ultrathin oxide films and interfaces for electronics and spintronics, *Adv. Phys.,* 60, p. 5–84 (2011).

95. F. Y. Bruno, et al., Insight into spin transport in oxide heterostructures from interface-resolved magnetic mapping, *Nat. Commun.,* 6, p. 6306 (2015).

96. Z. Sefrioui, et al., All-manganite tunnel junctions with interface-induced barrier magnetism. *Adv. Mater.,* 22, 5029 (2010).

97. M. Bowen, et al., Nearly total spin polarization in $La_{2/3}Sr_{1/3}MnO_3$ from tunneling experiments, *Appl. Phys. Lett.,* 82, p. 233 (2003).

98. J. S. Moodera, T. S. Santos, and T. Nagahama, The phenomena of spin-filter tunneling, *J. Phys.: Condens. Matter,* 19, 165202 (2007).

99. T. Nagahama, S. Yuasa, E. Tamura, and Y. Suzuki, Spin-dependent tunneling in magnetic tunnel junctions with a layered antiferromagnetic Cr(001) spacer: role of band structure and interface scattering, *Phys. Rev. Lett.,* 95, 086602 (2005).

100. S. Zhang, P. Levy, A. Marley, and S. Parkin, Quenching of magnetoresistance by hot electrons in magnetic tunnel junctions, *Phys. Rev. Lett.,* 79, pp. 3744–3747 (1997).

101. R. Y. Gu, L. Sheng, and C. S. Ting, Quantum spin assisted tunneling in half-metallic manganite tunnel junctions, *Phys. Rev. B,* 63, 220406(R) (2001).

102. F. Guinea, Spin-flip scattering in magnetic junctions, *Phys. Rev. B,* 58, pp. 9212–9216 (1998).

103. M. T. Fernandez-Diaz, J. L. Martinez, J. M. Alonso, and E. Herrero, Structural, thermal, transport, and magnetic properties of the charge-ordered $La_{1/3}Ca_{2/3}MnO_3$ oxide, *Phys. Rev. B,* 59, p. 1277 (1999).

104. S. Yunoki, A. Moreo, E. Dagotto, S. Okamoto, S. S. Kancharla, and A. Fujimori, Electron doping of cuprates via interfaces with manganites, *Phys. Rev. B,* 76, 064532 (2007).

105. L. Brey, Elementary electronic excitations in graphene nanoribbons, *Phys. Rev. B,* 75, 104423 (2007).

106. M. J. Calderón, J. Salafranca, and L. Brey, Electron gas at the interface between two antiferromagnetic insulating manganites, *Phys. Rev. B,* 78, 024415 (2008).

107. J. Salafranca, M. J. Calderón, and L. Brey, Magnetoresistance of an all-manganite spin valve: a thin antiferromagnetic insulator sandwiched between two ferromagnetic metallic electrodes, *Phys. Rev. B,* 77, 014441 (2008).

108. L. Brey, Continuous charge modulated diagonal phase in manganites, *Phys. Rev. Lett.,* 92, 127202 (2004).

109. D. Niebieskikwiat, M. B. Salamon, L. E. Hueso, N. D. Mathur, and J. A. Borchers, Nanoscale magnetic structure of ferromagnet/antiferromagnet manganite multilayers, *Phys. Rev. Lett.,* 99, 247207 (2007).

110. S. Dong, R. Yu, S. Yunoki, J. M. Liu, and E. Dagotto, Ferromagnetic tendency at the surface of CE-type charge-ordered manganites, *Phys. Rev. B,* 78, 064414 (2008).

Chapter 8

Orbital Symmetry and Electronic Properties of Two-Dimensional Electron Systems in Oxide Heterointerfaces

Gervasi Herranz
Laboratory of Multifunctional Oxides and Complex Structures, Institut de Ciència de Materials de Barcelona ICMAB-CSIC, Campus de la UAB, 08193 Bellaterra, Catalonia, Spain
gherranz@icmab.cat

8.1 Introduction

The discovery in 2004 of a conducting interface between the two wide bandgap insulators $LaAlO_3$ and $SrTiO_3$ grabbed a wide attention in the scientific community [1]. This result spurred the interest of studying physical properties that are absent in the parent compounds and could emerge within a nanometric region across the boundary of two different materials. Similar phenomena have been indeed observed in other systems, in which physical properties linked to, e.g., magnetic [2] or superconducting phases [3], are also found to arise just in the vicinity of the interface. In the particular

Oxide Spintronics
Edited by Tamalika Banerjee

ISBN 978-981-4774-99-4 (Hardcover), 978-0-429-46819-3 (eBook)
www.jennystanford.com

case of $LaAlO_3/SrTiO_3$, the two-dimensional character of the conductive region was demonstrated by a variety of experimental techniques, viz., analysis of Shubnikov de Haas oscillations[4, 5], real-space mapping with scanning proximity probes [6, 7], mid-infrared ellipsometry [8], anisotropy of superconducting critical magnetic fields [9], or photoemission spectroscopies [10, 11]. From all these experimental approaches, it has been inferred that the conductive layer extends just a few unit cells from the interface so that the electrons are confined inside nanometer scale quantum wells (QW) forming two-dimensional electron systems (2DESs) [12].

Since the first report of the conductive $LaAlO_3/SrTiO_3$ interface, a number of other transition metal oxide interfaces have been also found to host 2DESs, namely, $LaTiO_3/SrTiO_3$ [13], $LaVO_3/SrTiO_3$ [14], $YTiO_3/CaTiO_3$ [15], $LaGaO_3/SrTiO_3$ [16], $SmTiO_3/SrTiO_3$ [17], $GdTiO_3/SrTiO_3$ [18], $LaTiO_3/KTaO_3$ [19], $DyScO_3/SrTiO_3$ [20], $NdTiO_3/SrTiO_3$ [21], $RO/SrTiO_3$ (where R is La, Pr, Nd, Sm, or Y) [22], or γ-$Al_2O_3/SrTiO_3$ [23, 24]. In all these systems, the electrons participating in the transport are related to *d* orbitals derived from metal transition ions. This is in contrast with the orbital character of 2DESs in conventional QWs, e.g., III–V or II–VI semiconductors, where electrons are derived either from *s*- or *p*-states. The different nature of electrons in the QWs has important effects on their physical properties, since the density of states of *d*-bands is substantially larger than that of *s*- or *p*-bands, while the bandwidth of *d*-states is significantly narrower. Both features contrive to enhance correlations between states derived from *d*-bands and promote electronic phases that are usually not found in conventional semiconductors. For instance, the $LaAlO_3/SrTiO_3$ system has been shown to display magnetism [25] and superconductivity [26], and, as discussed below, the coexistence of both [27–29] might lead to interesting physical phenomena related to topological electronic states.

The discovery of all the aforementioned 2D systems based on transition metal oxides has been enabled by advances in deposition techniques, which have been crucial to grow the atomically sharp interfaces required to sustain high-mobility conduction in oxide heterostructures. Indeed, this is reminiscent of conventional semiconductors, where novel deposition methodologies perfected in the late 60s and 70s, such as molecular beam epitaxy (MBE) [30, 31], or metal organic chemical vapor deposition (MOCVD) [32] laid the ground for the growth of superlattice structures of alternate layers as thin as about 10 Å. At that time, this progress enabled the deep downscaling of semiconductors, especially of the III–V family. The reduced dimensionality led to new fundamental advances in physics, such as the integer and fraction quantum Hall effect [33, 34], and also paved the way to new technological developments, particularly in the domain of optoelectronics [35].

Similarly, the development of advanced growth methodologies, especially pulsed laser deposition (PLD) [36], has been also critical to allow the growth of oxide heterostructures, where atomically flat interfaces are indispensable to produce 2DESs. Thus, in the same way that engineered semiconductor superlattices opened up new prospects in fundamental science, oxide heterointerfaces may also provide insights into unexpected physical phenomena. In this respect, in addition to the coexistence of magnetism and superconductivity [37], the study of Rashba like spin–orbit interactions in $LaAlO_3/SrTiO_3$ interfaces [38, 39] has triggered a great interest, especially in relation of spin-injection into the 2DESs [40–43], or the possibility, in conjunction with superconductivity, to induce unconventional superconducting order parameters at the $LaAlO_3/SrTiO_3$ interface, such as the occurrence of triplet superconductivity [44, 45]. Similarly, the concurrence of large Rashba spin–orbit fields and superconductivity are key ingredients for the observation of pairs of Majorana fermions in topological superconductive states [46, 47]. Additionally, the 2D character of the superconductivity has led to phenomena not observed in the 3D regime, such as magnetic enhancement of superconductivity [48], or the violation of the paramagnetic Pauli limit for the upper critical fields [49]. The intricacy of all these complex phases and the evidence of the role of electron correlations have prompted the use of the concept of electron liquids to designate these electron systems [50], although we will refer them as 2DESs for the rest of the chapter.

While some of the above-mentioned interesting prospects have not been demonstrated yet in the archetypal $LaAlO_3/SrTiO_3$ interface, the latter system possesses a set of properties that make it singular with respect to other materials. In the first place, the 2D superconducting state can be largely modulated by gated electric fields, to the extent that the superconductivity can be switched on and off reversibly [51, 52]. Interestingly, this large modulation gives rise to an electrostatically tuned phase-transition that shows features of quantum transitions [53, 54]. Secondly, the spin–orbit coupling at the $LaAlO_3/SrTiO_3$ interface is found to be of the Rashba type, which in turn, is largely tunable with electric fields [38, 39]. This feature allows modulating the strength of the spin–orbit coupling felt by the electrons in the 2DES, opening up eventually the way towards new device concepts. Interestingly, it has been demonstrated that the interplay between orbital ordering and spin–orbit coupling leads to unconventional Rashba splitting, resulting in larger Rashba coefficients [55]. The tunability of the spin–orbit fields is, definitely, a comparative advantage over conventional semiconductor interfaces, where the intrinsic value of the spin–orbit coupling is weak and cannot be tuned.

The remarkable tunability of the 2D superconductivity and Rashba spin–orbit fields is intimately related to the electronic band structure of the 2DESs. More specifically, electric fields applied across the $LaAlO_3/SrTiO_3$ interface induce changes in the electron band filling and, as a result, changes of the Fermi level position in the QW energy landscape. Incidentally, the changes in the electron population imply, in the end, that states with different orbital symmetry can be occupied. This is relevant because the symmetry of the occupied orbitals often influences the physical properties. For instance, the possible multiband character of superconductivity at the $SrTiO_3$-quantum wells is expected to be linked to the detailed orbital structure of the t_{2g} states [56, 57]. On the other hand, the microscopic nature of the interface magnetism in $LaAlO_3/SrTiO_3$ has also been associated with the orbital energy hierarchy of t_{2g} and e_g levels in the $LaAlO_3/SrTiO_3$ QWs [58, 59], and even spectroscopic investigations emphasize the specific role of d_{xy} states regarding the emergence of magnetism in these interfaces [60]. For all these considerations, an incursion into the nature of the electronic band structure of the $LaAlO_3/SrTiO_3$ QWs is required.

8.2 The Electronic Band Structure at the $LaAlO_3/SrTiO_3$ Interface

8.2.1 Preamble: The Band Structure of Bulk $SrTiO_3$

As pointed out above, the electrons participating in the transport derive essentially from d-states, which in the case of $LaAlO_3/SrTiO_3$, arise from orbitals in the Ti ions. A rather intuitive way to understand the electronic band structure of QWs at the $LaAlO_3/SrTiO_3$ interface is to start from the bulk band structure of $SrTiO_3$ and derive the orbital reconstructions triggered by the quantum confinement when the electron motion is constrained within a narrow region across the interface.

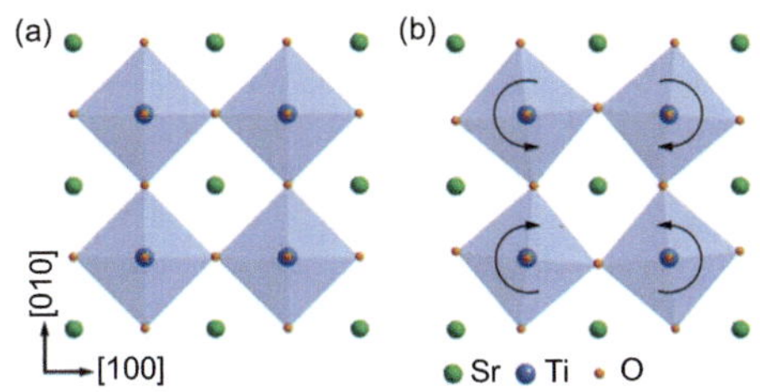

Figure 8.1 (a) Schematics of the structural cubic phase of $SrTiO_3$ with the structural TiO_6 octahedra units. (b) Schematics of the tetragonal structural phase (below ~105 K), in which an antiferrodistortive pattern of rotations of the TiO_6 octahedra is involved. Reprinted from Ref. [61], with permission from American Physical Society, Copyright 2010.

Before entering into details, a few words on the crystallographic properties are useful. At room temperature, bulk $SrTiO_3$ crystallizes in the ABO_3 cubic perovskite structure with space group Pm3m [62], where BO_6 octahedra are the basic structural units (Fig. 8.1a). The Goldschmidt tolerance factor is $t \approx 1$, close to the perfect cubic structure, with a lattice parameter of $c \approx 3.905$ Å and a density of $\rho \approx 5.12$ g/cm^3. Yet, at a temperature of $T \sim 105$ K, $SrTiO_3$ crystals undergo a transition to a tetragonal phase with a small distortion $c/a \approx 1.0005$ [63, 64] and with an I4/mcm space group. The transition to the tetragonal structure involves an antiferrodistortive pattern of rotations of the TiO_6 octahedra (Fig. 8.1b), where the pattern arises from counter-rotations of up to around 2° of neighboring octahedra along [001] axis. As a result, the unit cell volume changes from a^3 to $\sqrt{2}a \times \sqrt{2}a \times c$ [65] and the effect of tetragonality can be observed directly by the measuring the anisotropy of the dielectric constant [66], or the birefringence [67]. As discussed below, this subtle distortion lowers the symmetry of the crystal and breaks the degeneracy of the electronic states derived from the cubic structure at higher temperatures. Still, at lower temperatures, it has been also proposed that $SrTiO_3$ undergoes a further symmetry lowering to a triclinic structure created by small Sr displacements along the [111] direction convoluted with the oxygen-octahedron counter-rotations, which should induce additional splittings of the electronic band states [68]. We will restrict our discussion hereafter to the effects on the band structure of the cubic and tetragonal phases of $SrTiO_3$.

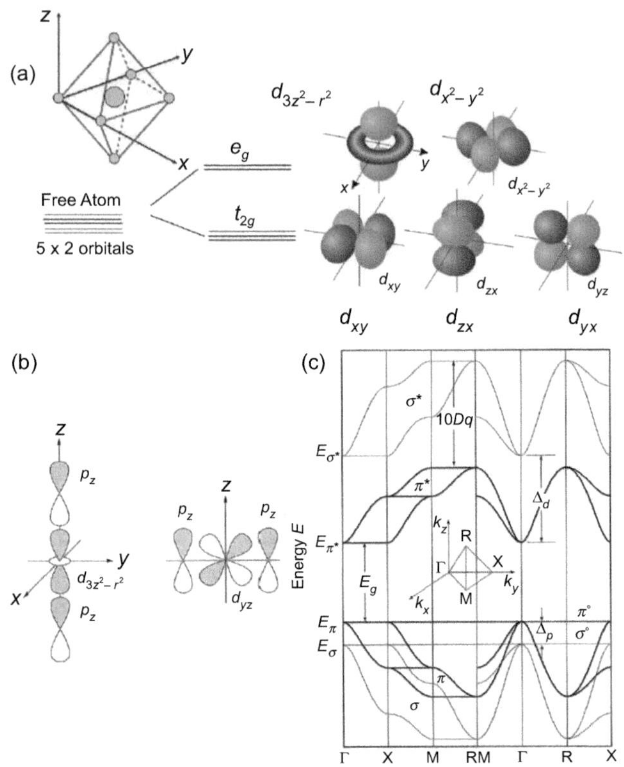

Figure 8.2 (a) Spatial distribution of t_{2g} (d_{xy}, d_{xz}, d_{yz}) and e_g ($d_{3z^2-r^2}$, $d_{x^2-y^2}$) orbitals. The 5 × 2 fold degeneracy (including the spin) is lifted by the octahedrally-coordinated environment (upper left corner). (b) Schematic representation of the σ and π bonds involving, respectively, the overlapping of e_g orbitals with $p_{||}$ orbitals, and the hybridization of t_{2g} orbitals with $p_{\perp}$ oxygen orbitals. (c) Energy bands for a typical perovskite. Lighter curves depict π-bands and darker ones σ-bands. Panels (b) and (c) are reprinted from Ref. [62], with permission from Cambridge University Press, Copyright 2009.

The digression on the structural properties of $SrTiO_3$ provides us with a way to a qualitative understanding of the bulk electronic band structure on the basis of simple chemical bonding considerations.

We start first the discussion of the cubic phase of $SrTiO_3$, which is the stable phase at room temperature and down to $T \sim 105$ K. The basic features of the band structure can be inferred by considering the TiO_6 octahedral unit, in which the Ti cations are coordinated to the six surrounding oxygens. Following this line of reasoning, and as a result of the crystal field, the five-fold degeneracy of the d-states of the free Ti ion is broken into three-fold t_{2g} and two-fold e_g states when the Ti ion is in the undistorted octahedral coordination of the cubic phase. An intuitive picture of the energy landscape is obtained by recalling that the $d_{x^2-y^2}$ and $d_{3z^2-r^2}$ orbitals with e_g symmetry are oriented towards the oxygen ligands, thus feeling a stronger electrostatic repulsion than the d_{xy}, d_{xz}, and d_{yz} orbitals with t_{2g} symmetry. Consequently, the energy of $d_{x^2-y^2}$ and $d_{3z^2-r^2}$ states is higher than that of d_{xy}, d_{xz}, and d_{yz} orbitals (Fig. 8.2a).

A more accurate approach to the band structure should account for the hybridization of d-states in Ti with p orbitals of the six oxygen ligands. It is indeed the overlapping of the wave functions of these states that enables the transfer of electrons back and forth between the ions and, eventually, the delocalization of electrons. The p-d hybridization can be resolved from calculations based on a linear combination of atomic orbitals (LCAO) approximations. From these calculations, it is inferred that the band structure is formed by sigma (σ) and pi (π) bands, where the former involve the overlapping of e_g orbitals with $p_{||}$ orbitals, while π-bands evolve from the hybridization of t_{2g} orbitals with $p_{\perp}$ oxygen orbitals (Fig. 8.2b). As a result of these orbital interactions, bonding (σ, π), and antibonding (σ^*, π^*) bands appear, with an admixture of p and d states that depends on the position of the wave vector $\vec{k}$ in the reciprocal space. For instance, in the particular case of the Γ-point ($\vec{k}=0$), the wave functions have either a pure d or p character, while departing from that point the bands acquire a certain degree of d or p character. The fundamental bandgap at Γ is given by the energy separation between the antibonding π^* bands (formed from t_{2g} states in the conduction band) and bonding π bands (formed from p orbitals in the valence band). Experimentally, the bandgap is found to be about 3.2 eV. Higher in energy by about 1 eV above t_{2g} states are the e_g levels that derive from antibonding σ^* bands. Figure 8.2c shows a tight-binding calculation displaying the electron bands in a typical perovskite. We note that, even though the overlapping between neighboring t_{2g} or e_g orbitals is mediated by *p–d* hybridization, the role of *p* orbitals is often neglected, and for the sake of simplicity, it is common to discuss in terms of the hybridization of *d* orbitals without mentioning *p* orbitals (see the discussion enclosed in next sections).

Semi-empirical approaches suffice to derive the basic features of the electronic band structure for the cubic phase of $SrTiO_3$ [69–72]. Yet, other models have been also applied to describe the properties across the cubic-to-tetragonal transition. Thus, in the 90s, studies used ab initio calculations, encompassing a range of different methods from the plane wave (PW) based on density function approximation-local density approximation (DFT-LDA) to advanced LCAO-based hybrid DFT schemes [73–76]. Due to the very small changes involved in the atomic and electronic structure across the transition, the use of ab initio calculations was particularly critical to interpret correctly the physical properties of $SrTiO_3$ around temperature where the cubic phase gives way to a tetragonal symmetry [73, 77, 78].

Apart from the dependence of the energy of the bands on the wave vector, the determination of the Fermi surface in the reciprocal space is also a basic aspect to calculate the physical properties. In the particular case of the cubic phase, band structure calculations predict that the Fermi surface of $SrTiO_3$ is formed by three orthogonal ellipsoids centered at the Γ-point, originated by the three degenerate t_{2g} orbitals (d_{xy}, d_{yz}, and d_{zx}), which as aforementioned, are lower in energy than the e_g orbitals [61, 70]. However, the Fermi surface undergoes important changes when the crystal undergoes the transition to the tetragonal structure. Across this transition, the structural TiO_6 octahedra experience antiphase octahedral rotations around the *c*-axis, so that the Ti-3*d* orbitals gradually rotate in the *ab* plain causing a change in the overlapping with the surrounding O-2*p* orbitals. As a result, the symmetry is lowered and the calculations predict a splitting of the three-fold degenerate t_{2g} states at the conduction band. More specifically, the t_{2g}-manifold is split into two subsets, namely, a two-fold degenerate band, which is lower in energy, and is formed by the d_{yz} and d_{zx} orbitals, and the d_{xy} band should locate higher. However, when angle-resolved photoemission spectroscopy (ARPES) experiments are used to scrutinize the electronic band structure, the energy landscape is different from these predictions. More specifically, what is found is that the d_{xy} band lies at a lower energy than the d_{yz} and d_{zx} orbitals [61]. The clue to this discrepancy is that, as we will describe below, ARPES is sensitive to a nanometric region at the surface, where the electronic structure may be changed significantly with respect to the bulk structure. This drives us into the formation of QWs at the surface of $SrTiO_3$, which is described in the following section.

8.2.2 The Band Structure of $SrTiO_3$ Quantum Wells

Stoichiometric $SrTiO_3$ is an insulator, but it can be doped n-type by substitution of La^{3+} for Sr^{2+}, Nb^{5+} for Ti^{4+}, or by reduction to $SrTiO_{3-\delta}$, where δ indicates the amount of oxygen deficiency. By carefully controlling the volume of dopants, $SrTiO_3$ can be rendered metallic with carrier densities spanning a wide range from about 10^{15} cm^{-3} to above 10^{20} cm^{-3} [79], with electronic mobilities well upwards of 2×10^4 cm^{-2}/Vs [80]. Taking advantage of this versatility, ARPES experiments were carried out recently in $SrTiO_3$ crystals in which the carrier densities covered a range of several orders of magnitude. The most striking outcome of these experiments is that in all cases strongly dispersive bands are found across the Fermi level, revealing the presence mobile electrons in surface quantum wells, even for nominally non-doped insulator crystals, for which no bulk bands are expected [81]. Furthermore, the bandwidths and Fermi momenta of the electronic bands are essentially unchanged, whatever the degree to which the crystal is doped (Fig. 8.3). The immediate conclusion is, thus, that the observed bands correspond to two-dimensional (2D) states in QWs [81].

The electronic structure revealed by ARPES is formed by lower parabolic bands with bandwidths typically above 100 meV, in conjunction with upper shallow bands with smaller bandwidths (<40 meV), see Fig. 8.4a for a pictorial depiction of the quantum well states. The more dispersive character of the former should correspond to light carriers with effective masses typically $m^* < 0.7\, m_e$ (m_e is the free-electron mass), while the shallow bands should be related to heavier bands ($m^* < 10\, m_e$) [81]. To understand qualitatively the electronic band structure observed by ARPES, we can reason in terms of the dispersive characters of the different bands. As aforementioned, the electronic states relevant for transport are the d_{xy}, d_{xz}, and d_{yz} bands with t_{2g} symmetry, and we will restrict our discussion to them. To lay out the argument, we come back briefly to the band structure of the bulk crystal. If we consider the conduction band of bulk $SrTiO_3$ at low temperatures (i.e., below the transition to the tetragonal unit cell), we see that the wave function overlapping depends strongly on the orbital symmetry. For instance, a large overlap exists for the d_{xy} states along the directions k_x and k_y of the reciprocal space. For d_{xz} and d_{yz} orbitals, the overlapping depends on the direction in the reciprocal space. Thus, along k_x, d_{xz} states overlap intensely, while the overlapping is weaker along k_y (Fig. 8.4c). Conversely, d_{yz} states overlap strongly along k_y and weakly along k_x, while the overlapping is much smaller for the same orbitals along the direction k_z (Fig. 8.4c). The resulting Fermi surfaces depicted in Fig. 8.4b are consequently, rather anisotropic. If, for instance, we cut the Fermi surface along the x–y plane, we observe two cigar-shaped curves, corresponding to d_{yz} and d_{xz} bands, and an isotropic d_{xy} band.

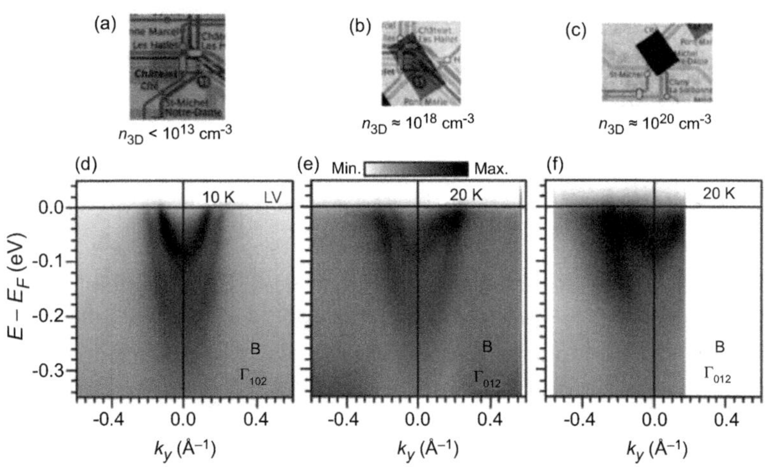

Figure 8.3 Experimental data extracted from ARPES experiments reported in Ref. [81] are displayed in this figure. (a–c) Show photographs and bulk doping of the studied samples. Samples with large electron doping, see panel (c), appear dark, whereas nominally undoped crystals, panel (a), are transparent. The electronic bands close to the Γ-point are shown in (d–f) for different bulk doping, revealing the presence of mobile electrons in surface quantum wells. Reprinted from Ref. [81], with permission from Springer Nature, Copyright 2011.

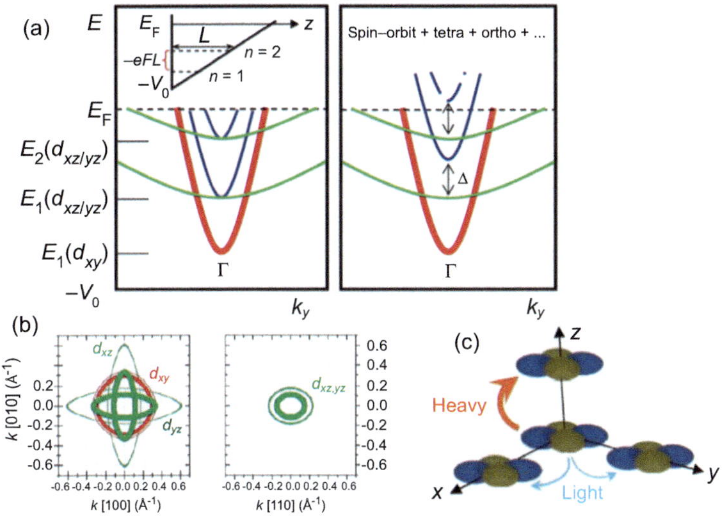

Figure 8.4 (a) The left figure shows a depiction of the quantum well states that arise from the confinement near the surface of $SrTiO_3$. The inset shows schematically the wedge-like potential that confines the electrons. The right figure of panel (a) shows how the degeneracy can be further lifted as a result of spin–orbit coupling or tetragonal distortions. (b) The Fermi surfaces of d_{xy} (red) and d_{xz}/d_{yz} bands (green) are shown, projected onto the (001) (left image) and (110) (right image) planes for different sheet carrier densities ($n_{2D} = 3 \times 10^{14}$ cm^2, thin lines, $n_{2D} = 3 \times 10^{14}$ cm^2, thick lines). The anisotropic dispersion observed in (b) is due to small and large overlaps of neighboring d_{xy}, d_{xz}, d_{yz} orbitals depending on the direction in space, as depicted in (c) for the d_{xy} orbitals. Figures (a) and (c) reprinted from Ref. [81], Copyright 2011, and figure (b) reprinted from Ref. [82], with permission from Springer Nature, Copyright 2014.

Now, we leave the bulk crystal and consider the case in which the electrons cannot move throughout the lattice so that their motion is restricted to travel in the plane of a QW, which we can designate as the x–y plane. From the fundamentals of quantum mechanics, we know that for particles confined in narrow wells, e.g., the simplest case of a particle in a box, the energy of the electron states depends on the inverse of the component of the effective mass along the normal to the confinement direction (that we can assume as the z-direction). Contrary to the strong overlapping in the plane x–y, the d_{xy} orbitals hybridize weakly along the z-direction, so that the effective mass of the d_{xy} band along z-direction is large (expected to be $m^*_{dxy,z} > 10\ m_e$) [83]. In contrast, the orbitals d_{xz} and d_{yz} overlap intensely along the z-direction, and therefore, along the direction normal to the quantum well, the band structure consists of a heavy d_{xz} band, and a doublet of light d_{xy}/d_{yz} bands. According to the above-mentioned quantum physics principles, heavier bands along z (d_{xy} states) are pushed to lower energies than lighter bands (d_{xz}/d_{yz} states), which correspondingly are located higher in energy. This is precisely what ARPES experiments indicate (see above) and gives an intuitive rationale of the energy landscape in which, contrary to the calculations of the bulk band structure, d_{xy} states appear at a lower in energy than d_{xz}/d_{yz} states [68].

The quantum well structure not only changes the energy landscape but also promotes the emergence of a series of quantized states or subbands for every orbital in the t_{2g} manifold. The subband structure as obtained from first-principles calculations [84] is shown in Fig. 8.5a. Three-dimensional views of the Fermi surface demonstrate that for typical sheet carrier densities in the $LaAlO_3$/$SrTiO_3$ system, i.e., around 10^{13} cm^{-3}, the electronic band structure obtained by ARPES shows the appearance of more than one QW subband. For instance, images reported in Ref. [83] show the complex anisotropic Fermi surface, in which in addition to elongated ellipsoidal d_{xz} and d_{yz} bands, the d_{xy} exhibit inner and outer rings corresponding to the first two d_{xy} subbands of the QW [Figs. 8.5(b–e)].

In the aforementioned picture, the hierarchy of electronic states is described as dominated by quantum confinement effects. Yet, a rigorous approach must consider also the effect of spin–orbit coupling and the tetragonal distortion on the electronic states. In the case of the bulk, tight-binding [70], and density functional theory calculations [85] show that in the cubic structure the spin–orbit coupling splits the degenerate t_{2g} bands by $\Delta_{SO} \approx 30$ meV at the Γ-point, giving way to a low-lying doublet involving both a light and heavy band, plus a higher energy singlet state with a light mass (Fig. 8.6). At temperatures below $T \sim 105$ K, where $SrTiO_3$ crystals undergo the antiferrodistortive cubic-to-tetragonal transformation, the degenerate low-lying doublet is further split into two strongly mixed bands separated by a few meV. Now, coming back to the case of the confined 2DESs and taking into account the arguments given in the previous paragraphs, the combined effect of spin–orbit coupling and tetragonal distortion is to lift the degeneracy between the d_{xz} and d_{yz} subbands, as illustrated in Fig. 8.6. All in all, the basic picture described above for the 2DES states, i.e., a bottom d_{xy} band,

and higher d_{xz} and d_{yz} bands, remains unaltered. However, recently Joshua et al. [86] have refined this picture, by proposing that spin–orbit coupling causes further changes near the bottom of the d_{xz} and d_{yz} bands that will be discussed in Section 8.3.2.

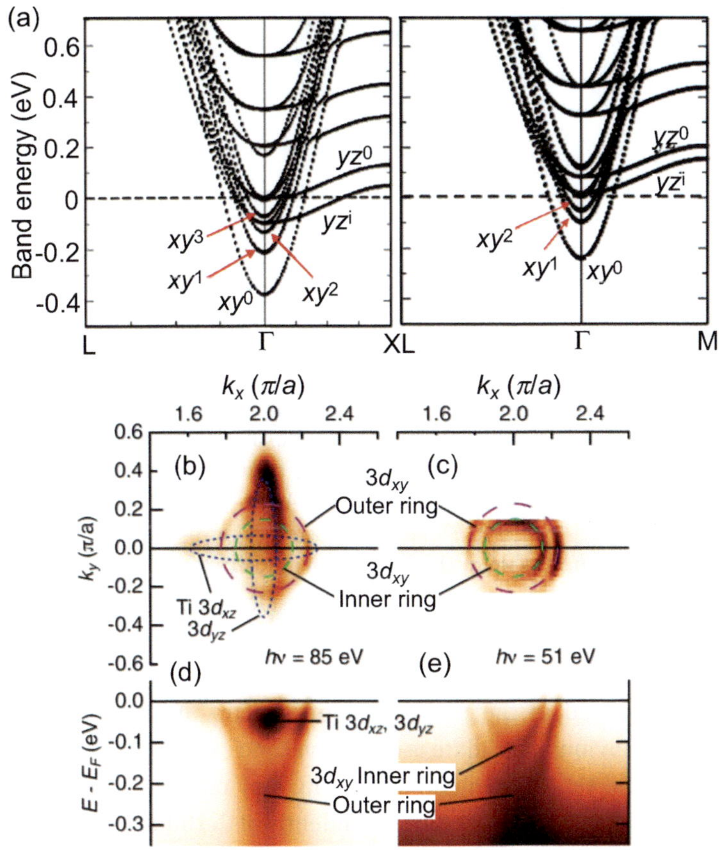

Figure 8.5 Band energies of $LaAlO_3/SrTiO_3$ quantum well states calculated from first-principles for sheet carrier densities $n_{2D} = 3.3 \times 10^{14}$ cm^2 and $n_{2D} = 1 \times 10^{14}$ cm^2. In this panel, d_{xy} and d_{yz} states are designated, respectively, as xy^n and yz^n, where n is the index of the subband. Panels (b) and (c) show the Fermi surface maps in the k_x-k_y plane measured, respectively, at photon energies $h\nu$ = 85 eV and $h\nu$ = 51 eV. The ellipsoidal d_{xz} and d_{yz} bands are illustrated by dotted blue lines, while the d_{xy} inner and outer rings are highlighted by short-dashed green lines and long-dashed magenta lines, respectively. The band dispersions along k_x at k_y= 0 are shown in (d) and (e) for the data depicted in panels (b) and (c). Figure (a) reprinted from Ref. [84], Copyright 2011, and figures (b–e) reprinted from Ref. [83], with permission from American Physical Society, Copyright 2014.

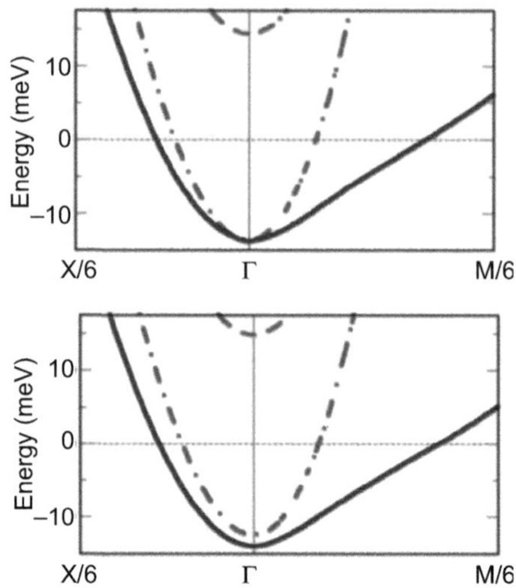

Figure 8.6 Bands calculated from density functional theory in Ref. [85] are shown for the cubic, and tetragonal structures in the upper and lower panels, respectively. The spin–orbit splits the bands into an upper singlet (dashed) and doublet with light (dashed-dotted), and heavy (solid) masses. Reprinted from Ref. [85], with permission from American Physical Society, Copyright 2012.

As mentioned above, the characteristics of the electronic band structure probed by ARPES on the surface of $SrTiO_3$ crystals is essentially unchanged, even in the case where the bulk of the crystal is insulating. This raises the question about the origin of the QW states observed in the photoemission spectra of cleaved crystals [81]. One obvious candidate is oxygen vacancies at the surface. It has been argued that cleaving or fracturing may produce a substantial depletion of surface oxygen, independently of the bulk concentration of vacancies [81]. The generation of these point defects provides two dopant electrons per created vacancy, which can delocalize within the potential wedge created by the positively charged layer of surface vacancies. Thus, oxygen vacancies have two roles: on the one hand, they provide carries; on the other hand, because vacancies are charged, they create a narrow wedge potential that confines the electrons. Additionally, controlled through exposure of the surface to intense ultraviolet light has also been claimed to be responsible for the generation of 2DESs at the surface of $SrTiO_3$ [87].

Now the question arises: Apart from the mechanisms described in the last paragraph, are there other alternative mechanisms by which 2DESs can be generated in $SrTiO_3$? This question brings us directly into the physics of the 2DESs at oxide non-polar/polar heterointerfaces, being the $LaAlO_3/SrTiO_3$ interface the archetypal system of this kind.

8.2.3 Quantum Wells at the $LaAlO_3/SrTiO_3$ Interface

Along the [001] direction of its pseudocubic unit cell, the ideal $LaAlO_3$ layer presents atomic planes that can be divided into alternating layers of LaO and AlO_2 planes. Taking into account the formal valences (O^{2-}, La^{3+}, Al^{3+}), the alternate planes have the formal charges $(LaO)^{+1}$ and $(AlO_2)^{-1}$, respectively. In contrast, bearing in mind the formal ionic valences of Ti^{4+} and Sr^{2+}, the atomic planes in $SrTiO_3$ along the direction [001] are all neutral, i.e., $(SrO)^0$ and $(TiO2)^0$, respectively. When the two materials are put in contact forming an atomically sharp interface, a polar discontinuity results from the mismatched electric boundary conditions set by the aforementioned different charges of the atomic planes (Fig. 8.7).

The interface can display two possible atomic compositions, i.e., it can be formed either by $AlO_2/LaO/TiO_2/SrO$ planes or by $LaO/AlO_2/SrO/TiO_2$ planes [1]. Due to the mismatched polarity at the interface, an electrostatic divergence occurs in both cases, although with opposite signs (Fig. 8.7) [88]. This divergence raises the electric potential as the thickness of the $LaAlO_3$ grows, by about 1 V per unit cell. As a consequence of the rising potential, the system becomes unstable, and in the case of the $AlO_2/LaO/TiO_2/SrO$ interface, it can curb this electrostatic divergence above a certain thickness, which is found to be 4 unit cells (uc) of $LaAlO_3$ [89]. Thus, $LaAlO_3$ films with a thickness above this critical value can avoid the ensuing "polar catastrophe" by adding half an electron to the last Ti layer. This process is thought to occur by transferring electrons from the uppermost layer of $LaAlO_3$ into the TiO_2 planes adjacent to the interface, i.e., it is a sort of Zener breakdown process [1, 88]. As a consequence, a 2DES is formed at the $LaAlO_3/SrTiO_3$ interface.

The behavior of the other interface, composed of $LaO/AlO_2/SrO/TiO_2$ atomic planes, is, however, radically different. In this case, the interface remains insulating and no 2DES is formed, whatever the thickness of the $LaAlO_3$ layer [1, 89]. It has been proposed that in this case the electrostatic mismatch at the interface is not compensated by any charge transfer across the interface, but by the introduction of oxygen vacancies. In consequence, the transport and electronic properties of the $LaAlO_3/SrTiO_3$ interfaces depend dramatically on its atomic chemical composition. When $LaAlO_3$ thin films are grown on TiO_2-terminated $SrTiO_3$ crystals, the 2DES is formed above a critical $LaAlO_3$ thickness (4 uc). In contrast, when the interface is formed at the boundary of $LaAlO_3$ with SrO-terminated $SrTiO_3$ crystals, it remains insulating, regardless of the $LaAlO_3$ thickness. The dramatic dependence on the interface atomic composition as well as the existence of a critical thickness for the formation of the 2DES is usually taken as strong arguments to support the "polar catastrophe" model as described above [89].

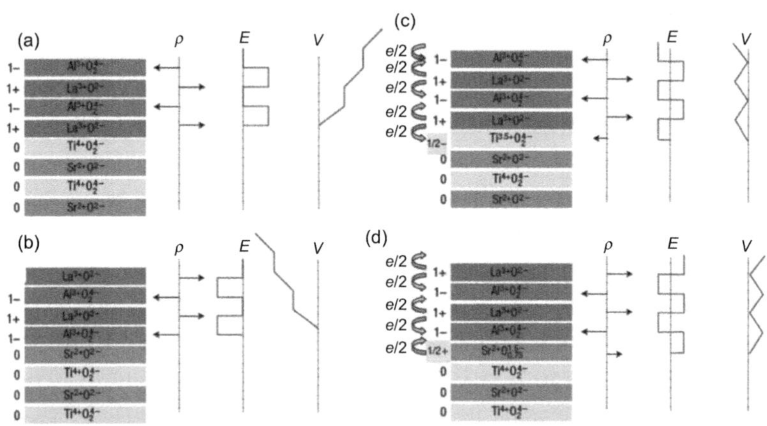

Figure 8.7 Schematic representation of the charge transfer based on the polar catastrophe scenario. (a) For (001)-oriented $LaAlO_3/SrTiO_3$ interfaces, the unreconstructed interface has neutral (001) planes in $SrTiO_3$ and alternating charges in $LaAlO_3$. In the case of the $AlO_2/LaO/TiO_2$ interface, this produces an electric field and an electric potential that diverges with thickness. (b) For the $AlO_2/SrO/TiO_2$ interface, the potential diverges with opposite sign. (c) To avoid the divergence catastrophe at the $AlO_2/LaO/TiO_2$ interface, an electron is transferred to the last Ti layer. Consequently, an electron is missing in the uppermost AlO_2 layer. (d) The divergence for the $AlO_2/SrO/TiO_2$ interface can also be avoided by removing half an electron from the SrO plane in the form of oxygen vacancies. Reprinted from Ref. [88], with permission from Springer Nature, Copyright 2006.

Yet, the origin of the 2DESs at the $LaAlO_3/SrTiO_3$ interface is still a debated issue. As seen in previous sections, oxygen deficiency generates 2DES at the surface of $SrTiO_3$ and it has been proposed that similar mechanisms may explain, at least partially, the conductivity at the $LaAlO_3/SrTiO_3$ interface [90–92]. Beyond that, cation intermixing, viz., interdiffusion of La/Sr across the interface [93, 94] has also been claimed as a contributor to the interface conduction. Furthermore, amorphous layers of $LaAlO_3$, where the polar catastrophe does not apply and even a handful of other amorphous oxides, have been shown to generate 2DESs when interfaced to $SrTiO_3$, also above a critical thickness that in this case may depend on the electron affinity of the overlying oxide [95–97]. Nevertheless, the model based on the polar discontinuity is generally accepted to be the main origin leading to the formation of the 2DES in epitaxial $LaAlO_3/SrTiO_3$ interfaces as it naturally explains the strong dependence of the electronic properties on the interface atomic composition, as well as other facts, such as the observed electrostriction [98] and ionic displacements [99] observed in the $LaAlO_3$ film, which are compatible to the existence of a sizable dipole electric field inside the layer [100]. On the other hand, although unintentional doping may lead to the conduction of $SrTiO_3$ layers adjacent to the interface, it cannot avoid the electric divergence generated by the polar catastrophe.

Beyond the simple polar catastrophe scheme outlined above, variant models have been proposed over the recent years, in which the polar discontinuity drives the spontaneous formation of certain defects, e.g., surface oxygen vacancies that generate mobile electrons at the interface [101–104]. Put in other words, in these models, the polar catastrophe is the factor driving the generation of 2DESs but in a more indirect way than in the simple direct charge transfer across the interface. Meanwhile, the role of defects is further substantiated in studies, which demonstrate that the cation stoichiometry of the nominal $LaAlO_3$ layer is an essential parameter to generate the 2DESs [50, 105]. Although the search for the ultimate reason for the generation of interface conduction at oxide heterostructures is undoubtedly of high interest, in the following we will focus on the electronic properties of the 2DESs, rather than the causes that create them. In particular, we will focus on what is known about two of the most conspicuous transport properties of these systems: the

2D superconductivity and the Rashba spin–orbit fields. Later on, we will show that the crystal orientation of the interface modifies the electronic subband structure of the 2DESs as well as their physical properties. For that reason, we first describe the properties of the (001)-oriented interface.

8.3 Low-Temperature Transport and Orbital Symmetries at the (001)-Oriented $LaAlO_3$/$SrTiO_3$ Interface

We have seen in Section 8.2 that the confinement of electrons at QW surface states of $SrTiO_3$ modifies substantially their electronic band structure. For similar reasons, the confinement of 2DESs at the $LaAlO_3$/$SrTiO_3$ interface also changes the electronic structure in a similar way. Thus, akin to the structure at the $SrTiO_3$ surface, it is found experimentally by X-ray spectroscopies [106], or by ARPES [107] that the states with minimum energy are those with d_{xy} symmetry, while the d_{yz}/d_{xz} orbitals lie higher in energy. This electronic structure, which is well reproduced by first-principles calculations [84, 108], is thus composed of states with different symmetry, which result in important consequences in the electronic transport, as shown below. And very importantly, the use of electric fields enables the variation of the electron band filling, so that the relative populations of electrons in d_{xy} or d_{yz}/d_{xz} states can be changed by applying fields in field-effect gated devices [109–111], and thus, modulate physical properties, like 2D superconductivity, or spin–orbital fields of the Rashba type.

8.3.1 2D Superconductivity

$SrTiO_3$ is notorious for being the most dilute known superconductor, with the onset of superconductivity at carrier densities as low as $\sim 10^{17}$ cm^3, orders of magnitude lower than any other superconductor [112]. These extreme low-carrier densities imply a particularly small Fermi surface, which in turn, implies other singular properties; for instance, the Fermi velocity, computed as $v_F = \frac{\hbar k_F}{m^*}$, where k_F is the Fermi wave vector, is as low as $v_F \approx 15$ km/s, i.e., almost as slow as sound velocity. Additionally, the Fermi temperature is an order of magnitude lower than the Debye temperature [112], a situation that is only encountered in non-conventional systems, e.g., heavy-fermion superconductors [113]. This situation has deep consequences in the interpretation of the superconductivity of bulk $SrTiO_3$ in terms of phonon–electron interactions, in particular regarding the relative weight of Coulomb repulsion with respect to the phonon-mediated attraction between electrons [114]. In view of this, the interpretation of superconductivity using the conventional Bardeen–Cooper–Schrieffer (BCS) theory is challenged and other more exotic pairing mechanisms have been proposed, e.g., exchange of plasmons between electrons [115], phonon-soft mode coupling [116], or the vicinity to the ferroelectric quantum critical point [117].

Taking into account all the peculiar properties of the bulk, it is understandable that reports of superconductivity at the $LaAlO_3$/$SrTiO_3$ interface (typically below $\approx$ 200 mK) [118–120] was followed with great expectation since the reduction of dimensionality was predicted to bring about new phenomena not observed in bulk. This was indeed the case; for instance, the 2-dimensionality of the superconductivity was shown to be dominated by the physics of Berezinskii–Kosterlitz–Thouless (BKT) theory, which implies a superconductive transition in which bound low-temperature vortex-antivortex pairs break up into unpaired vortices and antivortices at some higher critical temperature.* This physics is intrinsic to 2D systems, in which fluctuations control the establishment of phase coherence, and are expected to dominate the transport, as seen in the I–V characteristics [118, 121]. An additional effect of low dimensionality is the fact that the critical magnetic fields that destroy the superconductivity depend strongly on the orientation with respect to the plane of the interface. When the fields are parallel to the interface, the critical values are significantly larger than when the fields are applied normal to the interface by a factor order of larger magnitude [122]. Such strong anisotropy of the critical fields

*In 2016, the Nobel Prize in Physics was awarded to David J. Thouless, F. Duncan M. Haldane, and J. Michael Kosterlitz "for theoretical discoveries of topological phase transitions and topological phases of matter". In particular, the work of Kosterlitz and Thouless demonstrated that in an XY model, and at low enough temperatures, the lowest thermal fluctuations are formed by counter-circulating pairs of vortices. The breakup of these vortex–antivortex pairs at larger temperatures describes the physics of superfluid phases in 2-dimension.

is in good agreement with models of 2D superconductivity, and indeed, the thickness of the superconductive layer can be determined from such measurements, with typical values of ≈10 nm, which are significantly smaller than the estimated superconducting coherence length (≈ 70 nm) [122]. Although the value of the superconductive thickness obtained from the critical fields is almost one order of magnitude larger than that estimated from X-ray photoemission experiments [123], is still comparable to the value obtained conductive atomic force microscopy, which at low temperatures, gives an estimate of about 12 nm [124].

Soon after the discovery of the 2D superconductivity at the $LaAlO_3/SrTiO_3$ interface, it was found that the superconducting state was largely tunable by electric fields, which change the sheet carrier density and conductance to the point that the superconductivity can be switched on and off reversibly [51]. In these experiments, the superconductive transition temperature T_C is usually plotted either as a function of the applied gate voltage (V), the sheet conductance (σ_{2D}), or the sheet carrier density (n_{2D}) [38, 125]. A phase diagram can be constructed by plotting T_C *vs.* V, σ_{2D} , or n_{2D}. A salient feature of these phase diagrams is the appearance of a superconductive dome with a maximum $T_C \approx 300$ mK (Fig. 8.8). The dome-like curve, which separates regions with normal conductance from superconductive regions, consists of "under-doped" and "over-doped" regimes, as schematically depicted in Fig. 8.8. Although a superconductive T_C-n_{3D} dome (n_{3D} is the carrier density in the crystals) is well known to occur in the bulk, the superconductive state of the interface has characteristics that make it different from that of crystals. For one thing, the interfacial superconductivity develops only around a very narrow range of carrier densities—the sheet carrier density of the interfaces can be translated into an effective "bulk" density $n_{3D} \approx 10^{19}$ cm^{-3}, see, e.g., Ref. [125]. In contrast, the superconductive dome in bulk doped $SrTiO_3$ spans a much broader range of carrier densities ($n_{3D} \approx 10^{17}$– 5×10^{20} cm^{-3}) [125]. Another relevant difference is in the electronic structure; the first bands to be filled in bulk are heavy-electron d_{xz}/d_{yz} bands, whereas light-mass d_{xy} states are at the bottom in the $LaAlO_3/SrTiO_3$ interfaces (Section 8.2.2). As discussed next, the electrostatic control of the relative populations of the d_{xy}, d_{xz}, and d_{yz} bands play a pivotal role in the interfacial superconductivity.

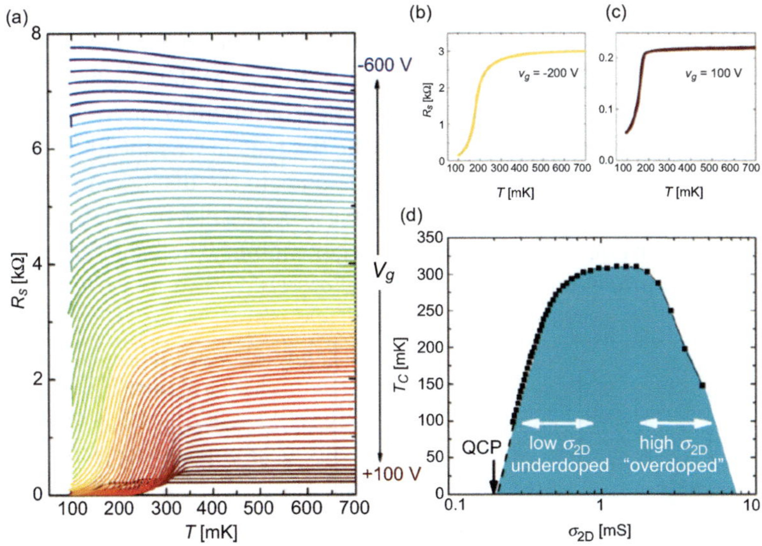

Figure 8.8 (a) Sheet resistance (R_S) *vs* temperature curves are shown for $LaAlO_3/SrTiO_3$ interfaces, in which the transport is modulated by electrostatic gating (extracted from Ref. [125]). By changing the gate voltage (V_g), the critical temperature for superconductivity is altered. Panels (b) and (c) show superconducting transitions for an under-doped, and an over-doped state, respectively. In panel (d), the superconducting dome is plotted *vs* the sheet conductance σ_{2D}. Reprinted from Ref. [125], with permission from Springer Nature, Copyright 2016.

Indeed, from transport experiments, there is evidence that electrostatic doping of the $LaAlO_3/SrTiO_3$ interface allows changing the orbital filling of t_{2g} states, with d_{xy} states being filled earlier than d_{xz}/d_{yz} states. For instance, the analysis of the low-temperature magnetoconductance suggests that as soon as d_{xz}/d_{yz} orbitals are occupied the superconductive state emerges [126], thus establishing a link between the orbital symmetry and superconductivity. In a related system, the $LaTiO_3/SrTiO_3$ interface, Biscaras et al. have also found multiple carriers with the superconductivity emerging as soon as high-mobility carriers are injected into the system [52]. This picture can be understood by considering that electrons with light mass ($m^* \approx 0.6\ m_e$) and heavy mass ($m^* \approx 2.2\ m_e$) participate in the transport [81, 126]. According to this scenario, light electrons should reside in d_{xy} states, which are extremely confined, extending over 1–2 uc from the interface [84, 127]. In contrast, heavy electrons are in d_{xz}/d_{yz} states, which are significantly more spread into the bulk [52] (Fig. 8.9). Although lighter in mass, d_{xy} electrons experience a stronger confinement, and feel large scattering at the interface due to increased disorder, suppressing superconductivity, and reducing the electron mobility [52]. This may be the reason why the depletion of carriers in the under-doped region leads to the disappearance of superconductivity (Fig. 8.8d). Conversely, d_{xz}/d_{yz} electrons, although with higher effective mass, have a larger mobility plausibly because they are more extended into the bulk.

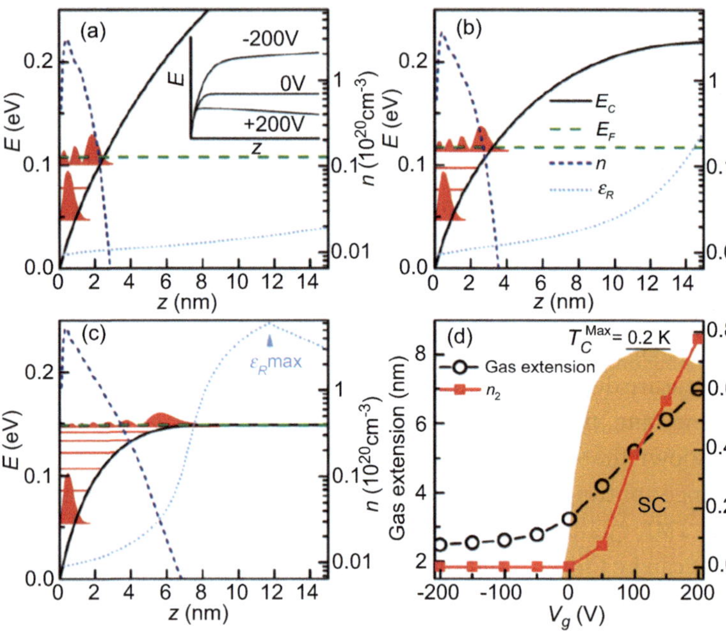

Figure 8.9 Panels (a–c) show the results of calculations of the band bending in the $SrTiO_3$ side of the interface for three different gate voltages (–200 V, 0 V, +200 V) obtained with the experimental sheet carrier densities of panel (c). In the figures, all magnitudes are plotted as a function of the depth *z* from the interface. The profiles of the conduction band E_C are shown in black, the Fermi energy E_F in dashed green lines, the subband energies in red and the carrier density *n* in deep blue. The dielectric constant ε_R is also plotted as a function of *z*. The red areas indicate the square modulus of the envelope function of the first and last filled subbands. The extension of the 2DES from the interface is plotted in panel (d) as a function of gate voltage. Reprinted from Ref. [52], with permission from American Physical Society, Copyright 2012.

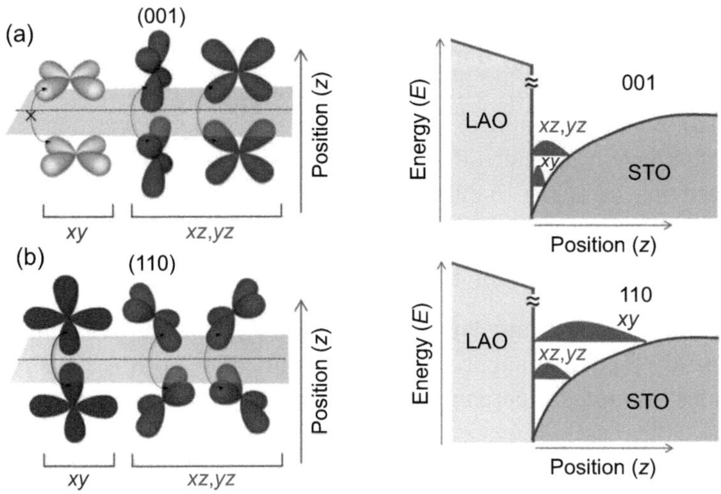

Figure 8.10 (Left panels) Schematic depiction of the overlapping of t_{2g} orbitals for (a) (001)-oriented and (b) (110)-oriented $LaAlO_3/SrTiO_3$ interfaces. Along (001), the overlapping of d_{xy} orbitals is very small, while it is moderately large for d_{xz}/d_{yz} states. However, along (110), the overlapping of d_{xy} orbitals is the largest. Since the effective mass along the confinement direction is inversely proportional to the orbital overlapping, it is expected that the largest effective mass along the out-of-plane direction z is for d_{xy} in (001)-interfaces; by the same token, the lightest masses along z are expected to be d_{xy} orbitals in the (110)-interface. d_{xz}/d_{yz} orbitals have intermediated masses either for (001)- or (110)-oriented interfaces. Based on these considerations, and applying the rules of quantum physics in solids, the energy landscape in the quantum wells is as follows (right panels in the figure): Orbitals with symmetry d_{xy} lie at the bottom of (001) quantum wells, while d_{xz}/d_{yz} are higher in energy; along (110), the d_{xz}/d_{yz} states have lower energy and the d_{xy} levels are at the top. The different energy hierarchy of orbitals determines a larger spatial extension for 2DESs along (110) as compared with (001). This figure is adapted from Ref. [128].

The above-mentioned experiments indicate a key role of d_{xz}/d_{yz} states in the emergence of 2D superconductivity at the $LaAlO_3/SrTiO_3$ interface. Indeed, a connection may be made between superconductivity and the orbital character of electrons. In the plane of the quantum well (x–y plane), where z is along the confinement, the band derived from d_{xy} states has a light mass. The reason is that d_{xy} orbitals have large overlap along the x and y directions (Fig. 8.10). On the contrary, for d_{yz} and d_{yz} bands, the overlap with neighboring orbitals is large or small, depending on the in-plane orientations. For instance, the d_{yz} band will be heavy along x and light along y, while the d_{xz} band will be heavy along y, and light along x (Fig. 8.10). As a consequence, as mentioned above, light electrons in d_{xy} states have a smaller effective mass $m^* \approx 0.6\ m_e$, whereas heavy electrons in d_{xz}/d_{yz} bands have larger masses $m^* \approx 2.2\ m_e$ [81, 126]. Recalling that in 2D systems, the density of states is given by $g(E)_{2D} = \frac{m^*}{\pi\hbar^2}$, we infer that the density of states of d_{xz}/d_{yz} bands should be significantly larger than that of d_{xy} bands. Interestingly, in the BCS theory of superconductivity, high values of the density of states promotes superconductivity, since the transition temperature goes as $T_C \sim \hbar\omega_c e^{-1/N(0)V}$, where ω_c is a cutoff frequency for phonon-mediated superconductivity, V is the electron-phonon coupling strength, and $N(0)$ is the density of states [129]. In view of this, as signaled by Gariglio et al. [125], the significantly higher density of states of d_{xz}/d_{yz} bands might be the reason why the transition to the superconductive state at $LaAlO_3/SrTiO_3$ and related interfaces seem to be linked to the occupancy of d_{xz}/d_{yz} states [52, 126].

8.3.2 Spin–Orbit Coupling and Rashba Spin Fields

In 2011, Joshua et al. suggested that the above-mentioned change of orbital symmetry from d_{xy} to d_{xz}/d_{yz} could be understood as a Lifshitz transition [86]. Lifshitz transitions were originally related to changes of the topology of the Fermi surfaces of metals as a function of the chemical potential [130]. In the case of $LaAlO_3/SrTiO_3$, Joshua et al. proposed that the changes in the orbital character of the occupied states and the subsequent changes in the Fermi surface were the manifestation of a Lifshitz transition occurring at a universal carrier density [86]. They also suggested that spin–orbit coupling significantly modifies the electronic band structure, especially at the electron band fillings where the Lifshitz transition occurs. More specifically, they argued that due to the spin–orbit induced admixture with d_{xy} orbitals, the bottom of the d_{xz} and d_{yz} bands is modified. Interestingly, the calculated spin–orbit strength was particularly large around the Lifshitz transition point [130]. This last observation is in agreement with the results of low-temperature magnetotransport reported by Caviglia et al., in which the spin–orbit coupling was observed to rise steeply precisely at the doping level at which the 2D superconductivity emerges at the interface, i.e., at the point of the change of the orbital character from d_{xy} to d_{xz}/d_{yz} [38].

Subsequently, it was also suggested that, in addition to spin–orbit coupling, electronic correlations were also important to understand the details of the evolution of the electronic properties with electrostatic doping [131]. In this work, the calculations included a Hubbard-type repulsion between electrons occupying different orbitals in the same unit cell. Owing to these interactions, the electron band structure is modified into the so-called "band 1" and "band 2" which are progressively occupied as the chemical potential that is increased (Fig. 8.11) [131]. At low energies, the first occupied states are in "band 1", which essentially has a d_{xy} character. However, as the orbital occupancy increases, "band 1" acquires a d_{xz}/d_{yz} character [131]. As the "band 1" is progressively occupied, the energy of "band 2" increases due to electronic correlations (Fig. 8.11a). At a given point, the chemical potential reaches the bottom of the shifted "band 2", where band density of states is large, and it begins to fill up (Fig. 8.11b). Finally, at higher orbital occupancy, the "band 1" becomes heavier than the "band 2", and it becomes preferable to increase the population of the former at the expense of the latter (Fig. 8.11c). Therefore, in this model, the population of "band 2", which is suggested to be responsible for superconductivity, is non-monotonic.

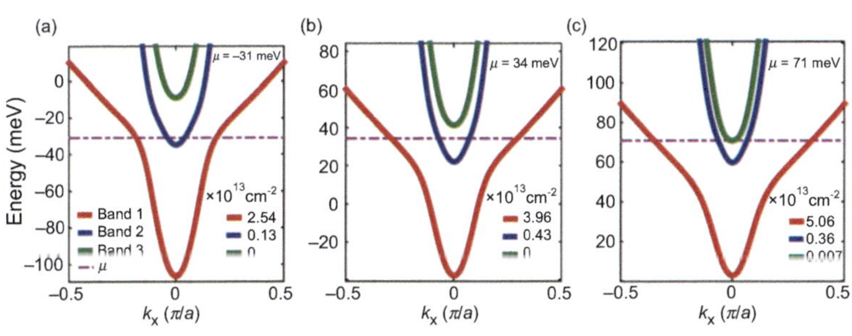

Figure 8.11 Panels (a–c) depict the electronic band structure of $LaAlO_3$/$SrTiO_3$ interfaces, in which the effects of electronic correlations and spin–orbit coupling are included for three different chemical potentials. Reprinted from Ref. [131], with permission from Springer Nature, Copyright 2015.

Summing up, the overall picture is that the orbital occupancy of 2DES states at the $LaAlO_3/SrTiO_3$ interface first proceeds by filling up levels with mostly d_{xy} character. As the doping level increases, the Fermi level reaches states with mostly d_{xz}/d_{yz} character. The critical point in which the orbital symmetry changes from d_{xy} to d_{xz}/d_{yz} is suggested to be described as a Lifshitz transition, in which two distinctive physical phenomena arise. First, 2D superconductivity emerges, and secondly, the spin–orbit coupling is peaked around the critical orbital occupancy. In view of this, it would be interesting to explore the physics of $LaAlO_3/SrTiO_3$ interfaces in which the orbital hierarchy is reversed, i.e., the order in which orbitals are filled up is inverted. As discussed in the next section, this change in the orbital hierarchy has been demonstrated in $LaAlO_3/SrTiO_3$ interfaces oriented along the [110] direction.

8.4 Orbital Reconstruction Induced by the Crystal Symmetry: the (110)-Oriented $LaAlO_3/SrTiO_3$ Interface

The description enclosed in the last section makes clear that the symmetry of the t_{2g} orbitals is essential to understand the physical properties of the 2DESs. A natural question arises: can we modulate the known physical properties of $LaAlO_3/SrTiO_3$ QWs by engineering their electron band structure? The answer to this question came when quantum wells were discovered at interfaces oriented along directions other than the conventional [001], more specifically, along (110) [132, 133], and (111) interfaces [132]. The transport characterization of these novel interfaces revealed values for the sheet resistance, carrier mobility, and sheet carrier density similar to those of the $LaAlO_3/SrTiO_3$ (001) interface [132]. Irrespective of their crystal orientation, a critical thickness for the metal–insulator transition was also found [132].

8.4.1 Anisotropy of the 2D Superconductivity

Soon, it was also shown that the electron band structure of (110)-oriented $LaAlO_3/SrTiO_3$ interfaces, and in particular, the orbital hierarchy of t_{2g} states was different from the one at the

(001) interface. For that purpose, Pesquera et al. used the element-sensitivity and symmetry-sensitivity of X-ray linear dichroism (XLD) spectroscopy to unveil the electron band structure [134]. In particular, they used photons with energy in the range of the Ti–$L_{2,3}$ edges, for which the fine structure of the XLD spectra corresponds to transitions to the t_{2g} (d_{xz}, d_{yx}, d_{xy}) and e_g (d_{z2}, d_{x2-y2}) levels of the Ti. In these experiments, the intensity resulting from the X-ray induced electronic transitions to the t_{2g} and e_g orbitals depend on their orbital symmetry, being the interaction strongest when light polarization is along the direction of the orbital lobes and minimal normal to them [135]. By exploiting this differential cross-section, Pesquera et al. found that the order in which the electronic states appear in the $LaAlO_3/SrTiO_3$ (110) interface as a function of energy is inverted with respect to the $LaAlO_3/SrTiO_3$ (001) system. More precisely, while d_{xy} states are the lowest in energy for the (001) oriented heterostructure (see Section 8.2.3), d_{xz}/d_{yx} states are instead at the bottom for the (110) interface [134] (see Fig. 8.10). As a consequence, crystal symmetry enabled the realization of different 2DES band reconstructions, thus allowing a selective occupancy of states of different symmetry.

This potentiality was subsequently used to probe the effects of orbital symmetry on the physical properties of the 2DESs. A first consequence is that in (110)-oriented samples the 2DESs spread significantly farther away from the interface [128]. The reason is that for the (001)-oriented interface the first d_{xy} subband is expected to be at the bottom of the well, with little spatial spread; on the contrary, along (110) the d_{xy} level raises its energy above the d_{xz}/d_{yz} states, and its spatial extent is considerably larger (Fig. 8.10). A consequence of the different spatial extension of the 2DESs is that the anisotropy of the 2D superconductivity is largely affected by crystal orientation. For both (001) and (110) interfaces, the coherence length (ξ), and the superconductive thickness (d) can be inferred from a Landau–Ginzburg analysis of the critical in-plane ($\mu_0 H_{c||}$) and out-of-plane ($\mu_0 H_{c\perp}$) magnetic fields as $\xi = \sqrt{\frac{\Phi_0}{2\pi\mu_{0H_{c\perp}}}}$ and $d = \frac{\Phi_0\sqrt{3}}{\pi\mu_{0H_{c||}}\xi}$, respectively, where Φ_0 is the quantum magnetic flux [136]. The results show that for the (001) interface the superconductive thickness is $d \approx 10$ nm [9, 122, 128], while for the (110) interface it is substantially larger, i.e., $d \approx 24 - 30$ nm [128]. In both crystal orientations, the superconductive thickness is considerably smaller than the coherence length, confirming the 2D character of the superconductivity [128].

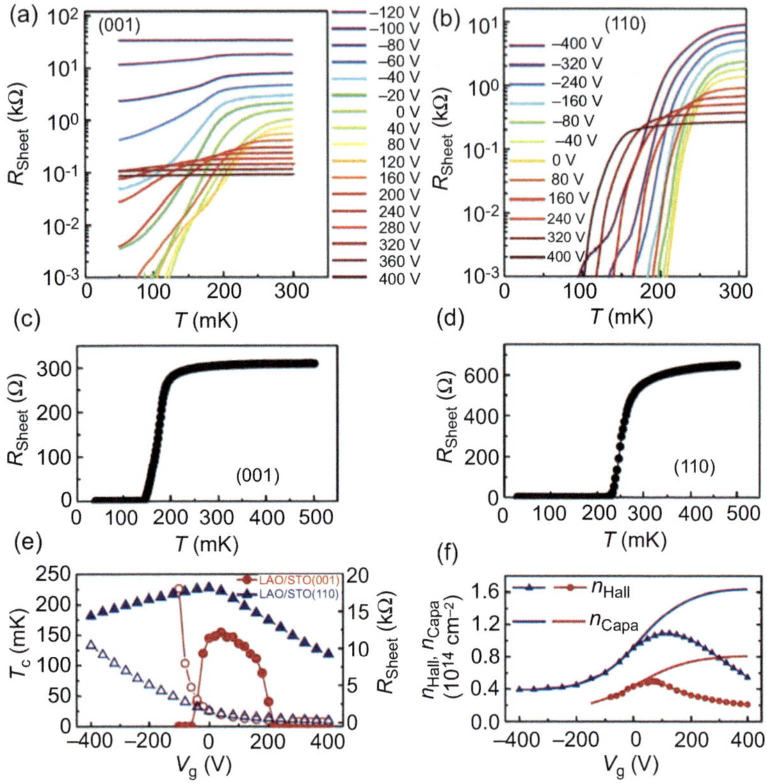

Figure 8.12 This figure shows the effects of electrostatic modulation of the 2D superconductivity at the (001) and (110) $LaAlO_3/SrTiO_3$ interfaces. Panels (a) and (b) show, respectively, the temperature dependence of the resistance as a function of gate voltage for (001)- and (110)-oriented interfaces. Panels (c) and (d) show the superconducting transitions of the (001) and (110) interfaces. Panel (e) displays the transition temperature T_C and the sheet resistance of (001) and (110) interfaces as a function of the gate voltage. In (f), the Hall carrier densities are shown. This figure is adapted from Ref. [128].

Electrostatic doping experiments give still more clues about the different evolution of the 2D superconductivity with the doping level. More specifically, in striking contrast to the case of the (001) oriented interface, the superconductivity in (110) samples is far more robust and the superconductive state cannot be suppressed for the same range of applied gate electric fields (Fig. 8.12). Two scenarios may explain this observation. On the one hand, the 2DES is considerably more extended for the (110) interface, so that the electric field profile is distributed over long distances, making it less efficient in modulating the chemical potential. On the other hand, an interesting perspective is the fact that the orbitals that are filled up in $LaAlO_3/SrTiO_3$ interfaces first are d_{xz}/d_{yz} states, which as mentioned in Section 8.3.1, have a significantly larger density of states, and therefore, are thought to promote superconductivity at the $LaAlO_3/SrTiO_3$ interface [121, 125].

8.4.2 Spin–Orbit Rashba Fields

In Section 8.3.2, the role of spin–orbit coupling was emphasized in relation to the electron band structure of $LaAlO_3/SrTiO_3$ interfaces. Indeed, the relevance of spin–orbit coupling is a consequence of the large confinement of the 2DESs, which breaks the inversion symmetry, and therefore, induces a Rashba spin–orbit field [38, 137]. The field B_{SO} arises because of the asymmetric electrostatic potential, caused by the asymmetric atomic environment at the interface, which induces an electric field $E_0 = -\nabla V(r)$. As a result, electrons moving relativistically under the influence of E_0 feel an effective magnetic field B_{SO}, i.e., the Rashba spin–orbit field [138]. At the atomic scale, Rashba fields can be understood as stemming from the polarization of atomic orbitals right at the interface, which break the spatial symmetry of t_{2g} states, allowing the hybridization within the t_{2g} manifold in the metal–oxygen network, which otherwise would be forbidden if $E_0 = 0$ [139, 140]. In a quantitative matter, the Rashba energy can be determined from $\Delta_{SO} \propto |k_{\|}| \frac{2}{c^2} \int dr \frac{\partial V}{\partial z} |\Psi_{k_{\|}}|^2$, where $k_{\|}$ is the in-plane component of the wave vector, $-\frac{\partial V}{\partial z}$ is the electric field felt by electrons and $\Psi_{k_{\|}} = \sum_l a_l \phi^l_{k_{\|}}$ is the wave function in which various angular moment (l) of the atomic orbitals $\phi^l_{k_{\|}}$ are mixed up by the spin–orbit coupling [141]. In particular, given that E_0 is along the out-of-plane z-direction, non-vanishing contributions come from cross terms between d_{xz} and p_x, and between d_{yz} and p_y, as well as d_{xy} via the admixture of d_{xz} with p_y and d_{yz} with p_x [141]. In this situation, and knowing that the spatial distribution of d_{xy}, d_{xz}, and d_{yx} orbitals changes with the crystal orientation (Fig. 8.10), one expects that B_{SO} should evolve differently as a function of the electrostatic doping for (010) and (110) interfaces. In particular, orbitals with large projections over the confinement direction are expected to be more sensitive to the inversion symmetry breaking fields E_0, giving larger Rashba effects [128, 139, 140]. Since the orbital hierarchy of (010) and (110) interfaces is different, B_{SO} is expected to change in a different way with the doping level.

To better understand this point, the atomic orbital polarization for (001) and (110) interfaces is graphically depicted in Fig. 8.13 in the form of spatially distorted orbitals. We see that for the (001)-oriented sample, d_{xy} orbitals are weakly polarized due to their minimal projection along the confinement direction, while d_{xz}/d_{yz} states have much stronger spatial asymmetry. As a consequence, the electrostatic modulation of the orbital occupancy is anticipated to give a significant variation of B_{SO} as a function of the orbital occupancy. More specifically, at low doping levels, only d_{xy} orbitals are populated; the orbital polarization is weak (Fig. 8.13) and B_{SO} is relatively small. When the chemical potential is raised, the population of d_{xz}/d_{yz} bands rises and B_{SO} is expected to increase significantly because of the significantly larger orbital polarization. These observations are in agreement with the experimental data, which show an asymmetric dependence of the spin–orbit field on gate fields: in the regime of carrier depletion, the values of the spin–orbit field are moderate ($B_{SO} < 0.5$ T), whereas for electron accumulation, the spin–orbit term rises up to $B_{SO} \approx 1.5$ T (Fig. 8.13a) [38, 126, 128]. The situation is radically different for the (110) interface. Now, the electrostatic modulation of B_{SO} is very weak and the spin–orbit field is largely unaffected by the electrostatic gating. This behavior can be well understood on the grounds of the similar atomic orbital polarizations of d_{xy} and d_{xz}/d_{yz} orbitals along the (110) direction, (Fig. 8.13b). Indeed, along (110), all t_{2g} orbitals are expected to undergo similarly strong polarizations (Fig. 8.5e), and therefore, the spin–orbit field B_{SO} is expected to have a rather weak dependence on the applied field, as confirmed by the experiments (Fig. 8.13a) [128].

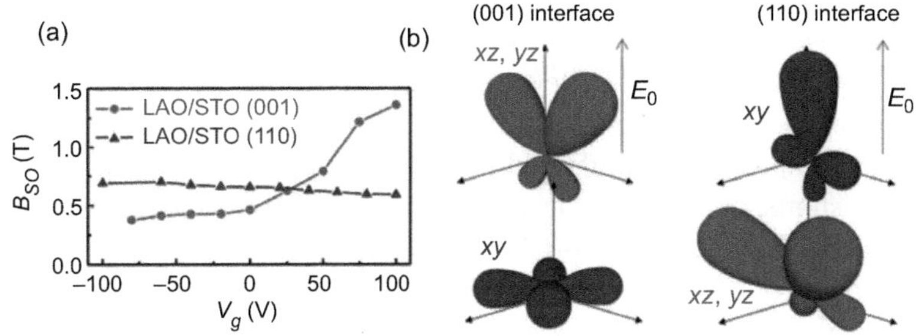

Figure 8.13 (a) Plot of the spin–orbit field B_{SO} as a function of the gate voltage V_g for the (001)- and (110)-oriented $LaAlO_3/SrTiO_3$ interface. (b) Sketches of the d_{xy} and d_{xz}/d_{yz} orbital polarizations at the interface induced by the inversion symmetry breaking field E_0. Along (001), d_{xy} states have a small projection along the out-of-plane direction and they are weakly polarized by E_0, whereas d_{xz}/d_{yz} states project along the direction of confinement, and are largely polarized. Instead, along (110), both kinds of orbitals are affected similarly by the interface fields. The differential orbital polarization across the interface explains the gate voltage dependence of B_{SO} shown in (a), i.e., a strong modulation of B_{SO} in (001) interfaces, but a very weak modulation for (110) interfaces. This figure is adapted from Ref. [128].

The comparative study of (001)- and (110)-oriented interfaces described in Sections 8.4.1 and 8.4.2 enriches considerably our perspective on the vital role that orbital symmetry plays in the physics of 2DESs at $LaAlO_3/SrTiO_3$ interfaces. The overall picture is that d_{xz}/d_{yz} orbitals have a significantly higher density of states than d_{xy} levels, which promotes superconductivity, and the symmetry of d_{xz}/d_{yz} states has a decisive role in increasing Rashba spin–orbit fields. Likewise, as discussed in Section 3.2, a Lifshitz transition is associated with the change of symmetry from d_{xy}-like to d_{xz}/d_{yz}-like states [86], which in turn, coincides with the position of the chemical potential where the spin–orbit term rises steeply [38, 142]. The combination of 2D superconductivity coexists with a sizable spin–orbit coupling may be the source of a rich physics in which the role of orbital symmetry, once again, may be prominent. Prospects about these physics are discussed in the next section.

8.5 Perspectives and Outlook on the Physics of 2DESs at the $LaAlO_3/SrTiO_3$ Interface

As previously noted in Section 8.3.1, the reduced dimensionality of the superconductivity gives rise to new physics, viz., the appearance of intrinsic fluctuations according to the XY-model described by the physics of Berezinskii–Kosterlitz–Thouless (BKT) transitions [118, 121]. Beyond this, it is fundamental to know how phase coherence is established through the superconductive transition, as this would shed light on the understanding of other 2D systems, particularly, high–transition temperature copper oxide superconductors [143, 144]. In the latter, in the under-doped region, the transition temperature drops, while the so-called *pseudogap*, i.e., a depression in the density of states around the Fermi level, increases. Understanding the origin of the pseudogap phase remains a challenge [145–147]. The pseudogap phenomenon has been attributed either to the emergence of competitive highly-correlated electronic phases [148], or the formation of Cooper pairs above the transition [149], where the system lacks the phase coherence to induce macroscopic superconductivity. Interestingly, Richter et al. have observed a pseudogap, which although analogous to that observed in high-T_C superconductors, it occurs for substantially much smaller carrier densities in the $LaAlO_3/SrTiO_3$ system [144] (Section 8.3.1). At the same time, the discovery of Richter et al. may imply that the pseudogap like behavior is a general property of 2D superconductivity [144]. Interestingly, transport experiments reveal electron pairing at temperatures as high as 900 mK (milli-Kelvin), well above the superconducting transition, demonstrating the existence electrons pairing without a macroscopic superconducting state [150]. In line with the BKT physics, the observation of the pseudogap behavior in $LaAlO_3/SrTiO_3$ may be related to the emergence of Cooper pairs without phase coherence above the transition [121]. Yet, alternative explanations have been proposed, based on the intrinsic electronic inhomogeneity of the system that may lead to a percolative behavior [151, 152]. An interesting perspective is offered by studies of nanostructured channels, which hint at a possible one-dimensional nature of superconductivity in $LaAlO_3/SrTiO_3$, with a possible role of ferroelastic domain boundaries [153]. It is relevant to note that the pseudogap phenomenon has been observed in the (001)-oriented interface; experiments in the (110) interface have not been done so far, which could reveal the effect of the change in the orbital hierarchy on the pseudogap physics.

Another interesting feature of the $LaAlO_3/SrTiO_3$ system and other similar interfaces is the simultaneous occurrence of 2D superconductivity, and large spin–orbit coupling [38, 54, 154]. Finally, signatures of spin-triplet pairing have been found in mesoscopic devices created within the LAO/STO interfaces [155]. Indeed, for the (001)-oriented interface, the Rashba energy rises sharply up to $\Delta_{SO} \approx 10$ meV as soon as d_{xz} and d_{yz} states are populated and becomes comparable to the kinetic energy of the d_{xz} and d_{yz} carriers [38, 121]. Certainly, there is experimental evidence supporting conventional pairing based on electron–phonon interactions [156, 157]. However, the large Rashba term has prompted suggestions about the promotion of unconventional superconductive states [141, 158–160]. In particular, it has been proposed that electron pairing may occur at a finite momentum, the so-called Fulde–Ferrell–Larkin–Ovchinikov (FFLO) state [141, 161, 162]. The peculiarity of the $LaAlO_3/SrTiO_3$ interface is that, contrary to other systems [161], the spin–orbit term is comparable to the Fermi energy. Bearing this in mind, Michaeli et al. extended the FFLO analysis to arbitrarily strong values of Δ_{SO} and found that spin–orbit coupling preserves electron pairing in the d_{xz} and d_{yz} bands [141]. According to this scenario, Josephson tunneling experiments should reveal Cooper pairs at moment $q \neq 0$. Interestingly, for the surface of $SrTiO_3$ crystals, quantum transport studies reveal signatures of a magnetically ordered ground state [163], in agreement with predictions [141, 164–166]. Yet, to date, this important observation remains to be tested experimentally in the $LaAlO_3/SrTiO_3$ interface. All in all, in this physical phenomenon or in other phenomena described throughout this chapter, the orbital symmetry plays a pivotal role and it is expected that over the next years other new insights will be gained revealing further aspects of symmetry in the physics of these 2DESs.

Acknowledgments

Gervasi Herranz acknowledges the support through Projects No. MAT2014-56063-C2-1-R and the Severo Ochoa SEV-2015-0496 grant and the Generalitat de Catalunya (2014SGR 734 project). Gervasi Herranz also acknowledges the scientific collaboration and priceless discussions with Jérôme Lesueur and Nicolas Bergeal from ESPCI-Paris Tech regarding the Physics of 2DESs in oxide heterointerfaces.

References

1. A. Ohtomo and H. Y. Hwang, A high-mobility electron gas at the $LaAlO_3$/$SrTiO_3$ heterointerface, *Nature,* **427**, pp. 423–426 (2004).
2. K. Ueda, H. Tabata, and T. Kawai, Ferromagnetism in $LaFeO_3$–$LaCrO_3$ Superlattices, *Science,* **280**, pp. 1064–1066 (1998).
3. A. Gozar, G. Logvenov, L. Fitting Kourkoutis, A. T. Bollinger, L. A. Giannuzzi, D. A. Muller, and I. Bozovic, High-temperature interface superconductivity between metallic and insulating copper oxides, *Nature,* **455**, pp. 782–785 (2008).
4. M. Ben Shalom, A. Ron, A. Palevski , and Y. Dagan, Shubnikov–De Haas oscillations in $SrTiO_3$/$LaAlO_3$ interface, *Phys. Rev. Lett.,* **105**, p. 206401 (2010).
5. A. D. Caviglia, S. Gariglio, C. Cancellieri, B. Sacépé, A. Fête, N. Reyren, M. Gabay, A. F. Morpurgo, and J.-M. Triscone, Two-dimensional quantum oscillations of the conductance at $LaAlO_3$/$SrTiO_3$ interfaces, *Phys. Rev. Lett.,* **105**, p. 236802 (2010).
6. M. Basletic, J. L. Maurice, C. Carrétéro, G. Herranz, O. Copie, M. Bibes, É. Jacquet, K. Bouzehouane, S. Fusil, A. Barthélémy, Mapping the spatial distribution of charge carriers in $LaAlO_3$/$SrTiO_3$ heterostructures, *Nat. Mater.,* **7**, pp. 621–625 (2008).
7. O. Copie, V. Garcia, C. Bödefeld, C. Carrétéro, M. Bibes, G. Herranz, E. Jacquet, J.-L. Maurice, B. Vinter, S. Fusil, K. Bouzehouane, H. Jaffrès, and A. Barthélémy, Towards two-dimensional metallic behavior at $LaAlO_3$/$SrTiO_3$ interfaces, *Phys. Rev. Lett.,* **102**, p. 216804 (2009).
8. A. Dubroka, M. Rössle, K. W. Kim, V. K. Malik, L. Schultz, S. Thiel, C. W. Schneider, J. Mannhart, G. Herranz, O. Copie, M. Bibes, A. Barthélémy, and C. Bernhard, Dynamical response and confinement of the electrons at the $LaAlO_3$/$SrTiO_3$ interface, *Phys. Rev. Lett.,* **104**, p. 156807 (2010).
9. N. Reyren, S. Gariglio, A.D. Caviglia, D. Jaccard, T. Schneider , and J. M. Triscone, Anisotropy of the superconducting transport properties of the $LaAlO_3$/$SrTiO_3$ interface, *Appl. Phys. Lett.,* **94,** p. 112506 (2009).
10. C. Cancellieri, M. L. Reinle-Schmitt, M. Kobayashi, V. N. Strocov, T. Schmitt, P. R. Willmott, S. Gariglio and J.-M. Triscone, Interface fermi states of $LaAlO_3$/$SrTiO_3$ and related heterostructures, *Phys. Rev. Lett.,* **110**, p. 137601 (2013).
11. G. Berner, M. Sing, H. Fujiwara, A. Yasui, Y. Saitoh, A. Yamasaki, Y. Nishitani, A. Sekiyama, N. Pavlenko, T. Kopp, C. Richter J. Mannhart, S. Suga, and R. Claessen, Direct *k*-space mapping of the electronic structure in an Oxide-Oxide interface, *Phys. Rev. Lett.,* **110,** p. 247601 (2013).

12. S. Gariglio, A. Fête , and J. M. Triscone, Electron confinement at the $LaAlO_3/SrTiO_3$ interface, *J. Phys.: Condens. Matter,* **27**, p. 283201 (2015).

13. J. Biscaras, N. Bergeal, A. Kushwaha, T. Wolf, A. Rastogi, R.C. Budhani, and J. Lesueur, Two-dimensional superconductivity at a Mott insulator/band insulator interface $LaTiO_3/SrTiO_3$, *Nat. Commun.*, **1**, p. 89 (2010).

14. Y. Hotta, T. Susaki, and H. Y. Hwang, Polar discontinuity doping of the $LaVO_3/SrTiO_3$ interface, *Phys. Rev. Lett.,* **99**, p. 236805 (2007).

15. X. Liu, D. Choudhury, Y. Cao, M. Kareev, S. Middey, and J. Chakhalian, Metallic interface in non-$SrTiO_3$ based titanate superlattice, *Appl. Phys. Lett.,* **107**, p. 191602 (2015).

16. P. Perna, D. Maccariello, M. Radovic, U. Scotti di Uccio, I. Pallecchi, M. Codda, D. Marré, C. Cantoni, J. Gazquez, M. Varela, S. J. Pennycook, and F. Miletto Granozio, Conducting interfaces between band insulating oxides: The $LaGaO_3/SrTiO_3$ heterostructure, *Appl. Phys. Lett.,* **97**, p. 152111 (2010).

17. S. Raghavan, J. Y. Zhang , and S. Stemmer, Two-dimensional electron liquid at the (111) $SmTiO_3/SrTiO_3$ interface, *Appl. Phys. Lett.,* **106**, p. 132104 (2015).

18. P. Moetakef, J. R. Williams, D. G. Ouellette, A. P. Kajdos, D. Goldhaber-Gordon, S. J. Allen , and S. Stemmer, Carrier-controlled ferromagnetism in $SrTiO_3$, *Phys. Rev. X,* **2**, p. 021014 (2012).

19. K. Zou, S. Ismail-Beigi, K. Kisslinger, X. Shen, D. Su, F. J. Walker, and C. H. Ahn, $LaTiO_3/KTaO_3$ interfaces: A new two-dimensional electron gas system, *APL Mater.,* **3**, p. 036104 (2015).

20. D. F. Li, Y. Wang , and J. Y. Dai, Tunable electronic transport properties of $DyScO_3/SrTiO_3$ polar heterointerface, *Appl. Phys. Lett.,* **98**, p. 122108 (2011).

21. P. Xu, T. C. Droubay, J. S. Jeong, K. A. Mkhoyan, P. V. Sushko, S. A. Chambers, and B. Jalan, Quasi 2D ultrahigh carrier density in a complex oxide broken-gap heterojunction, *Adv. Mater. Interfaces,* **3**, p. 1500432 (2016).

22. H. W. Jang, D. A. Felker, C. W. Bark, Y. Wang, M. K. Niranjan, C. T. Nelson, Y. Zhang, D. Su, C. M. Folkman, S. H. Baek, S. Lee, K. Janicka, Y. Zhu, X. Q. Pan, D. D. Fong, E. Y. Tsymbal, M. S. Rzchowski, C. B. Eom, Metallic and insulating oxide interfaces controlled by electronic correlations, *Science,* **331**, pp. 886–889 (2011).

23. Y. Z. Chen, N. Bovet, F. Trier, D. V. Christensen, F. M. Qu, N. H. Andersen, T. Kasama, W. Zhang, R. Giraud, J. Dufouleur, T.S. Jespersen, J.R. Sun, A. Smith, J. Nygård, L. Lu, B. Büchner, B.G. Shen, S. Linderoth, and N. Pryds,

A high-mobility two-dimensional electron gas at the spinel/perovskite interface of γ–Al_2O_3/$SrTiO_3$, *Nat. Commun.*, **4**, p. 1371 (2013).

24. T. Q. Ngo, N. J. Goble, A. Posadas, K. J. Kormondy, S. Lu, M. D. McDaniel, Jean Jordan-Sweet, David J. Smith, Xuan P. A. Gao, Alexander A. Demkov, and John G. Ekerdt, Quasi-two-dimensional electron gas at the interface of γ–Al_2O_3/$SrTiO_3$ heterostructures grown by atomic layer deposition, *J. App. Phys.*, **118**, p. 115303 (2015).
25. B. Kalisky, J. A. Bert, B. B. Klopfer, C. Bell, H. K. Sato, M. Hosoda, Y. Hikita, H. Y. Hwang, and K. A. Moler, Critical thickness for ferromagnetism in $LaAlO_3$/$SrTiO_3$ heterostructures, *Nat. Commun.*, **3**, p. 922 (2012).
26. N. Reyren, S. Thiel, A. D. Caviglia, L. Fitting Kourkoutis, G. Hammerl, C. Richter, C. W. Schneider, T. Kopp, A. S. Rüetschi, D. Jaccard, M. Gabay, D. A. Muller, J.-M. Triscone, J. Mannhart, Superconducting interfaces between insulating oxides, *Science*, **317**, pp. 1196–1199 (2007).
27. L. Li, C. Richter, J. Mannhart, and R. C. Ashoori, Coexistence of magnetic order and two-dimensional superconductivity at $LaAlO_3$/$SrTiO_3$ interfaces, *Nat. Phys.*, **7**, p. 762 (2011).
28. A. Ariando, X. Wang, G. Baskaran, Z. Q. Liu, J. Huijben, J. B. Yi, A. Annadi, A. Roy Barman, A. Rusydi, S. Dhar, Y. P. Feng, J. Ding, H. Hilgenkamp, and T. Venkatesan, Electronic phase separation at the $LaAlO_3$/$SrTiO_3$ interface, *Nat. Commun.*, **2**, p. 188 (2011).
29. J. A. Bert, B. Kalisky, C. Bell, M. Kim, Y. Hikita, H. Y. Hwang, and K. A. Moler, Direct imaging of the coexistence of ferromagnetism and superconductivity at the $LaAlO_3$/$SrTiO_3$ interface, *Nat. Phys.*, **7**, p. 767 (2011).
30. A. Y. Cho and J. R. Arthur, Molecular beam epitaxy, *Prog. Solid State Chem.*, **10**, pp. 157–191 (1975).
31. W. P. Maccray, MBE deserves a place in the history books, *Nat. Nanotechnol.*, **2**, pp. 259–261 (2007).
32. A. C Jones, M. L Hitchman, (eds.), *Chemical Vapor Deposition: Precursors, Processes and Applications*, RSC Publishing (2008).
33. K. von Kliyzing, G. Dorda, and M. Pepper, New method for high-accuracy determination of the fine-structure constant based on quantum Hall resistance, *Phys. Rev. Lett.*, **45**, pp. 494–407 (1980).
34. D. C. Tsui, H. L. Störmer, and A. C. Gossard, Two-dimensional magnetotransport in the extreme quantum limit, *Phys. Rev. Lett.*, **48**, pp. 1559–1562 (1982).
35. L. Esaki and R. Tsu, Superlattice and negative differential conductivity in semiconductors, *IBM J. Res. Dev.*, **14**, pp. 61–65 (1970).

36. M. Ohring, *The Materials Science of Thin Films: Deposition and Structure*, 2nd Ed., 2002, Academic Press, UK.

37. D. Stornaiuolo, C. Cantoni, G. M. De Luca, R. Di Capua, E. Di. Gennaro, G. Ghiringhelli, B. Jouault, D. Marrè, D. Massarotti, F. Miletto Granozio, I. Pallecchi, C. Piamonteze, S. Rusponi, F. Tafuri, and M. Salluzzo, Tunable spin polarization and superconductivity in engineered oxide interfaces, *Nat. Mater.*, **15**, pp. 278–284 (2016).

38. A. D. Caviglia, M. Gabay, S. Gariglio, N. Reyren, C. Cancellieri, and J. M. Triscone, Tunable Rashba spin–orbit interaction at oxide interfaces, *Phys. Rev. Lett.*, **104**, p. 126803 (2010).

39. M. Ben Shalom, M. Sachs, D. Rakhmilevitch, A. Palevski, and Y. Dagan, Tuning spin–orbit coupling and superconductivity at the $SrTiO_3$/$LaAlO_3$ interface: a magnetotransport study, *Phys. Rev. Lett.*, **104**, p. 126802 (2010).

40. N. Reyren, M. Bibes, E. Lesne, J. M. George, C. Deranlot, S. Collin, A. Barthélémy, and H. Jaffrès, Gate-controlled spin injection at $LaAlO_3$/$SrTiO_3$ interfaces, *Phys. Rev. Lett.*, **108**, p. 186802 (2012).

41. E. Lesne, Yu Fu, S. Oyarzun, J. C. Rojas-Sánchez, D. C. Vaz, H. Naganuma, G. Sicoli, J.-P. Attané, M. Jamet, E. Jacquet, J.-M. George, A. Barthélémy, H. Jaffrès, A. Fert, M. Bibes, and L. Vila, Highly efficient and tunable spin-to-charge conversion through Rashba coupling at oxide interfaces, *Nat. Mater.*, **15**, 1261–1266 (2016).

42. R. Ohshima, Y. Ando, K. Matsuzaki, T. Susaki, M. Weiler, S. Klingler, H. Huebl, E. Shikoh, T. Shinjo, S. T. B. Goennenwein, and M. Shiraishi, Strong evidence for d-electron spin transport at room temperature at a $LaAlO_3$/$SrTiO_3$ interface, *Nat. Mater.*, **16**, 609–614 (2017).

43. J. Varignon, L. Vila, A. Barthélémy, and M. Bibes, A new spin for oxide interfaces, *Nat. Phys.*, **14**, 322–325 (2018).

44. L. P. Gorkov and E. I. Rashba, Superconducting 2D system with lifted spin degeneracy: mixed singlet-triplet state, *Phys. Rev. Lett.*, **87**, p. 037004 (2001).

45. N. Mohanta and A. Taraphder, Multiband theory of superconductivity at the $LaAlO_3$/$SrTiO_3$ interface, *Phys. Rev. B*, **92**, p. 174531 (2015).

46. V. Mourik, et al., Signatures of majorana fermions in hybrid superconductor-semiconductor nanowire devices, *Science*, **336**, p. 1003 (2012).

47. L. P. Rokhinson, X. Liu, and J. K. Furdyna, The fractional A.C. Josephson effect in a semiconductor–superconductor nanowire as a signature of majorana particles, *Nat. Phys.*, **8**, p. 795 (2012).

48. H. J. Gardner, et al., Enhancement of superconductivity by a parallel magnetic field in two-dimensional superconductors, *Nat. Phys.*, **7**, pp. 895–900 (2011).

49. M. Kim, Y. Kozuka, C. Bell, Y. Hikita, and H. Y. Hwang, Intrinsic spin–orbit coupling in superconducting d-doped SrTiO3 heterostructures, *Phys. Rev. B*, **86**, p. 085121 (2012).

50. M. P. Warusawithana, et al., $LaAlO_3$ stoichiometry is key to electron liquid formation at $LaAlO_3/SrTiO_3$ interfaces, *Nat. Commun.*, **4**, p. 2351 (2013).

51. A. D. Caviglia, et al., Electric field control of the $LaAlO_3/SrTiO_3$ interface ground state, *Nature*, **456**, p. 624 (2008).

52. J. Biscaras, et al., Two-dimensional superconductivity induced by high-mobility carrier doping in $LaTiO_3/SrTiO_3$ heterostructures, *Phys. Rev. Lett.*, **108**, p. 247004 (2012).

53. T. Schneider, A. D. Caviglia, S. Gariglio, N. Reyren, and J. M. Triscone, Electrostatically-tuned superconductor-metal-insulator quantum transition at the $LaAlO_3/SrTiO_3$ interface, *Phys. Rev. B*, **79**, p. 184502 (2009).

54. J. Biscaras, et al., Multiple quantum criticality in a two-dimensional superconductor, *Nat. Mater.*, **12**, pp. 542–548 (2013).

55. P. D. C. King, S. McKeown Walker, A. Tamai, A. de la Torre, T. Eknapakul, P. Buaphet, S.-K. Mo, W. Meevasana, M. S. Bahramy, and F. Baumberger, Quasiparticle dynamics and spin -orbital texture of the SrTiO3 two-dimensional electron gas, *Nat. Commun.*, **5**, 3414 (2014).

56. R. M. Fernandes, J. T. Haraldsen, P. Wolfle, and A. V. Balatsky, *Phys. Rev. B*, **87**, p. 014510 (2013).

57. A. Joshua, S. Pecker, J. Ruhman, E. Altman, and S. Ilani, *Nat. Commun.*, **3**, p. 1129 (2012).

58. N. Pavlenko, T. Kopp, E. Y. Tsymbal, G. A. Sawatzky, and J. Mannhart, *Phys. Rev. B*, **85**, p. 020407(R) (2012).

59. N. Pavlenko, T. Kopp, and J. Mannhart, *Phys. Rev. B*, **88**, p. 201104(R) (2013).

60. J. S. Lee, Y. W. Xie, H. K. Sato, C. Bell, Y. Hikita, H. Y. Hwang, and C. C. Kao, *Nat. Mater.*, **12**, p. 703 (2013).

61. Y. J. Chang, A. Bostwick, Y. Su Kim, K. Horn, and Eli Rotenberg, Structure and correlation effects in semiconducting $SrTiO_3$, *Phys. Rev. B*, **81**, p. 235109 (2010).

62. T. Wolfram and S. Ellialtoglu, *Electronic and Optical Properties of d-Band Perovskites*, Cambridge University Press (2006).

63. F. W. Lytle, X-Ray Diffractometry of low-temperature phase transformations in strontium titanate, *J. Appl. Phys.*, **35**, pp. 2212–2215 (1964).

64. R. A. Cowley, The phase transition of strontium titanate, *Philos. Trans. A Math. Phys. Eng. Sci.*, **354**, pp. 2799–2814 (1996).

65. Q. Hui, M. G. Tucker, M. T Dove, S. A. Wells and D. A. Keen, Total scattering and reverse Monte Carlo study of the 105 K displacive phase transition in strontium titanate, *J. Phys.: Condens. Matter*, **17**, pp. S111–S124 (2005).

66. K. A. Müller and H. Burkard, An intrinsic quantum paraelectric below 4 K, *Phys. Rev. B*, **19**, p. 3593 (1979).

67. E. Courtens, Birefringence of $SrTiO_3$ produced by the 105 K structural phase transition, *Phys. Rev. Lett.*, **29**, p. 1380 (1972).

68. E. K. H. Salje, O. Aktas, M. A. Carpenter, V. V. Laguta, and J. F. Scott, Domains within domains and walls within walls: evidence for polar domains in cryogenic $SrTiO_3$, *Phys. Rev. Lett.*, **111**, p. 247603 (2013).

69. A. H. Kahn and J. Layendeccker, Electronic energy bands in strontium titanate, *Phys. Rev.*, **135**, pp. A1321 (1964).

70. L. F. Mattheiss, Energy Bands for $KNiF_3$, $SrTiO_3$, $KMoO_3$, and $KTaO_3$, *Phys. Rev. B*, **6**, p. 4718 (1972).

71. T. Wolfram, E. A. Krant, and F. J. Morin, d-Band surface states on transition-metal perovskite crystals: I, Qualitative features and application to $SrTiO_3$, *Phys. Rev. B*, **7**, p. 1677 (1973).

72. J. D. Zook and T. N. Casselman, *Surf. Sci.*, **37**, pp. 244 (1973).

73. N. Sai and D. Vanderbilt, First-principles study of ferroelectric and antiferrodistortive instabilities in tetragonal $SrTiO_3$, *Phys. Rev. B*, **62**, pp. 13942 (2000).

74. R. Wahl, D. Vogtenhuber, and G. Kresse, $SrTiO_3$ and $BaTiO_3$ revisited using the project or augmented wave method: Performance of hybrid and semi-local functional, *Phys. Rev. B*, **78**, p. 104116 (2008).

75. A. Lebedev, ab initio calculations of phonon spectra in $ATiO_3$ perovskite crystals (A = Ca, Sr, Ba, Ra, Cd, Zn, Mg, Ge, Sn, Pb), *Phys. Solid State*, **51**, pp. 362–372 (2009).

76. E. Blokhin, D. Gryaznov, E. Kotomin, R. Eevarestov, and J. Maier, A comparative hybrid DFT Study of Phonons in Several $SrTiO_3$ Phases, *Integr. Ferroelectr.*, **123**, pp. 18–25 (2011).

77. K. Uchida, S. Tsuneyuki, and T. Schimizu, First-principle calculations of carrier-doping effects in $SrTiO_3$, *Phys. Rev. B*, **68**, pp. 174107 (2003).

78. J. P. Buban, H. Iddir, and S. Öğüt, Structural and electronic properties of oxygen vacancies in cubic and antiferrodistortive phases of $SrTiO_3$, *Phys. Rev. B*, **69**, p. 180102(R) (2004).

79. A. Spinelli, M. A. Torija, C. Liu, C. Jan, and C. Leighton, Electronic transport in doped $SrTiO_3$: Conduction mechanisms and potential applications, *Phys. Rev. B*, **81**, p. 155110 (2010).

80. B. Jalan, S. J. Allen, G. E. Beltz, P. Moetakef, and S. Stemmer, Enhancing the electron mobility of $SrTiO_3$ with strain, *Appl. Phys. Lett.*, **98**, p. 132102 (2011).

81. A. F. Santander-Syro, O. Copie, T. Kondo, F. Fortuna, S. Pailhès, R. Weht, X. G. Qiu, F. Bertran, A. Nicolaou, A. Taleb-Ibrahimi, P. Le Fèvre, G. Herranz, M. Bibes, N. Reyren, Y. Apertet, P. Lecoeur, A. Barthélémy, and M. J. Rozenberg, Two-dimensional electron gas with universal subbands at the surface of $SrTiO_3$, *Nature*, **469**, pp. 189–193 (2011).

82. A. P. Petrovic, A. Pare, T. R. Paudel, K. Lee, S. Holmes, C. H. W. Barnes, A. David, T.Wu, E. Y. Tsymbal and C. Panagopoulos, Long-range electronic reconstruction to a $d_{xz,yz}$-dominated Fermi surface below the $LaAlO_3$/$SrTiO_3$ interface, *Sci. Rep.*, **4**, p. 5338 (2014).

83. N. C. Plumb, M. Salluzzo, E. Razzoli, M. Månsson, M. Falub, J. Krempasky, C. E. Matt, J. Chang, M. Schulte, J. Braun, H. Ebert, J. Minár, B. Delley, K.-J. Zhou, T. Schmitt, M. Shi, J. Mesot, L. Patthey, and M. Radović, Mixed dimensionality of confined conducting electrons in the surface region of $SrTiO_3$, *Phys. Rev. Lett.*, **113**, p. 086801 (2014).

84. P. Delugas, A. Filippetti, V. Fiorentini, D. I. Bilc, D. Fontaine, and P. Ghosez, Spontaneous 2-dimensional carrier confinement at the n-type $SrTiO_3$/$LaAlO_3$ interface, *Phys. Rev. Lett.*, **106**, p. 166807 (2011).

85. V. Laukhin, et al., Electronic subband reconfiguration in a d^0-perovskite induced by strain-driven structural transformations, *Phys. Rev. Lett.*, **109**, p. 226601 (2012)

86. Arjun Joshua, S. Pecker, J. Ruhman, E. Altman, and S. Ilani, A universal critical density underlying the physics of electrons at the $LaAlO_3$/$SrTiO_3$ interface, *Nat. Commun.*, **3**, p. 1129 (2012).

87. W. Meevasana, P. D. C. King, R. H. He, S. K. Mo, M. Hashimoto, A. Tamai, P. Songsiriritthigul, F. Baumberger, and Z-X. Shen, Creation and control of a two-dimensional electron liquid at the bare $SrTiO_3$ surface, *Nat. Mater.*, **10**, pp. 114–118 (2011).

88. N. Nakagawa, H. Y. Hwang, and D. A. Müller, Why some interfaces cannot be sharp, *Nat. Mater.*, **5**, pp. 204–209 (2006).

89. S. Thiel, G. Hammerl, A. Schmehl, C. W. Schneider, and J. Mannhart, Tunable Quasi-Two-dimensional electron gasses in oxide heterostructures, *Science*, p. **313**, pp. 1942–1945 (2006).

90. G. Herranz, M. Basletić, M. Bibes, C. Carrétéro, E. Tafra, E. Jacquet, K. Bouzehouane, C. Deranlot, A. Hamzić, J.-M. Broto, A. Barthélémy, and A. Fert, *Phys. Rev. Lett.*, **98**, p. 216803 (2007).

91. A. Kalabukhov, R. Gunnarsson, J. Börjesson, E. Olsson, T. Claeson, and D. Winkler, Effect of oxygen vacancies in the $SrTiO_3$ substrate on the electrical properties of the $LaAlO_3/SrTiO_3$ interface, *Phys. Rev. B*, **75**, p. 121404(R) (2007).

92. Wolter Siemons, Gertjan Koster, Hideki Yamamoto, Walter A. Harrison, Gerald Lucovsky, Theodore H. Geballe, Dave H. A. Blank, and Malcolm R. Beasley, Origin of charge density at LaAlO3 on SrTiO3 heterointerfaces: possibility of intrinsic doping, *Phys. Rev. Lett.*, **98**, p. 196802 (2007).

93. P. R. Willmott, S. A. Pauli, R. Herger, C. M. Schlepütz, D. Martoccia, B. D. Patterson, B. Delley, R. Clarke, D. Kumah, C. Cionca, and Y. Yacoby, Structural basis for the conducting interface between $LaAlO_3$ and $SrTiO_3$, *Phys. Rev. Lett.*, **99**, pp. 155502 (2007).

94. S. A. Chambers, M. H. Engelhard, V. Shutthanandan, Z. Zhu, T. C. Droubay, L. Qiao, P.V. Sushko, T. Feng, H.D. Lee, T. Gustafsson, E. Garfunkel, A.B. Shah, J.-M. Zuo, and Q.M. Ramasse, Instability, intermixing and electronic structure at the epitaxial view the MathML source heterojunction, *Surf. Sci. Rep.*, **65**, pp. 317–352 (2010).

95. Y. Chen, N. Pryds, J. E. Kleibeuker, G. Koster, J. Sun, E. Stamate, B. Shen, G. Rijnders, and S. Linderoth, Metallic and insulating interfaces of amorphous $SrTiO_3$-based oxide heterostructures, *Nano Lett.*, **11**, pp. 3774–3778 (2011).

96. E. Di Gennaro, U. Coscia, G. Ambrosone, A. Khare, F. Miletto Granozio, and U. Scotti di Uccio, Photoresponse dynamics in amorphous-$LaAlO_3/SrTiO_3$ interfaces, *Sci. Rep.*, **5**, p. 8393 (2015).

97. M. Scigaj, J. Gázquez, M. Varela, J. Fontcuberta, G. Herranz, and F. Sánchez, Photoresponse dynamics in amorphous-$LaAlO_3/SrTiO_3$ interfaces, Conducting interfaces between amorphous oxide layers and $SrTiO_3(110)$ and $SrTiO_3(111)$, *Solid State Ionics*, **281**, pp. 68–72 (2015).

98. C. Cancellieri, D. Fontaine, S. Gariglio, N. Reyren, A. D. Caviglia, A. Fête, S. J. Leake, S. A. Pauli, P. R. Willmott, M. Stengel, Ph. Ghosez, and J.-M. Triscone, Electrostriction at the $LaAlO_3/SrTiO_3$ interface, *Phys. Rev. Lett.*, **107**, p. 056102 (2011).

99. C. Cantoni, J. Gazquez, F. Miletto Granozio, M. P. Oxley, M. Varela, A. R. Lupini, S. J. Pennycook, C. Aruta, U. Scotti di Uccio, P. Perna, D. Maccariello, Electron transfer and ionic displacements at the origin of the 2D electron gas at the LaAO/STO interface: direct measurements with atomic-column spatial resolution, *Adv. Mater.*, **24**, pp. 3952–3957 (2012).

100. J. Gazquez, M. Stengel, R. Mishra, M. Scigaj, M. Varela, M. A. Roldan, J. Fontcuberta, F. Sánchez, and G. Herranz, Competition between polar and nonpolar lattice distortions in oxide quantum wells: new critical thickness at polar interfaces, *Phys. Rev. Lett.*, **119**, 106102 (2017).

101. L. Yu and A. Zunger, A polarity-induced defect mechanism for conductivity and magnetism at polar–nonpolar oxide interfaces, *Nat. Commun.*, **5**, p. 5118 (2014).

102. Z. Zhong, P. X. Xu, and P. J. Kelly, Polarity-induced oxygen vacancies at $LaAlO_3/SrTiO_3$ interfaces, *Phys. Rev. B*, **82**, p. 165127 (2010).

103. Y. Li, S. N. Phattalung, S. Limpijumnong, J. Kim, and J. Yu, Formation of oxygen vacancies and charge carriers induced in the n-type interface of a $LaAlO_3$ overlayer on $SrTiO_3$ (001), *Phys. Rev. B*, **84**, p. 245307 (2011).

104. N. C. Bristowe, P. B. Littlewood, and E. Artacho, Surface defects and conduction in polar oxide heterostructures, *Phys. Rev. B*, **83**, p. 205405 (2011).

105. E. Breckenfeld, N. Bronn, J. Karthik, A. R. Damodaran, S. Lee, N. Mason, and L. W. Martin, Effect of growth induced (non) stoichiometry on interfacial conductance in $LaAlO_3/SrTiO_3$, *Phys. Rev. Lett.*, **110**, p. 196804 (2013).

106. M. Salluzzo, J. C. Cezar, N. B. Brookes, V. Bisogni, G. M. De Luca, C. Richter, S. Thiel, J. Mannhart, M. Huijben, A. Brinkman, G. Rijnders, and G. Ghiringhelli, Orbital reconstruction and the two-dimensional electron gas at the $LaAlO_3/SrTiO_3$ interface, *Phys. Rev. Lett.*, **102**, p. 166804 (2009).

107. G. Berner, M. Sing, H. Fujiwara, A. Yasui, Y. Saitoh, A. Yamasaki, Y. Nishitani, A. Sekiyama, N. Pavlenko, T. Kopp, C. Richter, J. Mannhart, S. Suga, and R. Claessen, Direct k-space mapping of the electronic structure in an oxide-oxide interface, *Phys. Rev. Lett.*, **110**, p. 247601 (2013).

108. P. Delugas, A. Filippetti, V. Fiorentini, D. I. Bilc, D. Fontaine, and P. Ghosez, Spontaneous 2-dimensional carrier confinement at the n-Type $SrTiO_3/LaAlO_3$ interface, *Phys. Rev. Lett.*, **106**, p. 166807 (2011).

109. W. Liu, S. Gariglio, A. Fête, D. Li, M. Boselli, D. Stornaiuolo, and J. M. Triscone, Magneto-transport study of top- and back-gated $LaAlO_3$/$SrTiO_3$ heterostructures, *APL Mater.,* **3**, p. 062805 (2015).

110. S. Hurand, A. Jouan, C. Feuillet-Palma, G. Singh, E. Lesne, N. Reyren, A. Barthélémy, M. Bibes, J. E. Villegas, C. Ulysse, M. Pannetier-Lecoeur, M. Malnou, J. Lesueur, and N. Bergeal, Top-gated field-effect $LaAlO_3$/$SrTiO_3$ devices made by ion-irradiation, *Appl. Phys. Lett.,* **108**, p. 052602 (2016).

111. S. Hurand, A. Jouan, C. Feuillet-Palma, G. Singh, J. Biscaras, E. Lesne, N. Reyren, A. Barthélémy, M. Bibes, J. E. Villegas, C. Ulysse, X. Lafosse, M. Pannetier-Lecoeur, S. Caprara, M. Grilli, J. Lesueur, and N. Bergeal, Field-effect control of superconductivity and Rashba spin–orbit coupling in top-gated $LaAlO_3$/$SrTiO_3$ devices, *Sci. Rep.,* **5**, p. 12751 (2015).

112. X. Lin, Z. Zhu, B. Fauqué, and K. Behnia, Fermi surface of the most dilute superconductor, *Phys. Rev. X,* **3**, p. 021002 (2013).

113. Z. Fisk, H. R. Ott, T. M. Rice, and J. L. Smith, Heavy-electron metals. *Nature (London),* **320**, p. 124 (1986).

114. P. G. de Gennes, *Superconductivity of Metals and Alloys*, Benjamin, New York (1966).

115. Y. Takada, Theory of superconductivity in polar semiconductors and its application to N-type semiconducting $SrTiO_3$, *J. Phys. Soc. Jpn.,* **49**, p. 1267 (1980).

116. J. Appel, Soft-mode superconductivity in $SrTiO_{3-x}$, *Phys. Rev.,* **180**, p. 508 (1969).

117. S. E. Rowley, L. J. Spalek, R. P. Smith, M. P. M. Dean, M. Itoh, J. F. Scott, G. G. Lonzarich, and S. S. Saxena, *Nat. Phys.*, **10**, p. 367 (2014).

118. N. Reyren, S. Thiel, A. D. Caviglia, L. Fitting Kourkoutis, G. Hammerl, C. Richter, C. W. Schneider, T. Kopp, A.-S. Rüetschi, D. Jaccard, M. Gabay, D. A. Muller, J. M. Triscone, and J. Mannhart, Superconducting interfaces between insulating oxides, *Science,* **317**, pp. 1196–1199 (2007).

119. C. S. Koonce, M. L. Cohen, J. F. Schooley, W. R. Hosler, and E. R. Pfeiffer, Superconducting transition temperatures of semiconducting $SrTiO_3$, *Phys. Rev.,* **163**, pp. 380–390 (1967).

120. O. N. Tuft and P. W. Chapman, Electron mobility in semiconducting strontium titanate, *Phys. Rev.,* **155**, pp. 796–802 (1967).

121. S. Gariglio, M. Gabay, J. Mannhart, and J. M. Triscone, Interface superconductivity, *Physica C,* **514**, pp. 189–198 (2015).

122. N. Reyren, S. Gariglio, A. D. Caviglia, D. Jaccard, T. Schneider, and J. M. Triscone, Anisotropy of the superconducting transport properties of the $LaAlO_3$/$SrTiO_3$ interface, *Appl. Phys. Lett.,* **94**, p. 112506 (2009).

123. M. Sing, G. Berner, K. Goß, A. Müller, A. Ruff, A. Wetscherek, S. Thiel, J. Mannhart, S. A. Pauli, C. W. Schneider, P. R. Willmott, M. Gorgoi, F. Schäfers, and R. Claessen, Profiling the interface electron gas of $LaAlO_3$/$SrTiO_3$ heterostructures with hard X-ray photoelectron spectroscopy, *Phys. Rev. Lett.*, **102**, p. 176805 (2009).

124. O. Copie, V. Garcia, C. Bödefeld, C. Carrétéro, M. Bibes, G. Herranz, E. Jacquet, J.-L. Maurice, B. Vinter, S. Fusil, K. Bouzehouane, H. Jaffrès, and A. Barthélémy, Towards two-dimensional metallic behavior at $LaAlO_3$/$SrTiO_3$ interfaces, *Phys. Rev. Lett.*, **102**, p. 216804 (2009).

125. S. Gariglio, M. Gabay, and J. M. Triscone, Research update: conductivity and beyond at the $LaAlO_3$/$SrTiO_3$ interface, *APL Mater.*, **4**, p. 060701 (2016).

126. A. Fête, S. Gariglio, A. D. Caviglia, J. M. Triscone, and M. Gabay, Rashba induced magnetoconductance oscillations in the $LaAlO_3$–$SrTiO_3$ heterostructure, *Phys. Rev. B*, **86**, p. 201105(R) (2012).

127. Z. S. Popovic and S. Satpathy, Wedge-shaped potential and airy-function electron localization in oxide superlattices, *Phys. Rev. Lett.*, **94**, p. 176805 (2005).

128. Gervasi Herranz, Gyanendra Singh, Nicolas Bergeal, Alexis Jouan, Jérôme Lesueur, Jaume Gázquez, María Varela, Mateusz Scigaj, Nico Dix, Florencio Sánchez, and Josep Fontcuberta, Engineering two-dimensional superconductivity and Rashba spin–orbit coupling in $LaAlO_3$/$SrTiO_3$ quantum wells by selective orbital occupancy, *Nat. Commun.*, **6**, p. 6028 (2015).

129. J. Bardeen, L. N. Cooper, and J. R. Schrieffer, Theory of superconductivity, *Phys. Rev.*, **108**, p. 1175 (1957).

130. I. M. Lifshitz, Anomalies of electron characteristics of a metal in the high-pressure region, *JETP*, **11**, p. 1130 (1960).

131. E. Maniv, M. Ben Shalom, A. Ron , M. Mograbi, A. Palevski, M. Goldstein, and Y. Dagan, Strong correlations elucidate the electronic structure and phase diagram of $LaAlO_3$/$SrTiO_3$ interface, *Nat. Commun.*, **6**, p. 8239 (2015).

132. G. Herranz, F. Sánchez, N. Dix, M. Scigaj, and J. Fontcuberta, High mobility conduction at (110) and (111) $LaAlO_3$/$SrTiO_3$ interfaces, *Sci. Rep.*, **2**, p. 758 (2012).

133. A. Annadi, Q. Zhang, X. Renshaw Wang, N. Tuzla, K. Gopinadhan, W. M. Lü, A. Roy Barman, Z. Q. Liu, A. Srivastava, S. Saha, Y. L. Zhao, S. W. Zeng, S. Dhar, E. Olsson, B. Gu, S. Yunoki, S. Maekawa, H. Hilgenkamp, T. Venkatesan, Ariando, Anisotropic two-dimensional electron gas at the $LaAlO_3$/$SrTiO_3$ (110) interface, *Nat. Commun.*, **4**, p. 1838 (2013).

134. D. Pesquera, M. Scigaj, P. Gargiani, A. Barla, J. Herrero-Martín, E. Pellegrin, S. M. Valvidares, J. Gázquez, M. Varela, N. Dix, J. Fontcuberta, F. Sánchez, and G. Herranz, Two-dimensional electron gases at $LaAlO_3$/ $SrTiO_3$ interfaces: orbital symmetry and hierarchy engineered by crystal orientation, *Phys. Rev. Lett.*, **113**, p. 156802 (2014).

135. J. Stöhr, X-ray magnetic circular dichroism spectroscopy of transition metal thin films, *J. Electron Spectrosc. Relat. Phenom.*, **75**, p. 253 (1995).

136. M. Tinkham, *Introduction to Superconductivity*, 2nd ed., Dover (1996).

137. M. Ben Shalom, M. Sachs, M. Rakhmilevitch, D. Palevski, and A. Y. Dagan, Tuning spin–orbit coupling and superconductivity at the $SrTiO_3$/ $LaAlO_3$ interface: a magnetotransport study, *Phys. Rev. Lett.*, **104**, p. 126802 (2010).

138. M. I. Dyakonov (ed.), *Spin Physics in Semiconductors*, Spring Verlag (2008).

139. G. Khalsa, B. Lee, and A. H. Y. MacDonald, Theory of t_{2g} electron-gas Rashba interactions, *Phys. Rev. B*, **88**, p. 041302(R) (2013).

140. K. V. Shanavas and S. Satpathy, Electric field tuning of the Rashba effect in the polar perovskite structures, *Phys. Rev. Lett.*, **112**, p. 086802 (2014).

141. K. Michaeli, A. C. Potter, and P. A. Lee, Superconducting and Ferromagnetic Phases in $SrTiO_3$/$LaAlO_3$ Oxide Interface Structures: Possibility of Finite Momentum Pairing, *Phys. Rev. Lett.*, **108**, p. 117003 (2012).

142. H. Liang, L. Cheng, L. Wei, Z. Luo, G. Yu, C. Zeng, and Z. Zhang, Non-monotonically tuneable Rashba spin–orbit coupling by multiple-band filling control in $SrTiO_3$-based interfacial d-electron gasses, *Phys. Rev. B*, **92**, p. 075309 (2015).

143. J. Pereiro, A. Petrovic, C. Panagopoulos, and I. Bozovic, Interface superconductivity: history, development and prospects, *Phys. Express*, **1**, pp. 208–241 (2011).

144. C. Richter, H. Boschker, W. Dietsche, E. Fillis-Tsirakis, R. Jany, F. Loder, L. F. Kourkoutis, D. A. Muller, J. R. Kirtley, C. W. Schneider and J. Mannhart, Interface superconductor with gap behavior like a high temperature superconductor, *Nature*, **502**, p. 528 (2013).

145. W. W. Warren, et al., Cu spin dynamics and superconducting precursor effects in planes above T_C in YBa2Cu3O6.7, *Phys. Rev. Lett.*, **62**, pp. 1193–1196 (1989).

146. L. E. Hayward, D. G. Hawthorn, R. G. Melko, and S. Sachdev, Angular Fluctuations of a Multicomponent Order Describe the Pseudogap of YBa2Cu3O_{6+x}, *Science*, **343**, pp. 1336–1339 (2014).

147. H. Ding, et al., Spectroscopic evidence for a pseudogap in the normal state of underdoped high-T_C superconductors, *Nature,* **382**, pp. 51–54 (1996).
148. M. R. Norman, D. Pines, and C. Kallin, The pseudogap: friend or foe of high-T_C?, *Adv. Phys.,* **54**, pp. 715–733 (2005).
149. V. J. Emery and S. A. Kivelson, Importance of phase fluctuations in superconductors with small superfluid density, *Nature,* **374**, pp. 434–437 (1995).
150. Guanglei Cheng, Michelle Tomczyk, Shicheng Lu, Joshua P. Veazey, Mengchen Huang, Patrick Irvin, Sangwoo Ryu, Hyungwoo Lee, Chang-Beom Eom, C. Stephen Hellberg and Jeremy Levy, Electron pairing without superconductivity, *Nature,* **521**, p. 196 (2015).
151. L. Benfatto, C. Castellani, and T. Giamarchi, Broadening of the Berezinskii–Kosterlitz–Thouless superconducting transition by inhomogeneity and finite-size effects, *Phys. Rev. B,* **80,** p. 214506 (2009).
152. N. Scopigno, D. Bucheli, S. Caprara, J. Biscaras, N. Bergeal, J. Lesueur, and M. Grilli, Phase separation from electron confinement at oxide interfaces, *Phys. Rev. Lett.,* **116**, p. 026804 (2016).
153. Y.-Y. Pai, H. Lee, J.-W. Lee, A. Annadi, G. Cheng, S. Lu, M. Tomczyk, M. Huang, C.-B. Eom, P. Irvin, and J. Levy, One-dimensional nature of superconductivity at the $LaAlO_3/SrTiO_3$ interface, *Phys. Rev. Lett.,* **120,** 147001 (2018).
154. Z. Zhong, A. Tóth, and K. Held, Theory of spin–orbit coupling at $LaAlO_3$/$SrTiO_3$ interfaces and $SrTiO_3$ surfaces, *Phys. Rev. B,* **87**, p. 161102(R) (2013).
155. D. Stornaiuolo, D. Massarotti, R. Di Capua, P. Lucignano, G. P. Pepe, M. Salluzzo, and F. Tafuri, Signatures of unconventional superconductivity in the $LaAlO_3/SrTiO_3$ two-dimensional system, *Phys. Rev. B,* **95**, 140502 (2017).
156. H. Boschker, C. Richter, E. Fillis-Tsirakis, C. W. Schneider, and J. Mannhart, Electron-phonon coupling and the superconducting phase diagram of the $LaAlO_3/SrTiO_3$ interface, *Sci. Rep.,* **5**, p. 12309 (2015).
157. C. Shen, B. B. Chen, H. X. Xue, G. Cao, C. J. Li, X. X. Wang, Y. P. Hong, G. P. Guo, R. F. Dou, C. M. Xiong, L. He, and J. C. Nie, Gate dependence of upper critical field in superconducting (110) $LaAlO_3/SrTiO_3$ interface, *Sci. Rep.,* **6,** p. 28379 (2016).
158. N. Mohanta and A. Taraphder, Topological superconductivity and Majorana bound states at the $LaAlO_3/SrTiO_3$ interface, *Europhys. Lett.,* **108,** p. 60001 (2014).

159. M. S. Scheurer and J. Schmalian, Topological superconductivity and unconventional pairing in oxide interfaces, *Nat. Commun.*, **6**, p. 6005 (2015).

160. F. Loder, A. P. Kampf, and T. Kopp, Superconductivity with Rashba spin–orbit coupling and magnetic field, *J. Phys.: Condens. Matter*, **25**, p. 362201 (2013).

161. V. Barzykin and L. P. Gor'kov, Inhomogeneous stripe phase revisited for surface superconductivity, *Phys. Rev. Lett.*, **89**, p. 227002 (2002).

162. W. Zhang and W. Yi, Topological Fulde–Ferrell–Larkin–Ovchinnikov states in spin–orbit-coupled Fermi gasses, *Nat. Commun.*, **4**, p. 2711 (2013).

163. P. Gallagher, M. Lee, J. R. Williams, and D. Goldhaber-Gordon, Gate-tunable superconducting weak link and quantum point contact spectroscopy on a strontium titanate surface, *Nat. Phys.*, **10**, pp. 748–752 (2014).

164. N. Pavlenko, T. Kopp, E. Y. Tsymbal, G. A. Sawatzky, and J. Mannhart, Magnetic and superconducting phases at the $LaAlO_3/SrTiO_3$ interface: The role of interfacial Ti 3d electrons, *Phys. Rev. B*, **85**, p. 020407 (2012).

165. L. Fidkowski, H. C. Jiang, R. M. Lutchyn, and C. Nayak, Magnetic and superconducting ordering in one-dimensional nanostructures at the $LaAlO_3/SrTiO_3$ interface, *Phys. Rev. B*, **87**, p. 014436 (2013).

166. S. Banerjee, O. Erten, and M. Randeria, Ferromagnetic exchange, spin–orbit coupling and spiral magnetism at the $LaAlO_3/SrTiO_3$ interface, *Nat. Phys.*, **9**, pp. 626–630 (2013).

Index